Electrochemistry in
Research and Development

Electrochemistry in Research and Development

Edited by
R. Kalvoda
J. Heyrovsky Institute of Physical Chemistry and Electrochemistry
Prague, Czechoslovakia

and
Roger Parsons
The University of Southampton
Southampton, England

Plenum Press • New York and London

Library of Congress Cataloging in Publication Data

UNESCO Forum on Electrochemistry in Research and Development (1984: Paris, France)
  Electrochemistry in research and development.

  "Proceedings of the UNESCO Forum on Electrochemistry in Research and Development, held June 3–6, 1984, in Paris, France" – T.p. verso.
  Includes bibliographies and index.
  1. Electrochemistry – Congresses. I. Kaldova, Robert. II. Parsons, Roger. III. Title.
QD551.U54 1984   541.3′7   85-30067
ISBN-13: 978-1-4684-5100-9   e-ISBN-13: 978-1-4684-5098-9
DOI: 10.1007/978-1-4684-5098-9

Proceedings of the UNESCO Forum on Electrochemistry in Research and Development, held June 3–6, 1984, in Paris, France

© 1985 Plenum Press, New York
Softcover reprint of the hardcover 1st edition 1985

A Division of Plenum Publishing Corporation
233 Spring Street, New York, N.Y. 10013

PREFACE

This volume contains the papers presented at the UNESCO Scientific Forum on Chemistry in the Service of Mankind - Electrochemistry in Research and Development, held in Paris, June 4-6, 1984.

Electrochemistry is concerned with the way electricity produces chemical changes and in turn chemical changes result in the production of electricity. This interaction forms the basis for an enormous variety of processes ranging from heavy industry through batteries to biological phenomena. Although there are many established applications, modern research has led to a great expansion in the possibilities for using electrochemistry in exciting future developments. To encourage this progress, UNESCO has set up an Expert Committee on Electrochemistry and its Applications in the European and North American region, which has already held a number of meetings devoted to specific topics. To achieve a synthesis of the main directions of development and to demonstrate the importance of these for the needs of our modern society, the Expert Committee organized a Forum on Electrochemistry in Research and Development.

The object of this was to assess the future trends in research and development and to establish a dialogue between experts in electrochemistry and their colleagues in the many other disciplines which can make use of electrochemistry. The Forum was also intended to present electrochemistry and its applications in a form accessible to non-specialists so that science policy-makers will be aware of the potentialities of this subject for the future needs of mankind.

The program of the Forum included four sections with plenary lectures followed by discussion on the following topics:

Electrochemistry and

    - Energy   - Solar Energy Conversion
                         Energy Conversion and Storage
                         Hydrogen Economy

    - The Environment - Analysis and Removal of Pollutants
                         Trace Metal and Analysis
                         Food and Drug Control

- Biosciences   - In Vivo Applications
                          Membranes
                          Genetic Engineering
                          Polymer Electrodes

- Technology   - Potentialities in New Technologies
                          Processes in Chemical Industry

Prof. A. A. Vlček, the Chairman of the UNESCO Expert Committee on Electrochemistry for the European and North American Region, was in the chair: about 140 visitors attended the meeting.

R. Kalvoda
R. Parsons

CONTENTS

A. A. Vlček

J. Heyrovský Institute of Physical Chemistry
and Electrochemistry
Czechoslovak Academy of Science, Prague

Electrochemistry is an interdisciplinary science, combining mainly chemistry, physics, solid state science and electronics, of immense theoretical and practical importance.

This special interdisciplinary position of electrochemistry led, on one side, to its development as an almost independent branch of science, on the other hand has caused a certain isolation of electrochemistry from chemistry, where its roots come from. This isolation can be for example demonstrated at universities as well as on the position of electrochemistry in industry. Very few universities teach a proper course of electrochemistry, and if so, then mainly oriented toward electroanalytical applications or as a minor part of physical chemistry.

In industry there is, on one side, a huge application (batteries, galvanoplating, heavy inorganic technology) with its specific technologies, on the other side; there are very few examples of electrochemical procedures used in integrated technologies. As the most recent example of the latter, use of electrochemistry in the industry of semiconductor devices can be quoted. However, the potentialities of electrochemistry are much wider but the R and D people, and especially those designing new technologies, are rarely aware of them.

Electrochemistry as science per se formulates very fundamental questions and problems and develops a basis for interpretation of many phenomena of nature. The basis electrochemistry has developed is such that electrochemical methods can be used as a standard method in chemical research when investigating redox properties or structure of inorganic and organic compounds. Electrochemical methods of preparation can provide unusual species hardly accessible by other techniques. Again, due to the specific nature of electrochemistry not always are research people in other branches aware of electrochemical results and possibilities of their application in other scientific discipline. As an exception, electroanalytical chemistry can be mentioned. Methods of electroanalytical chemistry are widely used in the control of industrial processes, monitoring of environment, biology, clinical chemistry etc.

All these reasons led in 1977 to the decision to establish an international cooperation in electrochemistry on a UNESCO basis. The main goal of this cooperation has been to bring electrochemists into closer cooperation, to specify the most urgent problems and future lines of development of electrochemistry and last but not least to make scientists and R and D people more aware of the possibilities electrochemistry offers, or might offer in future, in solving scientists tasks as well as problems for the development of new technologies.

This main idea of cooperation, the results of which, will be mentioned in the contribution by Prof. Gierst, represents also the chief motive which led to the decision to organize this Scientific Forum on Electrochemistry in Research and Development. At the same time we have hoped to increase the awareness of electrochemists of the necessity to make a greater offer in getting their results more widely known and used.

GENERAL INTRODUCTION II

L. Gierst

Université Libre de Bruxelles
Belgium

UNESCO works in a world-wide framework aimed at promoting education, science and culture. In this it has quickly become aware of the fact that Electrochemistry is a particularly fertile domain with far-reaching potentialities; they extend well beyond its academic frontier and have important implications in economic and social matters.

As a first step, a meeting of experts, working on the general theme "Perspectives in Electrochemistry" was held in Prague in March 1977. The multifarious subdisciplines of Electrochemistry were appraised critically and a limited number of fields were selected as more likely than others to lead to significant and fruitful developments.

The delineation of priority fields came as the logical result of a convergence between criteria of purely scientific nature and the fundamental background of the UNESCO philosophy. One of the major concerns of this philosophy is to promote the exchange of information (cultural, educational as well as scientific) into the whole world, irrespective of artificial barriers of any possible nature. This involves giving help in disseminating knowledge as soon as it is acquired, and taking the decisive steps that could answer the most pressing needs of society, perceived locally as well as on any broader scale. This necessity implies a totally new approach in the scientific, technological and educational ways of defining domains of action, through the creation of new types of relationships between pure science, its educational vehicles, the research laboratories, the electrochemical industry and society itself, with its growing awareness of the need of upgrading the level of knowledge, in its ever-lasting struggle for better material conditions and increasing quality of life. For this to work smoothly and efficiently, the free flow of information and the dissamination of new findings would profit by being channelled through adequate facilities, in order to coordinate the efforts on a regional or a world-wide base, to stimulate the rate of mutual exchanges, with the additional bonus of lowered running costs.

But let us return to Electrochemistry. The expert committee of UNESCO, on the basis of thorough critical exploratory work, reached the conclusion that three main areas for research and training should be retained as the major lines of force of the program.

Project one deals with electrochemical energy conversion and storage, with special emphasis on photoelectrochemistry, on the methodology of electrochemical storage and on the synthesis of energy-rich compounds. This project has been vigorously pursued in recent years, with three international workshops, successively dedicated to Electrochemical Energy Conversion, Photoelectrochemistry and High Energy Density Light Metals. Three other meetings, already scheduled, will focus on Photoelectrochemical Processes, Modified Electrodes, Solar Energy Conversion and New Developments in Conversion and Storage.

The second project considers electrochemistry in its relations with environmental problems. The two major topics are the development of electrochemical monitors specially adapted to pollution, and the electrochemical removal of environmental pollutants, these processes can be made economic and highly selective. Three successive workshops have already been devoted to these topics, with special emphasis on electrochemical sensors in water analysis.

The third project is concerned with the electrochemical aspects of materials, taken in their broadest sense. It includes electrode interfaces, adsorption layers, bioelectrochemistry and the huge field of electro-organic synthesis. This project has resulted up to now in five workshops, the accent being put on organic electrochemistry and the behavior of interfaces.

These three series of workshops have led to significant results, judged not only from a strictly scientific point of view: a number of laboratory networks have been established, informal cooperative research has started and most of the available research facilities have been surveyed. Rapid exchange of information has been enhanced by facilitation of short visits, microsymposia and contacts with young scientists from developing countries. Most of these activities, including the workshops, have been strongly catalyzed by the action of the UNESCO coordinating group, with the help of UNESCO funds supplemented by local contributions from governmental and private organizations.

It is perhaps still somewhat premature to assess the full impact of the UNESCO program, as it is felt by the electrochemical community and society. It is clear, nevertheless, that the direct and fast flow of information, the formation and strengthening of flexible scientific networks and the exchange of collaborators have resulted 1) in stimulating ideas, 2) in enhancing the pace of dedicated research, 3) in extending the use of underemployed equipment and 4) in conducing to a better based selection of specific research topics. These positive efforts are leading to a better use of all available means, human as well as technical. This will contribute to more efficient fund distribution and utilization by discouraging too-redundant programs and by easing a part of the routine work.

The organizing committee, from the very start of its activities, has been fully aware that science in general, and Electrochemistry in the present case, is a matter which concerns the whole planet. In order to promote dissemination of Electrochemical Science, taking account of the basic needs of the Society. It has been decided to launch an additional project, devoted to the publication of a book on "Teaching in Electrochemistry". The major objective here is to allow any teacher working far from a suitable educational center to construct his own course of electrochemistry, with minimal demands in equipment. This book is organized in such a way that full coverage of the Electrochemical Science will be presented at three different levels of qualification. This structure will allow the teacher to adjust his training course to the average background of his class, taking into account the local availibility in materials and equipment.

Besides these various activities (coordination work, specialized workshops, joint projects, educational aspects, etc.) the organizing committee felt compelled to set up the present meeting. It has been appropriately called a "FORUM" in the sense that its major objective is to trigger off an effective interaction between Science and Society, by bringing together specialists in Electrochemistry and non-specialists who are well aware of the pressing need to raise the level of knowledge, but are uncertain about the routes to follow. They know that enhancement of this knowledge constitutes one of the firmest bases of action for improving living conditions.

Human implications will thus be constantly present among us, particularly perhaps in relation with the developing countries where Science and Technology, after suitable adaptation, remain the best hope for economic as well as social improvements. UNESCO successfully manages to bring the scientists of the developing countries into closer and closer contact with the laboratories of the more industrialized countries, in order to afford, in this way, a simpler and more economical access to scientific know-how than was the case in the past and still is at present. Its constitutional bylaws allow UNESCO to assume a privileged role among other international organizations, in its capability of helping scientific progress by direct aid, coordination, stimulation and crossfertilization between adjacent fields. Electrochemists are not automatically fully aware of the problems and needs of the Society, either at local and regional or world-wide level. Conversely, people who have a better perception of these problems but are not dedicated to scientific research may lack the degree of expertise required for taking appropriate decisions - which may sometimes involve millions of individuals. By asking specialists to assess the state-of-the-art and the perspectives in their respective fields, and by offering ample time for debating with the audience, the organizing committee is convinced that a new type of communication bridge will be built and that a far-reaching dialogue will develop and become firmly established. It will be an exciting and unique experience.

We are particularly glad to express our thanks to all the participants, who have shown their dedication to the problems which will contribute the core of this FORUM. More particularly, we express our gratitude to the UNESCO authorities, to the lecturers for their care and help and to the chairmen who have reacted enthusiastically to the perspective of sharing the efforts of the organizers and are ready to make this meeting as relevant and fruitful for as many people as possible.

We are also convinced that those of us who will attend the three days of this FORUM will not, afterwards, disperse to the four cardinal points, without seizing the present opportunity to confirm durable and enriching contacts.

Whether in Science or in Social development, the main tasks are still ahead of us. Working closely together will increase our force and strengthen our determination.

INTRODUCTORY TALKS

WHAT IS ELECTROCHEMISTRY ?

R. Parsons

Laboratoire d'Electrochimie
Interfaciale du CNRS

INTRODUCTION

The science of electrochemistry was born at the end of the eighteenth century following intense interest in the production of electricity by animals such as the torpedo and the electric eel. The key phenomena were observed just at the turn of the century. Volta invented his pile in which chemical species were consumed with the production of electricity and a few months afterwards Nicholson and Carlisle used this electricity to decompose water into hydrogen and oxygen. It is this two-way process which lies at the heart of electrochemistry; it may be represented by a very simple diagram

Electricity $\overset{\leftarrow}{\underset{\rightarrow}{}}$ chemical compounds

which hides the many complexities of this interchange. One might imagine that study of a branch of science over nearly two centuries would exhaust its interest and its potentialities but this is far from being so. Electrochemistry continues to find new applications in wider fields as well as providing scope for greater understanding of its fundamentals. A brief and rather selective outline of the history of the subject should illustrate this as well as the range of its interest.

NINETEENTH CENTURY

The first half of the century was dominated by the work of Davy and Faraday. Although they both made many fundamental discoveries, they were always interested in practical applications. Thus Davy used electrolysis to discover the alkali metals but also applied his electrochemical knowledge to suggest cathodic protection to the Royal Navy as a way of preventing the corrosion of their copper-sheathed wooden ships. Corrosion remains a major problem today. It was estimated that losses due to corrosion cost the United States $4\frac{1}{2}\%$ of the GNP or about $70,000 million per year in 1975.

Faraday showed that all known forms of electricity were identical and then discovered the quantitative relation between the electric charge passed through a cell and the amount of chemical change which occurs. These observations are nowadays easily understood because an electrical

9

current in a metal is known to be a stream of electrons. He laid the basis for electrochemical terminology and also for the understanding of passivation which plays such an important role in the protection of materials.

Not long after Faraday's electrochemical work, Sir George Grove in 1842 showed that Nicholson and Carlisle's experiment of the decomposition of water could be reversed, that is, hydrogen and oxygen could be combined to form water, in an electrochemical cell, with the production of electricity. Besides giving a complete and simple demonstration of the primary electrochemical process, this was important because it led the way to a fuel cell in which a fuel, hydrogen, could be continuously oxidized electrochemically in contrast to the cells derived from Volta's pile which have a capacity limited by the amount of active chemical contained in them. Grove's fuel cell was the forerunner of the hydrogen-oxygen fuel cells used in the Gemini and Apollo space capsules. In fact Grove used his cell already in 1842 with primitive incandescent lamps to light the lecture theatre of the London Institution.

Developments of Volta's pile which are known as primary batteries, were made during the nineteenth century. The most enduring is the cell invented by Leclanché in 1867 in which zinc dissolves at the anode and manganese dioxide is reduced at the cathode. This basic principle is still used in many of the dry cells which are the familiar power sources for portable electronic apparatus like radios. These cells are essentially irreversible; one cannot recharge them, so once the electroactive material is consumed they are usually thrown away. Rechargeable cells or secondary batteries are more economical and the best known of these is the lead accumulator invented by Planté in 1860. It is now used in every automobile to power the starter motor and as an energy reservoir as well as being the sole power source in the majority of electric vehicles. Electricity is produced from the reduction of lead dioxide to lead sulphate and the oxidation of lead to lead sulphate and these processes are reversed on charging the battery. Fundamentally the working of this battery depends on the properties of lead which can exist in three states of oxidation each step requiring the transfer of two electrons and also on the fact that suitable solid compounds of lead are available. This simplifies the practical achievement of the battery although not the detailed explanation of its operation.

Humphry Davy's use of electrolysis using molten electrolytes to produce active metals culminated near the end of the century in the industrial production of aluminium on the basis of the independent discoveries by Hall and Héroult in 1886 that alumina could be electrolyzed in a molten cryolite bath.

The intense activity in electrochemistry in the nineteenth century covered both fundamental and practical aspects, but although it was established early that the process itself was one which occurred at the junction between a metallic and a non-metallic conductor. Little progress was made in the basic understanding of this reaction. To a large extent this was because the nature of the conduction of electricity itself was poorly understood and such understanding developed in parallel with the development of electrochemistry itself. In fact, the idea that an electric current always consists of a flow of charged particles replaced the idea of an "electric fluid" gradually during the century and was only firmly established by the discovery of the electron by J. J. Thomson in 1897 and the proposition of the ionic theory of electrolytic conduction by S. Arrhenius in 1187. Thus it became clear that the basic electrochemical event was a result of the change in the mode of conduction of electricity, from the flow of the minute, negatively charged, electrons as in a metal to the much larger and heavier charged particles which could be identified with the species Faraday called an "ion", a term derived from Greek with the aid of William Whewell and signifying "that which goes". The important difference is that when conduction is by electrons there is no movement of matter, while when there is conduction by ions, matter is moved from one place to another in the system. When there is a change in the mechanism of conduction, the production of electrons from the ions or the consumption of electrons by the ions is inevitably accompanied by reduction or oxidation of the ions concerned, that is, a change in their chemical nature. This is the beginning of a clear understanding of the nature of the electrochemical process.

Electrons are usually the charge carriers in metals and semi-conductors while ions carry the charge in electrolytic solutions like battery acid or sea water where the ions are a minor species dissolved in the solvent (water) or like solid or liquid salts where the ions are the only constituents. Both ions and electrons may coexist for example in plasmas or ionized gases but the rapid movement of the small, light electron tends to dominate the conduction and a smaller proportion of the current is due to the larger, heavier ions.

TWENTIETH CENTURY

Towards the end of the nineteenth century Walther Nernst had used the powerful system of energy relations known as thermodynamics to show how one could calculate the maximum amount of electrical energy that could be obtained from a chemical reaction occurring in an electrochemical cell. These predictions can often be achieved in practice but only if the current flow is very small, that is the process is carried out very slowly. This observation shows that electrochemical reactions like all chemical reactions occur at a finite rate. The study of the rates of reaction is called kinetics and the understanding of the kinetics of electrochemical reactions has been the central preoccupation of electrochemists this century.

Although basic observations were made by Tafel in 1905 the development of a clear understanding of the kinetics has been slow. This was initially because it was difficult to make reliable observations. This is due to the interfacial nature of the electrode reaction mentioned above. The exchange of electrons and the rearrangement of chemical species actually occurs in a very small region of space. Although some work on this was done earlier, that done by Georges Gouy in the first decade of this century first showed clearly the structure of this region and that it is frequently only a few molecular diameters thick. This means that the environment of the electrochemical reaction can be markedly changed by the replacement of a full or even a partial monomolecular layer at the interface. Such a change can easily occur as a result of the presence of minute quantities of substances in the system which are not evident in the general properties of the system as a whole. Thus, reliable results can be obtained only if a scrupulous control of the purity of the system is maintained. Techniques for achieving this have developed slowly and only in the last few decades have they been satisfactorily mastered. (A similar problem has occured in the closely related field of heterogeneous catalysis). The first steps towards a satisfactorily controlled environment for the study of electrode reactions were made in Frumkin's laboratoroy in Moscow in the 1930's but an important contribution was also made by Heyrovsky who introduced the renewable surface of the dropping mercury electrode. Fully reliable experiments on solid metal surfaces of known composition are a quite recent development in parallel with the explosion of surface science techniques of the last decade or two.

Mercury, however, has an advantage other than that of providing a clean surface easily; that surface is uniform and atomically smooth. It is not surprising then that much of the early progress in understanding the kinetics and mechanism of electrode reactions has arisen from experiments with mercury electrodes. During the twenties and thirties of this century the idea emerged clearly that electrode reactions were heterogeneous chemical reactions which obeyed the same sort of kinetic laws as ordinary chemical reactions; that is, the rate was proportional to a product of concentrations at the reacting species:

Rate = k[A][B]

where the square brackets indicate concentration of the species A, B etc. This arises because the reactant must collide to react. The proportionality constant k is known as the rate constant of the reaction. The unique feature of electrochemical reactions is that this "constant" can be adjusted by changing the electrode potential. In other words there is a simple electrical way of controlling the rate of this type of chemical transformation. Since the relation between the rate constant and the electrode potential is exponential, this control can be exercised over a very wide range; in the extreme example it is possible to change the rate by a factor of 10^{10} in this way. This opens up an enormous range of possibilities for the control of chemical reactions and emphasises the facility with which electrochemistry can be coupled with modern electronic systems, for the selective preparation of desired products, for example.

It should be emphasized that there is a limit to which an electrochemical reaction can be accelerated. This is because it is an interfacial reaction and the reacting species must be brought to the interface to react. As the interfacial reaction is accelerated a point is always reached when it is this transport of matter to the electrode which ultimately determines the rate of the whole process. Such behavior is of great practical importance because the rate of the transport process is directly proportional to the concentration of the reacting species in the solution. Hence this the basis for a whole series of powerful methods of quantitative analysis. This principle was first exploited in the polarograph invented by Heyrovsky in 1925 was the first automated analytical instrument. Now the range is enormously expanded and concentration of the order of parts per billion can be determined. Of great importance for example is the trace analysis of seawater is the fact that these advanced electrochemical methods of analysis are sensitive not only to the amount of the species but also to its nature and to the way it is combined with the other chemicals present. With the aid of modern digital electronics these analysis may be made routinely and on-line.

Of the many new developments in electrochemistry, one of the most interesting is that in which electrochemical processes are combined with photo-excitation; that is the use of light energy. The basic principle is that the light energy is used to form a positive charge and an electron in a region where an electric field can separate these charges and lead them to produce useful work. In the semiconductor p-n junction the region of electric field is produced by a very slight change in the composition of the semiconductor which is brought about by very carefully controlled diffusion of a dopant using very sophisticated furnaces. A similar type of field can be produced at the surface of a semiconductor simply by dipping it into an electrolyte. This semiconductor electrode also has the advantage that the electron driven towards the electrolyte can be used either to produce electricity or to carry out a useful chemical reaction, whereas the p-n junction can only do the former. This holds out the promise of a solar energy based chemical production or of the possibility of storing solar energy in chemical form. This exciting development is still in the laboratory stage but progress is rapid.

It should be remarked, in conclusion, that the natural sources of energy on which mankind has so far depended are the result of a process which is essentially photo-electrochemical:photosynthesis. Light is absorbed by an array of chlorophyll molecules resulting in the production of a pair of charges which are separated in the natural analogue of the field described previously and a series of oxidation-reduction reactions ensure resulting ultimately in the production of carbohydrates from carbon dioxide and water. This and other biological processes such as nerve conduction involve the transfer of charged ions across membranes and the ideas of electrochemical reactions are proving of great help in their understanding.

It is remarkable that the subject of electrochemistry after almost two hundred years of development is still of enormous interest and its field of application is continually expanding. The forum is an illustration of the present state of this fascinating and widely practical subject.

WHAT CAN ELECTROCHEMISTRY DO ?

A. R. Despić

Faculty of Technology and Metallurgy
University of Beograd
Beograd, Yugoslavia

The use of electrochemistry seems to be older than electrochemistry itself. In the valley of Mesopotamia, between the Euphrates and the Tigris, in the ruins of some Parthian villages, archeologists have found some 2200 years old remains of what is today popularly called a battery or, technically, a chemical power source of the primary cell type. As seen in Figure 1, it consisted of an iron rod and a copper sheet bent into a cylinder around it, both placed in a ceramic jar filled possibly with grape juice. Its likely use was that of a mysterious driving power source for the process of plating metal objects with gold or silver[1].

This skill, however, was then lost for two millenia, and electroplating and the development of other electrochemical practices had to await the rediscovery of sources of continuous flow of electricity within the last two hundred years. It is these, relatively recent inventions, that have made electrochemistry boom and cover vast areas of practical use.

ELECTROPLATING AND SURFACE FINISHING

Plating objects, both metallic and non-metallic, with metals or alloys or other types of surface finishing has become a widespread activity that major metal-working industries, above all the automobile industry and the manufacture of domestic appliances, cannot do without. The major objective is to obtain coatings with a dual purpose: that of decoration, to impart a lustrous appearance and that of protection against corrosion.

There are, however, many more subtle uses of plating and surface finishing such as the hardening of surface parts of objects subjected to especially heavy wear e.g. engine shafts which are electroplated nowadays with "hard-chromium", the plating of plastic reflectors to obtain mirror-like surfaces of high reflectivity, the electrolytic etching and electroplating of microelectronic components to produce patterns in the micrometer range with a precision unattainable by any other process, etc.[2].

All this is based on considerable fundamental knowledge of metal deposition and dissolution processes acquired during the boom of science of the last few decades[3]. Thus, the reasons for the formation of metal crystals of widely different appearance are well understood. They do not lie in the properties of the metals themselves (e.g. zinc and copper).

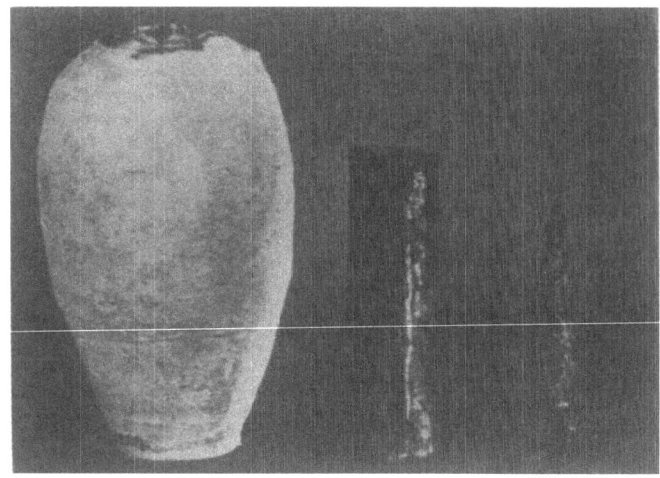

Fig. 1. Parthian jar cell from 2000 B.C.

Semi-quantitative theories available at the present time relate morphology, texture and physical properties of the deposits to the conditions of deposition. They enable us to produce the kind of metal deposits required.

Relatively recently electrochemistry has turned its attention to colloidal solutions of organic polymers. A new line of processes of industrial importance has thereby been opened up, that of electrophoretic deposition of paints and plastic coatings, which provide improved adherence and has technological advantages over classical spraying processes[4].

It is difficult to assess the volume of the plating and surface finishing industry with any precision, because it is either incorporated into larger production complexes of scattered in small workshops with a few craftsmen and manual operation. Still, rough estimates maintain that e.g. in the USA alone, it runs into billions of dollars.

ELECTROCHEMICAL MACHINING

The expertise in electrochemical dissolution of metals has grown to the extent that this process can be used for shaping metal objects into shapes difficult or impossible to achieve by other metal-working processes. This has resulted in the development of a new technology - electrochemical machining[5]. As shown in Figure 2, this technique could be used in a number of machining operations as are external shaping, cavity sinking, plunge cutting, turning, trepanning and internal grooving.

The cathode is made as the negative of the form to be achieved and is impressed into the material to be machined, erroding the latter by anodic dissolution. A very fast stream of electrolyte flows in between the cathode and the anode, serving manifold purposes carrying electricity through, carrying away the dissolution products and cooling the set-up by carrying away large amounts of heat evolved in the process. Considerable technical difficulties (arising from the use of very high dissolution currents, mass - and heat-transfer) had to be overcome in order to attain satisfactory shaping up and drilling or trepanning rates, required by efficient contemporary machining processes.

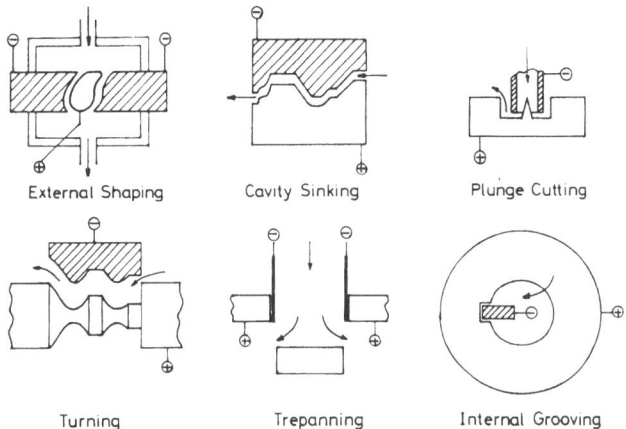

Fig. 2.   Schematic representation of different electrochemical machining
          operations.

ELECTROCHEMICAL POWER SOURCES

     As already mentioned, all the above processes could only be set-up
after mastering ways of providing a continuous flow of electrons or an
electric current. It is interesting to note that two entirely independent
ways of doing this were discovered within about 30 years of each other.
The first was found by Alessandro Volta in 1800 in an investigation of the
causes of the well known Galvani frog-leg phenomenon, and the second was
the discovery of Michael Faraday in 1831 that a magnet moving near a metal
wire induces motion of electrons it it. The latter discovery resulted in
the present day production of electrical energy in power plants, while the
former enabled us to obtain electricity independently of the mains supply,
from chemical power sources, commonly known as batteries. It may not be
common knowledge however, that these two power production methods have
developed to a similar level of use. The electric power delivered at the
present time by automobile batteries alone amounts to about 1.5 TW and is
of the same order as that delivered by all the electric power-plants in the
world.

     Two chemical power sources have dominated general use for over a
century: The "Leclanché cell" or "dry-battery" and the "lead-acid battery",
invented by Gaston Planté. They divided the field of chemical power
sources into two general areas: that of the primary cells and that of
electrically rechargeable secondary cells, or - from the point of view of
the user - those of electricity producing and those electricity storing
devices[6]. Their size and power has been determined by two major (and for
many years almost exclusive) uses: that of the dry-cell for torchlights and
of lead-acid batteries for starting automobiles and other vehicles.

     In recent years, however, demand for autonomous electric power has not
only increased, but also significantly broadened in the areas of use and
hence in requirements. It ranges from large megawatt power installations
for energy storage for load-levelling and reserve (emergency) power in
electric networks, down to microwatt sources for the stimulation of the
human heart beat.

     The amount of electricity which can be produced from a limited amount
of substances in the primary batteries, or from the limited storing
capacity of the secondary ones, could not satisfy some needs where continuous supply was necessary. This prompted interest in cells to which

active chemical substances could be fed continuously and the energy of their reaction converted into electrical energy. Since from the very early attempts these substances have been a fuel and oxygen from the air, this type of cell was called a "fuel cell". With the development of the fuel cell system, the concept of "fuel" has shifted from the conventional ones, such as coal, hydrocarbons and hydrogen, to the non-conventional, but more reactive ones, such as hydrazine or electrochemically active metals. Thus the category of metal-air batteries was introduced. Since metals cannot be fed continuously as gases or liquids, but have to be replaced periodically, these batteries are also known as mechanically rechargeable batteries and are placed in the secondary battery group.

Metal-air batteries have proved easier to develop to a state of simplicity and reliability required from a power source than the fuel cells with gaseous or liquid fuels. Hence, although a type of hydrogen-oxygen fuel cell has reached such a state of perfection that it could be assigned the task of the main power supply in the cabin of the Apollo spacecraft, more recently the zinc-air battery seems to have taken its place.

In the secondary-battery area, it is the drive to develop electric vehicles competitive with gasoline driven ones that has stimulated invest- ment in research and development. This has covered a broad spectrum from improved lead-acid battery to a large selection of new chemical power sources.

In any optimization between the weight of energy reservoirs and the vehicle range, it is the energy density, i.e. the number of kWh obtainable per kg, which is the critical parameter. That of the classical leaf-acid battery is only about 20 Wh/kg, so that it can do no more than power electric carts for internal factory-transport or the slow milk-distribution floats. A two-fold to ten-fold increase in the electric energy density was needed in order for electric drive to have any chance of penetrating the broad area of passenger, utility and commercial vehicles. This goal has not yet been achieved. Among the many candidates of the electrically rechargeable battery type, such as the sodium/sulphur, lithium/ferrous sulfide, zinc/chlorine, etc. the zinc/bromine battery seems to come closest to satisfying the requirements both from the point of view of energy density and the cost of production.

Electrically rechargeable batteries as energy source for electric vehicles are, however, the obsession of the developed part of the world, where the electricity supply is ample, the distribution network wide spread and the possesion of one's own garage with electric plugs a commonplace.

In the larger part of the world such an energy supply for electric vehicles is scarcely available. Fuel cells and mechanically rechargeable batteries of the metal-air type will for a long time be the only viable possibilities there. (The best example of the latter type will be dis- cussed towards the end of this review.)

Finally, the fast development of electronics and its penetration into everyday life and military technology has stimulated the development of a series of low-power-long-operation and high-power-short-operation batter- ies. Requirements for high reliability and low cost are basic for both. The range of systems used and types produced is so large that it is impos- sible even to enumerate them within this brief review. One typical example of a low-power-long-operation battery is the button type cell used for powering a watch, which is shown in Figure 3. Other uses of this or similar cells cover photographic equipment, pacemakers, electronic instru- mentation, pocket-computers, etc. Somewhat bigger cells have a large market in audio-and video-equipment.

Fig. 3. Mercury 1.5 V button cell for driving an electric watch.

It is worth mentioning the ingeniously constructed polaroid-camera battery (as an example of the second type). This can be as thin as a sheet of paper and able to deliver very high power for a very short time of milliseconds to parts of a second. Others, which are especially interesting to the military, are the "thermal-batteries", a combination of an electrochemical cell and a chemical charge which, when ignited, could heat up the cell within parts of a second and bring it into a state of operation. They could be stored indefinitely in torpedoes, rockets and similar gadgets, because at ambient temperature no corrosive degradation or "self-discharge" takes place. Yet, at elevated temperature, they are powerful power-sources, capable of delivering high power for a reasonably long time.

The area of chemical power sources is far from exhausted. New combinations are being tested all the time and it is difficult to forsee in which direction new breakthroughs will appear.

ELECTROCHEMICAL ANALYSIS

Electrochemcial processes are especially suitable for quantitative chemical analysis[7]. This is due to the fact that the electrical response of an electrochemical system to the substance whose presence is to be assessed, can be made quantitative and can easily be converted into signals acceptable to electronic instrumentation and computer processing. One good example is the "polarographic" method, for which its inventor, Jaroslav Heyrovsky, was awarded the Nobel Prize in 1964. In its classical form, the record of the current as a function of voltage across a polarographic cell, contains both qualitative and quantitative information on the analyzed species, in terms of the so-called "half-wave potential" and the "diffusion limiting current", respectively. This method has advanced since then, both in precision and automation, as have numerous other electroanalytical methods based on measurements of current, potential, quantity of electricity or impedance of the electrochemical systems. A record obtained in one of these methods - linear sweep voltammetry, shown in Figure 4, gives a quantitative estimate, in terms of the surface area under the curve, of as little as one hundredth part of one atomic layer of a metal (e.g. lead) on the surface of another (e.g. gold) while the peak-potentials reflect the energetics of their interaction. Such a level of detection has not been achieved in analytical chemistry by any other method known so far.

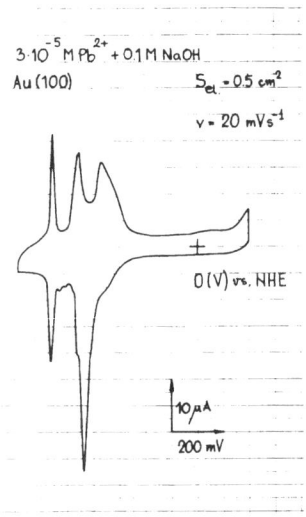

$3 \cdot 10^{-5}$ M Pb^{2+} + 0.1 M NaOH

Au (100) $S_{el} = 0.5$ cm^2

v = 20 mVs^{-1}

0 (V) vs. NHE

10 μA

200 mV

Fig. 4. Potential-sweep voltamogram showing the presence of a monoatomic
layer of lead on a gold surface.

Thus, electroanalytical chemistry is particularly suited for assessing
trace amount of elements or chemical compounds. Its other characteristic,
the sensitivity of particular electrochemical systems to specific substances, gives the possibility of building electroanalytical tools of
extraordinary selectivity. The range of such tools has recently been
greatly extended by the invention of the so-called "chemically modified"
electrodes. Special treatment of electrode surfaces, including e.g.
implication of specific enzymes, imparts sensitivity to substances no other
analytical tool could detect, at the same time converting the signals due
to their presence into electrical impulses.

The ease of automation, the selectivity and sensitivity of electrochemical systems, have led to a wide use of electrochemical sensors for the
majority of pollutants in the air and in natural waters, as well as in
other media affecting human life and environment.

All these characteristics have placed electroanalytical methods in the
forefront of chemical analysis. The equivalence between the amount of
electricity, which can be measured by electrical and electronic instrumentation and the results processed by most sophisticated computer-systems,
and the amount of chemical substance demonstrated by Michael Faraday as
long ago as in 1838, has resulted in a proposal that the unit of electricity, the Coulomb, be taken as the basic standard in in chemical analysis.

ELECTROCHEMICAL SYNTHESIS AND WINNING OF MATERIALS

Since the very early days of electrochemistry, electrolysis has been
used as a method of preparing substances, especially elements. For the
synthesis of many substances there have been two parallel routes available:
the chemical and the electrochemical one. The electrochemical route won,
and was developed to industrial scale, in the preparation of substances
having a very high tendency to react with an appropriate counterpart to
regenerate the state from which they were produced. These are the highly
oxidative or highly reductive substances, such as chloride, hypochlorite,
chlorate, perchlorate, hydrogen peroxide and a number of other per-salts,
manganese dioxide, chromium trioxide etc. on the one hand and hydrogen,

metallic sodium and other alkali and alkaline-earth metals, magnesium, aluminium and zinc on the other[8].

In recent years electro-organic synthesis has been gaining in importance[9]. The first to be carried out on an industrial scale was that of adiponitrile, the raw material for large scale production of nylon. However, the production of many more substances, in particular those of pharmaceutical interest, has been set up at many points around the world in limited volumes achievable with small-scale electrochemical reactors. A review of these products is given in Table 1.

Two electrochemical production processes, that of chlorine and caustic alkalis, and that of aluminium, exceed in volume and value all others by orders of magnitude. These processes indeed belong to a small group of major industrial processes of production of basic chemical materials.

The world production of chlorine amounted in 1983 between 28 and 30 million tons, with an equivalent amount (similar) of caustics. The chlorine-alkali electrolysis consumed in that year 90 billion kWh of electric energy and produced about 10 billion dollars in market value. It represents a striking example of the effect electrochemical science has on technology, since three breakthroughs in recent years have resulted in major improvements in its environmental acceptability and major energy saving.

The classical salt-brine electrolysis was based mainly on the mercury-cell, producing sodium amalgam and chlorine, the former being subsequently decomposed into sodium-hydroxide and hydrogen. The process represented a serious environmental hazard since tons of mercury were being released with the effluents of the plants into rivers and seas, accumulating in fish and thus endangering humans who use the fish as food. The alternative "diaphragm" process, in which chlorine and hydrogen are obtained by electrolysis gave poor quality hydroxide and had therefore, never achieved widespread use.

It was the development of ion-exchange membranes which opened a new era in brine electrolysis. These membranes have the extraordinary property of allowing the passage of electrolytic current and yet almost completely preventing the mixing of the hydroxide formed in one compartment of the electrolytic cell with brine introduced as raw material into the other compartment. Hence, a "membrane cell" could be developed for large scale industrial electrolysis, giving satisfactory quality of products without environmental hazard. This is likely to make the "mercury cell" vanish as the enormous investment into new installations pays off.

Table 1. Organic Compounds Synthesized Electrochemically on Semi-Industrial or
Industrial Scale

Tetraalkyllead
Sebacic Acid
Silicylaldehyde
Propylene Oxide
Cyanogen Bromide
Tetraethylammonium Hydroxide
Sorbit
Isoindole
Cyclohexanedicarboxylic Acid
Adiponitrile

Figure 5 represents brine electrolysis plants of the same capacity using the two different technologies.

Two other breakthroughs relate to energy saving in the process of electrolysis. The first has come in the form of dimensionally stable anodes (DSA) made of titanium, replacing graphite anodes. The latter are consumed during the process, thus requiring readjustments of the inter-electrode spacing. Titanium anodes not only do not change shape, but when coated with some catalyst (titanium oxide/ruthenium oxide mixture or a platinum/iridium alloy) enable electrolysis to be carried out with significantly lower voltage or higher current density (amounting to increased productivity of the electrochemical reactor) or both. The improvement is so significant that within the last ten years most electrolytic chlorine plants around the world have undergone reconstruction and have introduced DSA.

The second breakthrough is at this moment at the pilot-plant stage. It involves the replacement of the hydrogen evolving cathode with an air-consuming one, as shown in Figure 6. The difference in potential between the two, promises an energy saving as high as 30%.

(a)

(b)

Fig. 5. Brine-electrolysis installations (a) with mercury cells and (b) with membrane cells.

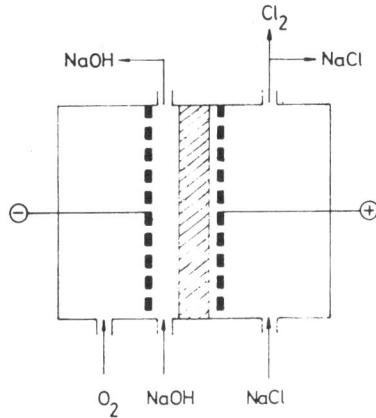

Fig. 6. Air-breathing membrane cell for electrolytic production of
chlorine and caustics.

Hence, the air-breathing membrane cell is the future of brine electrolysis and is likely to promote further significant increase in the
volume of production around the world.

ALUMINIUM

Aqueous solution and molten salt electrolysis from the bases of the
bulk of production and refining of non-ferrous metals. Thus winning zinc
and refining copper alone render 5.9 and 9.5 million tons per year respectively of these important metals.

Yet by far the most important electrometallurgical production is that
of aluminium amounting in 1982 to 14 million tons in the USA alone.

Aluminium is one the marvels of nature[10]. It has a very large
affinity for oxygen and hence can be regarded as a potent fuel. Indeed,
it is being currently used for special welding requiring very high temperatures, as well as for propulsion of some rockets for specific mission.

Yet it is commonly known to be a very good and stable construction
material as well, and so far this is its overwhelming use.

Figures 7 and 8 show two examples where aluminium is used in building
industry: The Place Ville Marie building in Montreal and a barn with
painted aluminium siding at some ranch in the USA. The lustrous appearance
and high resistance to corrosion is achieved by a special electrolytic
process of anodization. The automobile industry is making extensive use of
anodized aluminium, as exemplified by the automotive trim parts. In one of
its major uses, that for cans containing various beverages, it has already
penetrated every home practically all over the world.

Its resources on the earth are inexhaustible, for it is the most
abundant element after oxygen and silicon, accounting for 7.5% of the earth
crust. With the present technology only one mineral - bauxite, represents
a feasible raw material. This problem, however, is trivial in view of the
expected life of our civilization. So far we have had only one hundred and
seventy five years to develop the technology for its winning from the first
experiment made by Sir Humphry Davy.

Fig. 7. Use of anodized aluminium for panneling outer surfaces of
buildings: the Place Ville Marie building in Montreal.

Fig. 8. Barn with painted aluminium siding.

Even the technology of its extraction from bauxite is far from perfect. It was developed in a world unaware of the value of energy. Hence, it is anything but energy efficient. Yet, Figure 9 shows that significant advances have been made in the course of time and there is no reason to fear that further reduction in energy consumption will not be achieved in the future in the same or alternative processes for its winning.

The stability of aluminium in an oxygen containing atmosphere is due to a particularly tough, protective skin of its oxide, which it acquires at the very moment it comes into contact with air.

However, the protection by this skin can be overcome in at least two ways: by the action of alkalis and by the application of chloride-containing solutions. In both cases aluminium becomes an electrochemically active substance. Hence, it can be used e.g. for corrosion protection of sea-bound objects - ships, oil platforms, marine installations, - in the form of sacrificed anodes, in the same way as zinc has been used so far. Yet, it can also be used as an electrochemical fuel to build a fuel-cell type installation for obtaining electric power and this is likely to become one of its important uses[11]. Since aluminium is a solid fuel, the battery based on it belongs to the category of "mechanically rechargeable" batteries. One solution to the problem of refuelling such batteries is shown schematically in Figure 10, for the example of an aluminium-air cell with a saline electrolyte. The wedge shaped anode provides for the atomatic maintenance of a constant interelectrode spacing. The anode sinks as it is consumed and new aluminium-fuel is applied in the form of flat plates to the reserve anode compartment. Low frequency pumping of the electrolyte up and down the anodes results in even and smooth dissolution, with the reaction product, aluminium oxide, accumulating as a sand-like precipitate at the bottom of the electrolyte sump which is periodically discharged.

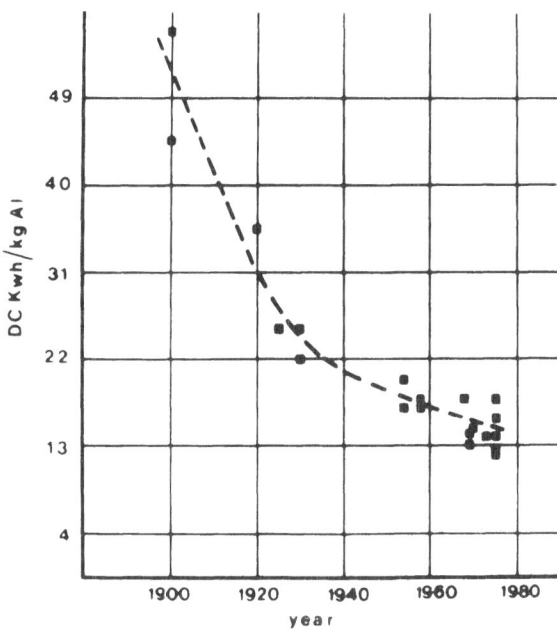

Fig. 9. Energy consumption in the Bayer-Hall process of winning of aluminium as a function of development.

The saline-electrolyte aluminium-air battery represents the most benign and environmentally acceptable electrical power source ever made. The cost of the metal consumed per kWh of energy is given in Table 2, compared to other candidates for metal-air batteries. It is seen to be significantly larger than the cost per kWh drawn from the network mains supply for domestic use. However, it is quite acceptable for any use where the electricity mains are not accessible e.g. for lights and communication equipment in sailing-boats, television relays on mountain tops, light-buoys at sea, remote camping-sites, and even as a main source of electric energy in households in underdeveloped parts of the world, where electricity supply is scarce. When calculated on mileage of an electric vehicle versus a standard gasoline driven automobile, the cost of energy from an aluminium-air battery compares acceptably with that obtained from high-octane fuel.

The penetration of the aluminium-air battery into general use depends on ingenuity of design and construction of reliable and cheap hardware, including the air-electrodes as the major component. One of the first publicly disclosed models is shown in Figure 11.

Electrochemistry is both a science and a craft. One aspect can hardly be separated from the other.

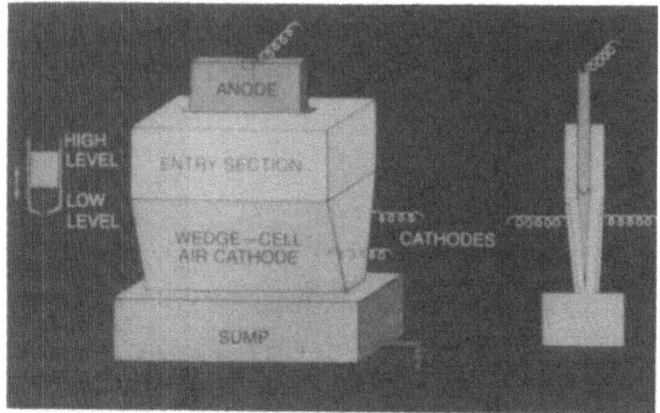

Fig. 10. Schematic representation of an aluminium-air cell with wedge shaped anodes.

Table 2. Cost of Metal Consumption per kWh of Energy Produced

Metal	$M/kg\ mol^{-1}$	$Q_M/\$\ kg^{-1}$	$\Delta G/kWh\ mol^{-1}$	$Q_W^o/\$\ kWh^{-1}$	η_v	$Q_E/\$\ kWh^{-1}$
Li	$6.94 \cdot 10^{-3}$	44.15*[1]	0.125	2.45	0.65	3.77
Na	$23.0 \cdot 10^{-3}$	2.052*[2]	0.116	0.407	0.58	0.70
Mg	$24.32 \cdot 10^{-3}$	2.958*[3]	0.158	0.455	0.51	0.89
Al	$26.97 \cdot 10^{-3}$	1.66*[4]	0.219	0.204	0.44	0.46
Zn	$65.38 \cdot 10^{-3}$	1.324*[5]	0.0884	0.979	0.60	1.63

*[1](99.9%) Metal Bull., 6852 (10.1.1984)
*[2]Chem.Marketing Reporter (2.1.1984)
*[3]Metalstatistik 1972-82, Metallges.Ag., Frankfurt a.M., quotation for 1982
*[4](99.7%) Metal Bull. 6850 (30.12.1983)
*[5](99.95%) Metal Bull. 6850 (30.12.1983).

Fig. 11. A model of an aluminium-air battery developed by the Alcan Corp.

This brief review of the achievements of electrochemistry in building up our technical civilization could give little more than a hint regarding its true potential. When the spectrum of its application has broadened so much from its emergence 150 years ago until today, there is no reason to be sceptical concerning the birth of new developments in the future, hidden at present in the yet unborn minds of future generations. Even within our life-time we have witnessed the appearance of unthinkable new discoveries and inventions. There is no justification for believing that our life-time is in any way exceptional, and that future generations will not be equally ingenious for further developments on a similar scale.

Acknowledgement

The author is indebted to Mr V. Jović for his help in the technical preparations of this contribution.

REFERENCES

1. A. C. Clarke, "Mysterious World," W. Collins & Sons Co. Ltd., London (1980).
2. C. J. Raub, Electroplating, in: "Comprehensive Treatise in Electrochemistry," J. O'M. Bockris, B. E. Conway, E. Yeager and R. E. White, eds., Vol.2, Ch.7, Plenum Press, New York (1981).
3. E. Budevski, Metal deposition and dissolution, Part A: Electrocrystalization, and A. R. Despić, Part B: Mechanism, kinetics, morphology and texture, in: "Comprehensive Treatise in Electrochemistry," Vol.7, Ch.7, Plenum Press, New York (1983).
4. F. Beck, Electrodeposition of paint, in: "Comprehensive Treatise in Electrochemistry," J. O'M. Bockris, B. E. Conway, E. Yeager and R. E. White, eds., Vol.2, Ch.10, Plenum Press, New York (1981).
5. I. P. Hoare and M. A. Laboda, Electrochemical machining, in: "Comprehensive Treatise in Electrochemistry," J. O'M. Bockris, B. E. Conway, E. Yeager and R. E. White, eds., Vol.2, Ch.8, Plenum Press, New York (1981).
6. M. Barak, ed., "Electrochemical Power Sources," Institution of Electrical Engineers, London (1980).
7. A. J. Bard, ed., "Electroanalytical Chemistry," Marcel Dekker, New York (1969).

8. A. Kuhn, ed., "Industrial Electrochemical Processes," Elsevier, Amsterdam (1971).

9. M. M. Baizer, ed., "Organic Electrochemistry," Marcel Dekker, New York (1973).

10. W. E. Haupin and W. B. Franks, Electrometallurgy of aluminum, in: "Comprehensive Treatise in Electrochemistry," J. O'M. Bockris, B. E. Conway, E. Yeager and R. E. White, eds., Vol.2, Ch.5, Plenum Press, New York (1981).

11. A. R. Despić, Electrochemical Energy Conversion, in Proc.29th Cong. IUPAC, Köln (1983).

ELECTROCHEMISTRY AS TRANSFER AGENT TO TECHNOLOGY

N. Hackerman

Rice University
Houston
Texas

Starting with Galvani there has been a tendency for quick application of the effects of moving charged particles across interfaces to practical use. Indeed, this tendency has from earlier times been a boon to technology. The latter is defined in Webster's Unabridged Dictionary as "the science of the application of knowledge to practical purposes". The dictionary correctly recognizes what we call science as our understanding of nature. Thus the field of electrochemistry even in its narrowest sense has been long active in quickly turning its science to use.

An early direct use by a distinguished scientist appears in a series of papers in the Philosophical Transaction of the Royal Society of London in 1824-25. Sir Humphry Davy, in papers read on January 22 and June 17, 1824 and June 9, 1825, described what is probably the earliest version of catholic protection. He records his experiment in the use of zinc to help maintain the integrity of copper in seawater of copper-sheathed hulls of British warships of the day. To quote from one of these papers:

> "The Lords Commissioners of the Admiralty, with their usual zeal for promoting the interests of the Navy by the application of science, have given me permission to ascertain the practical value of these results by experiments upon ships of war; and there seems every reason to expect (unless causes should interfere of which our present knowledge gives no indications) that small quantities of zinc, or which is much cheaper, of malleable or cast iron, placed in contact with the copper sheeting of ships, which is all in electrical connection, will entirely prevent its corrosion. And as negative electricity cannot be supposed favourable to animal or vegetable life: and as it occasions the deposition of magnesia, a substance exceedingly noxious to land vegetables, upon the copper surface: and as it must assist in preserving its polish, there is considerable ground for hoping that the same application will keep the bottom of ships clean, a circumstance of great importance both in trade and naval war".

Without going into a fully referenced work on the history of electrochemistry, it is easily possible to describe numerous other applications.

The ubiquitous dry cell in its long-standing form used zinc, ammonium chloride, manganese dioxide and carbon. Some more exotic recent systems of this kind involve very high power density, short life packages for special military purposes. Similarly, the widely used rechargeable systems using lead or nickel are also examples of the art of electrochemistry in everyday use.

To the list there needs to be added the very important contribution to the conversion of natural inorganic materials to useful metals by electrochemical means. Aluminium, magnesium, copper, sodium, and lithium are among the metals obtained by electrowinning. Add to this the many metals applied to solid - usually metal - surfaces by electroplating. Until the more recent vapor deposition processes, electroplating competed only with hot dipping as the means to produce duplex systems.

The production of several vital, inorganic chemicals has long been carried out via electrochemical processes. These include chlorine and sodium hydroxide as well as related products like sodium hypochlorite. A very few organic chemicals have been produced commercially by electrochemical means. In recent years many more have been synthesized in laboratories, in large part because of improved control of potential plus better insight into electrodes as catalysts. Controlled electrochemical steps in commercial organic synthesis are now clearly in prospect.

Electroanalytical procedures have long been in use but recent instrument developments make for new procedure plus very helpful automation. This provides for effective and beneficial use of these methods both for programing and for rapid, accurate quality control.

There still remain other examples of the science of electrochemistry deeply involved in technological functions. Electrochemical machining was a key to turbine bucket manufacturing for aircraft engines, here the science elicited in the study of corrosion processes in conducting media paid off handsomely. More recently applications in the health field of our experience with ion transport through membranes has been fruitful. This line may well have started with the electrochemists' long time interest in how the electric eel functions.

The implication being made here is that electrochemical science transfers readily into technology, i.e. is an effective agent for this often difficult process. In fact, there is a long history of technology in the field, and it may well have inhibited earlier insight into a basic understanding of moving charged particles, especially across interfaces between conductors of electrons and conductors of ions.

Certainly conduction in less facile conducting elements and in solid compounds was not fully appreciated until just recently. Indeed, the ease with which a little science could be turned into industrially useful items - batteries, metals, etc. is what has made the field so fruitful. That is technology followed small insights readily. So this has been a good transfer agent which may well have retarded electrochemical science until just a few years ago.

In fact, though, happily the science has progressed remarkably in the last twenty years. It is arguable that electrochemistry-based technology has not developed nearly as rapidly in this period as compared to earlier times.

The latter is pure conjecture but a general - quite simple - model for a system which connects ignorance to use may be helpful here. It involves three sequential letters of the alphabet:

S → T → U

Namely, S is science, the ever increasing understanding of nature's forces and process; T is technology, the application of this understanding to lever our human capability – both physical and mental; and U is ultimate use, these items which take root in the marketplace. There are, of course, numerous feedback systems and interweaving but basically the model says S is basic to the remainder.

The coupling between the phases is all-important in understanding the system. The first one on the left is very fluid and frequently shows little direct connection except on looking backwards. Electrochemical science seems to have been out of line in this regard. The second coupling is direct and quite solid. It is market driven and this in turn is need driven. Where needs are well met, of course, the drive becomes that of demand which can include a good deal of non-necessity.

The feedback to S is not overly complex. Everything done to the right of S exposes some ignorance (knowledge gaps). Thus science is called on to reduce our recognized ignorance as well as to delve into pure ignorance.

It is interesting to note once again that in the early days of electrochemistry a little science went a long way. Indeed, the profound scientific advances of the last two decades in the field have yielded quite a bit of technology but proportionately not nearly as much as in the early years. This, of course, may be characteristic of any field. Still, the large technological impact of early electrochemists may have predisposed electrochemists to think practically to a greater extent than have scientists in other fields of chemistry.

The close follow-on of a technology to the science appears clearly in some modern salients into our ignorance. Two good examples are very high speed intergrated circuits and genetic engineering. In any event, the field of electrochemistry has been and continues to be a fertile area for good scientists and engineers to work in and have impact both on human affairs generally and on science more narrowly.

SECTION I
ELECTROCHEMISTRY AND ENERGY

SOLAR ENERGY CONVERSION THROUGH PHOTOELECTROCHEMISTRY

A. J. Bard

Department of Chemistry
The University of Texas
Austin

INTRODUCTION

Of the many problems facing mankind, those concerning the availability and distribution of energy will ultimately be the most important. As fossil fuels become depleted, we will turn more and more to alternative sources and eventually depend upon energy technologies based on solar energy and perhaps fusion. Moreover, an important standard of living in undeveloped nations will require an abundant, inexpensive, and decentralized energy source, solar energy seems ideally suited for this application. It is the thesis of this paper that electrochemical methods will be useful in the conversion of solar energy to electricity or useful chemical products. Indeed, photoelectrochemical cells utilizing semiconductor materials have demonstrated the highest efficiencies in systems for the sustained conversion of solar energy to useful chemicals.

Let us first try to put the energy problem in perspective. While it is difficult to project exactly when fossil fuels will be depleted to the extent that they will no longer be a viable energy source, it is clear that eventually this depletion will occur. Consider the time scales involved in the production of fossil fuels and in their usage. This can be illustrated by a "world calendar" such as that shown in Figure 1. In this calendar, the time scale from the origin of the solar system to the present day is represented as one calendar year[1]. On this calendar, the earth is formed on January 1 (4.5×10^9 years ago) and the present time is midnight on December 31. On this time scale, each one second tick of the clock represents about one hundred and forty years and each month, about 375 million years. On this calendar, Homo sapiens appears at about 11:50 pm on December 31, and the Cambrian period, when land plants and animals began to flourish, starts on November 15. The major deposition of the materials which produced fossil fuels, which depended upon photosynthesis and the production of biomass, occurred over millions of years or, on the scale of the calendar, during the month of December. The industrial revolution, when large scale consumption of fossil fuels occurred, began only two seconds ago. Thus, representing the world history as shown, we have essentially consumed one month's worth of solar energy stored as chemical energy in the form of fossil fuels in a couple of seconds.

It is difficult to predict exactly when fossil fuels will be depleted, but it is clear that as the world population swells to six billion persons

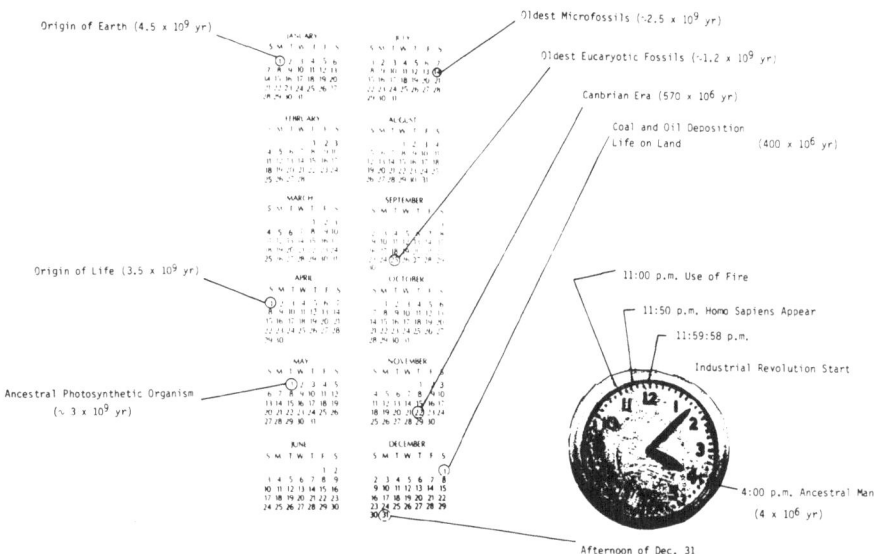

Fig. 1. The World Calendar representing different times in the history of the earth.

by the end of this century and to between nine and twelve billion persons about one century from now, energy usage will grow beyond current rates. Estimates have been made for the total amounts of oil and coal available[2], and most estimates, even assuming a low annual growth rate in consumption and optimism about available reserves and recovery, make it unlikely that our fossil fuels will last beyond the end of the next century (see, for example, Figure 2)[3,4]. As Bartlett[4] has said, "The rise and fall of the world's rate of consumption of fossil fuel resources is like the flame of one match in the long night, a delta function in darkness." Actually, long before natural gas, oil, and coal are depleted, they will cease being used as fuels in energy conversion because of their high value in the synthesis of useful chemicals, such as plastics, fibers, and pharmaceuticals. What sources of energy will be available in the post-fossil period? The only known ones that seem possible of making a significant contribution are nuclear fission, nuclear fusion, and solar energy in its various forms. Nuclear fission also relies on finite resources and involves significant problems in its widespread utilization. While it can delay our need for other energy sources, it will not be a long term solution. Controlled fusion has not yet been demonstrated, and its utilization as an energy source will undoubtedly involve major materials and engineering problems[5]. Solar energy, either through available natural sources such as biomass and wind or via devices for the production of heat, electricity, or chemicals, appears to be the best option for the long term. The widespread utilization of a solar technology, now and in the near future, primarily depends upon the construction of inexpensive systems based on materials that can be used to cover large areas. The most effective systems for the production of electricity and the promotion of chemical reactions which produce fuels (e.g. hydrogen) or other useful products by the utilization of solar energy are based on semiconductor materials. During the past fourteen years, such materials have been investigated in photoelectrochemical (PEC) cells. The principles of utilizing semiconductors as electrodes in such cells have been established and many systems

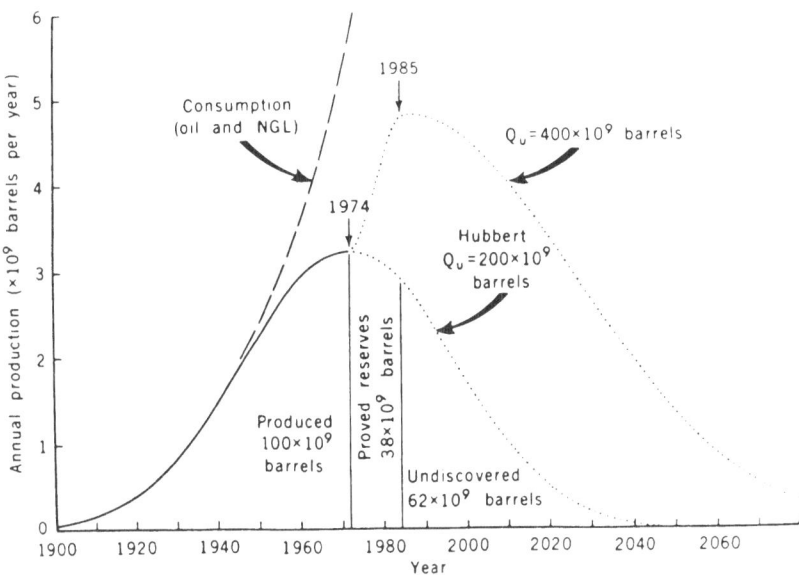

Fig. 2. Total oil production history for the United Stated assuming an ultimate recovery Q = 200 x 10^9 barrels and possible recovery of 400 x 10^9 barrels. The consumptive curve includes crude oil and natural gas liquids (NGL). From R. R. Berg, J. C. Calhoun, Jr., and R. L. Whiting, Science, 184:331 (1974).

have been described for carrying out different chemical reactions. In addition, rather simple particulate systems based on the principles of these semiconductor PEC cells have been described. The remainder of this paper deals with a brief introduction to semiconductors and the basic principle of photoelectrochemical cells. An overview of the types of reactions that have been carried out with these systems and the state-of-the-art will be given. Finally, I will discuss the future tends in research and development in this area.

SEMICONDUCTORS

Before discussing the photoelectrochemical cells, a brief review of the nature of semiconductors is perhaps desirable. The electronic properties of solids can be described in terms of a band model which treats the behavior of electrons moving in the field of the atomic nuclei. When a solid is formed, the isolated atoms, which are characterized by filled and vacant atomic orbitals are assembled into a lattice containing about 5×10^{22} atoms cm^{-3}. This leads to the formation of new molecular orbitals which are so closely spaced that they form essentially continuous bands. The valence band involves the bonding electrons and the filled bonding orbitals, while the conduction band is made up of the vacant antibonding orbitals (Figure 3). These bands are separated by a forbidden region, or bandgap, with an energy, E_g. The electronic properties of a solid depend upon the size of this gap. When E_g is small compared to kT or when the conduction and valence bands overlap, the material will be an electronic conductor (e.g. metals like copper and platinum.) When the gap is very large, for example 6 or 7 eV, the material is an insulator (e.g. SiO_2 Al_2O_3). Materials which have intermediate gaps, for example, in the range of 0.7 to 3.5 eV, can be considered as semiconductors. The resistance of these materials depends upon E_g and also upon the extent to which these

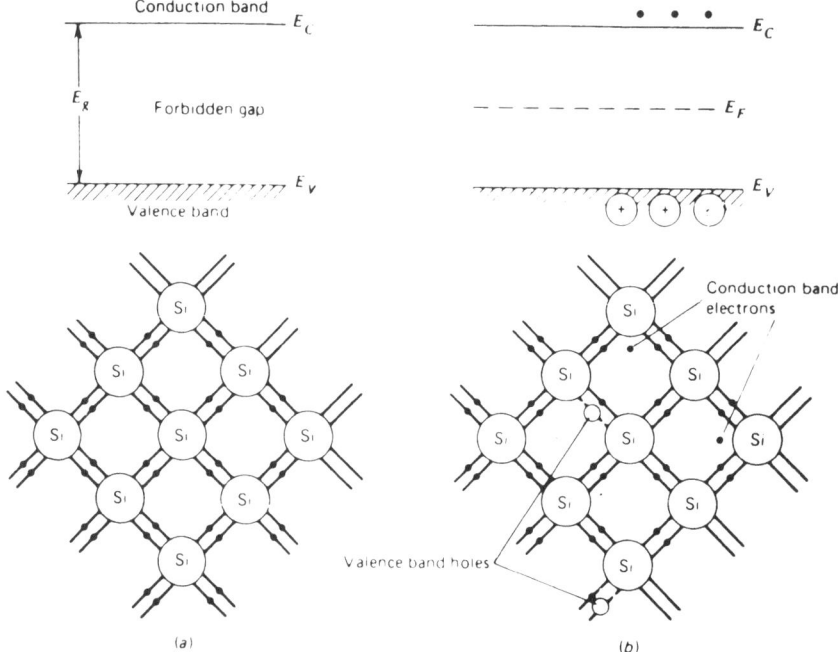

Fig. 3. Energy bands and two-dimensional representation of a semiconductor lattice.

are doped with various impurities. Typical semiconductors that have been
employed in photoelectrochemical cells are listed in Table 1. The color of
the material, which indicates what portion of the solar energy spectrum is
absorbed, is determined by E_g. The visible region of the spectrum ranges
from about 1.5 eV (red) to about 3 eV (violet.) Semiconductors with E_g
below 1.5 eV, such as Si and GaAs, appear black, since essentially all of
the visible light is absorbed in them, while those with E_g's above 3 eV,
such as TiO_2 and $SrTiO_3$, appear white because they reflect the visible
spectrum and only absorb in the ultraviolet. For solar energy conversion,
the optimum E_g is about 1.1 to 1.3 eV. For practical utilization on a
large scale, the semiconductor must be composed of abundant and inexpensive
elements and be readily prepared in a form and at a level of purity consistent with good efficiencies. Silicon, for example, meets many of these
requirements and has been widely used for electronic devices such as
integrated circuit chips. However, in the latter applications, only tiny
amounts of silicon are required for each chip, so the cost of the needed
high purity silicon and the expense in growing large single crystals are
not major factors. For solar devices, large areas are needed, since solar
radiation is diffuse. For example, in the southern United States, the
insolation or solar flux at the earth's surface (averaged over 24 years for
the whole year) is about 230 W m^{-2} so that at a ten percent conversion
efficiency, only 23 watts of useful energy can be obtained per square meter
of semiconductor surface. Considered in another way, the total solar
influx in Paris is about 1000 kW-h m^{-2} y^{-1}. Since the per capita usage of
all energy in Western Europe is about 2.9×10^4 kW-h y^{-1}, it would require
about 290 m^2 of semiconductor material to take care of the needs of each
person via solar energy devices, assuming again a ten percent conversion
efficiency. This need for large areas will almost certainly require that
the semiconductors be in the form of thin films or small particles. While
some devices of this sort have been described, most of the fundamental work
on photoelectrochemical cells have utilized expensive, single crystal
materials. These have shown good efficiencies, but they are clearly not
practical for large scale utilization.

Table 1.   Some Semiconductors used in Photoelectrochemical Cells and their Bandgaps

	E_g/eV
Si	1.2
WSe_2	∿1.1
$MoSe_2$	∿1.1
GaAs	1.4
InP	1.4
CdTe	1.4
CdSe	1.8
GaP	2.2
Fe_2O_3	∿2.3
CdS	2.5
WO_3	2.8
TiO_2 (rutile)	3.0
$SrTiO_3$	3.2
SnO_2	3.5

SEMICONDUCTOR PHOTOELECTROCHEMICAL CELLS

There have been a number of reviews of PEC cells based on semiconductors[7-13] and only a brief outline of the operating principles will be
given here. When light is absorbed in a semiconductor, it gives rise to an
electron-hole (e^-h^+) pair. This electron-hole pair represents reducing and
oxidizing power which can be utilized to carry out chemical reactions or to
cause a flow of current through an external circuit to produce electricity.
When a semiconductor is immersed in a liquid containing a redox couple, D,
D^+, the establishment of electrostatic equilibrium causes the formation of
an electric field at the interface. This electric field extends for
several hundred Å into the semiconductor and is called the space charge
region. This field is important because it serves to separate the photogenerated electrons and holes before they recombine (Figure 4). For
example, a photoelectrochemical cell can be prepared by immersing an n-type
semiconductor in a liquid and connecting it via an external circuit to a
metal electrode. When the semiconductor is irradiated, the holes produced
in the space charge region will move towards the surface of the semiconductor where they will oxidize D to D^+. The electrons move to the bulk
semiconductor and through an external circuit, creating a current flow and
causing a reduction reaction at the metal electrode. Thus, the photoelectrochemical cell can be used to transduce the radiant energy from the sun
into chemical energy in the form of products generated at the electrodes
and electricity. The separation of the e^-h^+ pair in the electric field at
the semiconductor-solution interface is similar to the process that occurs
at the p-n junction of a solid-state photovoltaic cell (for example, the
silicon solar photovoltaic cell). However, in the solid-state cells, light
pumps electrons through an external circuit, but no actual chemistry
occurs. In the liquid junction cells, chemical reactions occur at the
interfaces and charge is transported, as in most electrochemical cells,
by ionic movement in the liquid. One of the earliest cells for energy
conversion based on these principles was that described by Fujishima and
Honda[14] which utilized a single crystal of n-type TiO_2 as the semiconductor. Irradiation of this material with light in the ultraviolet
region of the spectrum resulted in oxygen evolution via the oxidation
water. TiO_2 has remained an interesting and widely investigated material,
since it is readily available and very stable. However, its bandgap is so
large that only about 5% of the solar spectrum is available with this
material, and thus, the practical efficiencies obtainable are probably too

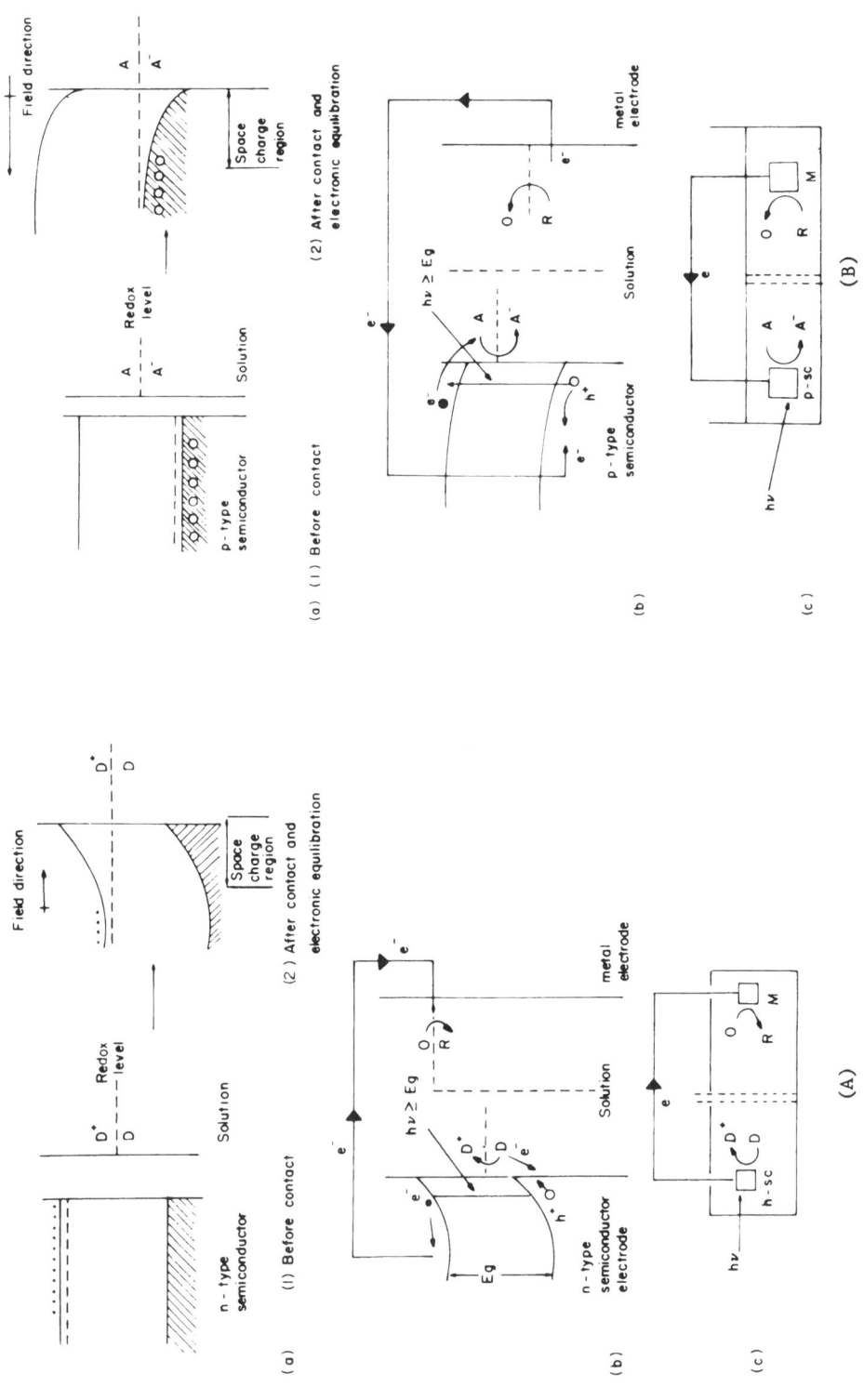

Fig. 4. Schematic representation of (A) n-type semiconductor photoelectrochemical cell. (B) p-type semiconductor cell.

low for many applications. A number of other semiconductors with smaller bandgaps have been investigated and techniques for producing photovoltaic cells which show reasonably stable operation have been devised. A representative sampling of these, along with their efficiencies, is given in Table 2. Note that the efficiency of a photovoltaic cell is defined as the amount of electrical power (watts) produced by the cell divided by the radiant power impinging on it. While these cells show good efficiencies and stabilities, most still utilize only small pieces of semiconductor material and at this time, are probably not competitive with solid-state photovoltaic cells in terms of cost and availability.

A more promising application of photoelectrochemical cells involves their use in the generation of desirable chemical species. While conventional electrochemical cells powered by solid-state photovoltaic devices can be employed in similar applications, the direct utilization of semiconductor immersed in the solution simplifies the construction of the device and, if good stabilities can be attained, probably represents the more desirable configuration. Cells of this type for the production of hydrogen and bromine or the evolution of chlorine have been described[15]. To determine what chemical reactions are possible in a semiconductor PEG cell, it is necessary to know the energies of the photogenerated electrons and holes. These are determined by the location of the conduction and valence bands as represented, for example, by potentials vs a reference electrode. These band locations can be obtained by electrochemical measurements, and along with the bandgaps, are useful in determining the effectiveness of a given semiconductor for carrying out a desired chemical process. It is possible to utilize an external power source, such as a battery, to supplement the radiant energy falling on the semiconductor and promote reactions which would not occur without external power. However, the use of external power sources complicates the design of the system and will probably not be important in practical devices.

Table 2.   Representative Liquid Junction Photovoltaic Cells

Semiconductor	E_g/eV	Redox System	Efficiency %	Reference
n-GaAs (xyl)	1.4	Se_2^{2-}, Se^{2-}	12 (solar)	1, 2
n-GaAs (poly)	1.4	Se_2^{2-}, Se^{2-}	7.8 (solar)	1, 3
n-CdTe (xyl)	1.4	Te_2^{2-}, Te^{2-}	10 (632.8 nm)	4
n-Si (xyl)	1.1	$Fc^{+/o}$ (MeOH)	10 (solar)	5
p-WS$_2$ (xyl)	1.3	$Fc^{+1/o}$ (MeCN)	7 (652.8)	6
p-InP (xyl)	1.4	$V^{3+/2+}$	9.4 (solar)	7

1. A. Heller, H. J. Lewerenz and B. Miller, Ber.Bunsenges.Phys.Chem., 84:592 (1980).
2. R. Noufi and D. Tench, J.Electrochem.Soc., 127:188 (1980).
3. A. Heller, B. Miller, S. S. Chu and Y. T. Less, J.Am.Chem.Soc., 101:7633 (1979).
4. A. B. Ellis, S. W. Kaiser and M. S. Wrighton, J.Am.Chem.Soc., 98:6418 (1976).
5. C. M. Gronet, N. S. Lewis, G. Cogan and J. Gibbons, Proc.Natl.Acad. Sci., 80:1152 (1983).
6. A. J. Ricco, M. S. Wrighton and G. P. Zosla, J.Am.Chem.Soc., 105:2246 (1983).
7. A. Heller, B. Miller, H. J. Lewerenz and K. J. Bachmann, J.Am.Chem. Soc., 102:6556 (1980).

The basic concepts that resulted from studies of photoelectrochemical cells with semiconductor electrodes have been applied to the design of systems in which semiconductor particles are used for similar purposes. Consider, for example, a particle of TiO_2 with platinum dispersed on its surface. The particle can be pictured as a "short-circuled" photoelectrochemical cell where photo-driven oxidation and reductions can still occur without any external current flow. These particles are obviously much simpler and potentially much less expensive than the semiconductors used in photoelectrochemical cells. However, they have the disadvantage that the oxidation and reduction sites formed upon absorption of a photon are now in relatively close proximity so that the probability of recombination is higher and the conversion efficiency lower; moreover, both oxidized and reduced products are produced together and come off in a mixture rather than being produced in separate product streams as in a PEC cell. A relatively early application of these particulate systems was the photo-oxidation of cyanide at TiO_2 particles[16] (Figure 5). While CN^- will react with oxygen in the dark, the reaction rate is very slow and in processes where cyanide is produced as a waste, oxidants such as chlorine must be used for cyanide removal. However, when a suspension of TiO_2 in the anatase (or pigment grade) form, is placed in a cyanide solution and irradiated with sunlight, a photocatalytic reaction occurs. The photo-generated holes oxidize CN^-, while the photogenerated electrons react with oxygen. In this case, light has been used to catalyze the reaction of CN^- and O_2. In this case, no metallization of the TiO_2 particles is required, since the reduction of oxygen occurs quite favorably on the TiO_2 surface. In other applications, however, platinization of the particle is necessary. For example, platinized TiO_2 can be used to photodecompose carboxylic acids in the so-called photo-Kolbe reaction. A typical reaction would be the photodecomposition of acetic acid to methane and CO_2. Platinization of the particle is necessary in this case because oxygen is not a reactant in the process. A number of other semiconductor materials, e.g. CdS and WO_3, have been used for such processes. Reactions carried out under these conditions include the oxidation of many organics with the production of hydrogen, the deposition of metals such as platinum and copper, photo-initiated polymerizations, and several interesting syntheses, including the reduction of amino acids in very small amounts from methane, ammonia, and water. A number of representative reactions of this type which occur at particles are given in Table 3. It is important to recognize that these particulate systems can be designed based on data obtained from photoelectrochemical cells with semiconductor electrodes. This concept of a particle as a microelectrochemical cell has been rather widely adopted and is also useful in other areas of heterogeneous catalysis[17].

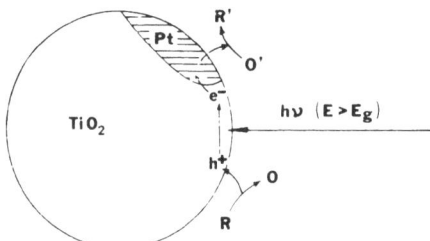

Fig. 5. Schematic representation of a photocatalytic reaction on a platinized semiconductor particle.

Table 3. Representative Heterogeneous Particulate Photoreactions

Reactants	Products	Semiconductor	Refs.
CN^-, O_2	CNO^-	TiO_2, ZnO, CdS	1
CO_2, H_2O	CH_3OH, HCHO	TiO_2, CdS, GaP	2, 3
H_2O	H_2O, O_2	Pt/TiO_2, $Pt/SrTiO_3$ $Pt/CdS/RuO_2$	4–6
CH_4, NH_3, H_2O	amino acids	Pt/TiO_2	7
many organics, H_2O	H_2, CO_2	Pt/TiO_2	8, 9
$\overset{Ph}{\underset{Ph}{>}}=$, O_2	$\overset{Ph}{\underset{Ph}{>}}=O$ (MeCN)	TiO_2	10
CH_3COOH	CH_4, CO_2	Pt/TiO_2	11

1. S. N. Frank and A. J. Bard, J.Phys.Chem., 81:1484 (1977).
2. T. Inoue, A. Fujishima, S. Konishi and K. Honda, Nature, 277:637 (1979).
3. M. Halmann and B. Ourian-Blajeni, Nature, 275:115 (1978).
4. M. S. Wrighton, A. B. Ellis, P. T. Wolczanski, D. L. Morse, H. B. Abrahamson and D. S. Ginley, J.Am.Chem.Soc., 98:2774 (1978).
5. A. V. Bulatov and M. C. Khidekel, Izr.Akad.Nauk.SSSR, Ser.Khim., 8:1902 (1976).
6. K. Kalyanasundaram, K. Borgarello, E. Grätzel, Helv.Chim.Acta., 64:362 (1981).
7. W. W. Dunn, Y. Aikawa and A. J. Bard, J.Am.Chem.Soc., 103:6893 (1981).
8. I. Izumi, W. W. Dunn, K. Wilbourn, F. Fan and A. J. Bard, J.Phys.Chem., 84:3207 (1980).
9. T. Kawai and T. Sakata, Chem.Phys.Letters, 80:341 (1981); Nouv.J.Chim., 5:279 (1981).
10. M. A. Fox and C. C. Chen, J.Am.Chem.Soc., 103:6757 (1981).
11. B. Kraeutler and A. J. Bard, J.Am.Chem.Soc., 100:5985 (1978).

INTEGRATED CHEMICAL SYSTEMS

While systems that show reasonable stabilities and efficiencies have been produced, practical photoelectrochemical systems will require less expensive materials and new structures. A concept that has emerged in recent years is the integrated chemical system[18]. By this we mean a designed heterogeous system involving several components (e.g. semiconductor, polymer, metal, catalyst) to carry out a specific process and often showing synergistic effects among the components. Fabrication of integrated systems is necessary because the semiconductor, which serves as the light absorbing vehicle, does not have good catalytic properties to promote desired chemical reactions at its surface. Moreover, good utilization and dispersal of the semiconductor will often require the use of another support, such as a polymer, with means of interconnection between the different semiconductor particles. Fabrication of these systems will often utilize techniques that are not generally familiar in chemical synthesis, such as spin coating, chemical vapor deposition, vacuum evaporation, and molecular beam epitaxy. Among the earlier integrated chemical systems described for photoelectrochemical cells were those involving polymer films containing the viologen moiety on a p-type semiconductor electrode surface and containing dispersed platinum[19]. Such systems were shown to be quite efficient for the evolution of hydrogen via photoreduction of the viologen to the cation radical and then catalysis of the reaction of the cation radical with protons by the platinum contained within the polymer layer.

Perhaps the most advanced integrated chemical system for solar energy conversion is the Texas Instruments Solar Energy System (TISES)[20]. In this system, small (0.25 mm) silicon spheres (n-Si photocathodes and p-Si on n-Si photoanodes) are embedded in a glass matrix and connected to a conductive backing; the basic concepts of this system are shown schematically in Figure 6. The glass matrix serves as a support for the small silicon spheres and also allows a better utilization of the silicon, since light reflected from the back plane as well as light directly impinging on the silicon sphere is useful in photogeneration of carriers. The microspheres are coated with noble metal films which act as electrolytic contacts to the solution, stabilize the substrate silicon from decomposition, and catalyze the electrode reaction. Electronic contact is made by the conductive back plane, and the spheres are exposed to a solution of HBr. Upon irradiation, hydrogen is produced at the photocathode and bromine is produced at the photoanode. The products are kept separated by an ion exchange membrane between the portions of the array containing photocathodes and those containing photoanodes. Thus, the arrays can be used to photodecompose HBr to hydrogen and bromine. In the TISES system, these are stored and subsequently recombined in a fuel cell to produce electricity. The arrays have the advantage of being potentially much less expensive than single crystal solar devices, since only small spheres of silicon are used and large single crystals are not required. Such arrays have shown efficiencies for generation of H_2 and Br_2 of about 8 to 10 percent. These arrays have another important feature which will probably be incorporated into many future photoelectrochemical devices. The maximum driving force for a single photo-active junction (e.g. semiconductor/solution or semiconductor/metal) is rarely more than 0.8 V and is often below 0.6 V. This is also true of the p-Si/n-Si junction in the solar photovoltaic cell. To cause energetic reactions (such as the decomposition of HCl to H_2 and Cl_2) to occur without an external bias, multiple photo-active junctions must be employed[21]. In the TISES array, two junctions are connected in series to produce an overall voltage of 1.1 V. In future systems, it would be desirable to produce multilayer devices with a number of photoactive junctions in series. These could involve, for example, an internal Schottky barrier involving a metal/semiconductor junction in series with a semiconductor/solution junction. It may be possible to learn how to couple together a number of junctions in series to produce much larger voltages. To take an example from biology, consider the electric eel. In this animal, individual nerve cells, which produce only about 100 ml each, are coupled in series to produce hundreds of volts.

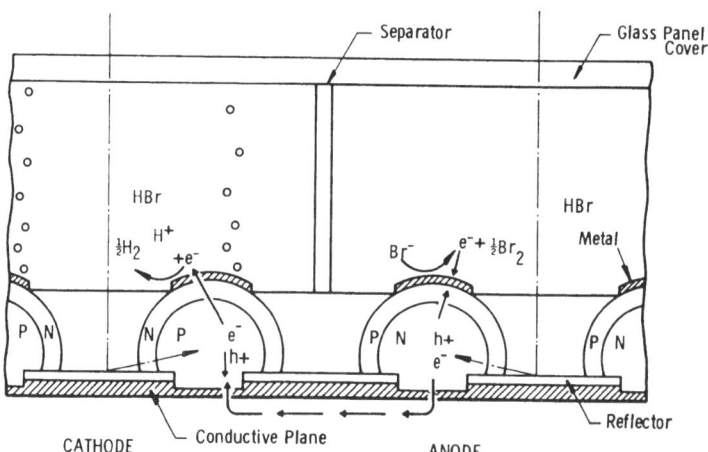

Fig. 6.   Texas Instruments solar array based on p-Si/n-Si spheres for photodecomposition of H_2 and Br_2.

Integrated chemical systems based on particle photocatalysis are also possible. The particle systems have several disadvantages in large scale utilization. Because they are dispersed in the solvent, they are inconvenient to use in continuous flow systems. Those with small dimensions, such as colloids, tend to flocculate and settle out with time. A group at the University of Texas has been working on means of immobilizing particles in a polymer matrix and incorporating appropriate catalysts for a desired reaction such as hydrogen evolution. For example, CdS can be immobilized rather easily in a polymer (Nafion) matrix[22,23]. Nafion, a perfluorinated cation exchange membrane produced by Du Pont, serves as the support. When this film is soaked in a solution containing Cd^{2+}, this ion is strongly bound at the anion sites in Nafion. When these films are then treated with H_2S gas, CdS precipitates inside the polymer matrix.

$$\begin{matrix} SO_3^- \, Na^+ \\ SO_3^- \, Na^+ \end{matrix} \; + \; Cd^{2+} \longrightarrow \begin{matrix} SO_3^- \\ SO_3^- \end{matrix} Cd^{2+} \; + \; 2Na^+ \tag{1}$$

$$\begin{matrix} SO_3^- \\ SO_3^- \end{matrix} \; + \; Cd^{2+} + H_2S \longrightarrow \begin{matrix} SO_3^- H^+ \\ SO_3^- H^+ \end{matrix} CdS \tag{2}$$

By appropriate regulation of the precipitation conditions and the use of multiple treatments with Cd^{2+} and H_2S, the amount of CdS and its location within the membrane can be controlled. It is also possible to incorporate redox couples, which serve as mediators or relays, within the membrane. For example, methyl viologen (MV^{2+}) can be incorporated by cation exchange after the CdS production to produce the system Nafion/CdS/MV^{2+}. In this system when the initial yellow membrane is irradiated, it turns blue, because electrons that are photogenerated in the CdS transfer to the MV^{2+} reducing it to MV^+. In this case, the photogenerated holes will oxidize the CdS material unless suitable sacrificial donors, such as sulfide ion, are also included in the solution. It is also possible to incorporate Pt into the membranes, for example, by soaking the Nation films in $Pt(NH_3)_2I_2$ and then treating with a $NaBH_4$ solution[23]. Such films when immersed in a 0.1 M Na_2S solution (pH=13) and irradiated show the evolution of hydrogen (Figure 7) for an overall reaction.

$$2S^{2-} \; + \; 2H_2O \; \xrightarrow[2h\nu]{CdS} \; S_2^{2-} + H_2 + 2OH^- \tag{3}$$

The method of deposition of the CdS and platinum and the crystalline form of the CdS material (α or cubic vs β or hexagonal) are important factors in the efficiency of such systmes. Future work on related integrated systems involving new supports (other polymers, clay or zeolites, glass and other ceramics) and new catalysts and semiconductors may lead to greatly improved and inexpensive systems for solar energy utilization.

FUTURE TRENDS IN RESEARCH AND DEVELOPMENT

The photoelectrochemical systems continue to be interesting ones from the scientific standpoint and potentially practical for the utilization of solar energy. While such systems may be used for direct electricity generation, the cost of solid-state photovoltaic devices utilizing thin films and amorphous materials has decreased rapidly in recent years and it seems likely that these will prove most useful for direct electricity generation. However, photoelectrochemical systems are particularly useful

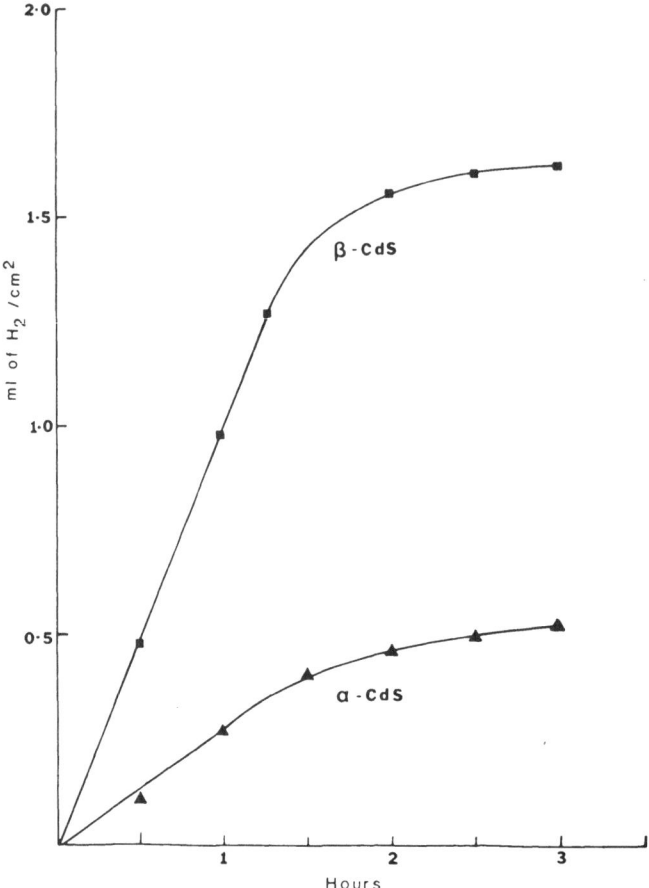

Fig. 7. Time dependence of H_2 production on a 1 cm^2 Nafion film containing CdS and Pt (0.02 mg cm^{-2}). (From ref. 23).

for carrying out chemical reactions. Those of particular interest for practical utilization are listed below:

1. Fuel production such as the splitting of water to produce H_2 or the reduction of CO_2 to produce formaldehyde and methanol. Solar energy utilized in this way has the advantage of yielding chemical energy in a form that is easily stored. For example, H_2 produced by this method can be used in place of natural gas or later utilized in fuel cells with a suitable oxidant such as oxygen to produce electricity. Products arising from the reduction of CO_2 will be liquid fuels which can be used in transportation.

2. Water purification. Semiconductor systems, such as particulate TiO_2, have already demonstrated the possibility of oxidation of contaminants such as C^-, SO_2, and organics in aqeous solutions in the presence of oxygen. Irradiated semiconductors are very potent oxidizing agents, since hydroxyl radical is produced in some systems. Semiconductor dispersions should also be useful for disinfection and the oxidation of even very resistant organic molecules.

3. Photofertilization via N_2 reduction. The reduction of N_2 to NH_3 has been demonstrated at semiconductors, although the efficiencies are very

46

low. This reaction, of course, is of enormous importance in the production of fertilizers and in intensive cultivation of crops. The current synthesis utilizes hydrogen and requires high temperatures and pressures. One can envisage the use of semiconductor systems for similar purposes. If a very inexpensive semiconductor which can carry out this reaction efficiently can be discovered, it might even be possible to utilize the semiconductor directly as a fertilizer and have the photofertilization (e.g. $N_2 \rightarrow NH_3$) occur in the field. In this way the semiconductor would replace nitrogen-fixing bacteria. A possible alternative is the oxidation of N_2 to NO_3^-.

4. Generation of Cl_2 from seawater. Chlorine is a widely used industrial chemical and is currently produced in the chloralkali process, which involves electrochemical cells and the consumption of large amounts of electrical energy. Photogeneration of chlorine at semiconductors at semiconductors has already been demonstrated and the efficient generation of Cl_2 (e.g. with the reduction of O_2) with an inexpensive semiconductor system would be very useful. One can even envisage small units for the production of Cl_2 for bleaching, disinfection, and chemical production.

5. New photosynthetic and photocatalytic reactions. As shown in Table 3, a number of reactions have been carried out at irradiated semiconductors. As energy costs increase, production of useful chemicals such as H_2O_2 and various organic molecules by this route may become competitive. It is probably not too far fetched to consider the possibility of production of amino acids and proteins by artificial photosynthetic systems. The solar efficiency for the growth of most crop plants is only about 1% in the field. The swelling world population may make it necessary to have chemical systems for food production that operate at higher efficiencies and semiconductor materials may well play a role in such systems.

What are the important areas of research in the field of photoelectrochemistry that may lead to practical systems? New semiconductor materials will certainly be important. The bulk of the research carried out so far has involved relatively few materials, such as Si, GaAs, TiO_2, and CdS. However, a number of binary and ternary semiconductor compounds exist which have never been investigated in photoelectrochemical cells. To aid in the search for new semiconductors, theoretical guidelines which can be used to predict E_g and carrier energetics will be particularly useful. Theoretical chemistry has reached the stage where calculations of these properties, at least for the bulk semiconductor, appear possible. Even semi-empirical guidelines or correlations would be useful in predicting what materials are worth more detailed investigation. Theoretical models which could describe surface properties of the semiconductors and predict their stability under irradiation would be particularly useful. New methods for the physical characterization of particle systems should be sought. A number of methods such as electron spin resonance with spin trapping, spectroscopy, electrochemistry, and electrophoresis have been used to characterize the particles. New methods which could probe more directly the photophoresis at the surface and the nature of the electron transfer reaction are needed. Methods for separating the particles and obtaining more monodispersed suspension would also be useful. Finally, the discovery of new catalysts, preferably ones not based on noble metals, for such reactions as oxygen reduction, hydrogen evolution, and CO_2 and nitrogen reduction would be of interest not only in the design of semiconductor PEC system, but also in electrochemistry in general. It is possible that homogeneous catalysts in solution or attached to the semiconductor surface could also be found. Advances in inorganic chemistry, such as the production of new cluster compounds, has lead to a much better understanding of homogeneous catalysis by transition metal species. Such species might also play the role as mediators and relays in photoelectrochemical systems.

Acknowledgements

The support of the National Science Foundation (CHE8304666), the Robert A. Welch Foundation and the Gas Research Institute (5082-260-0756) of some of the research described in this paper is gratefully acknowledged.

REFERENCES

1. C. Sagan, "The Dragons of Eden," Random House, 1977, used a similar calendar with the origin of the universe (15 billion years ago) set on Jan. 1.
2. M. K. Hubbert, "Resources and Man," Freeman, San Francisco, pp.157-242 (1969); "US Energy Resources," a Review as of 1972, Part I, Ser.No.93-40, (92-75), US Government Printing Office (1974).
3. R. R. Berg, J. C. Calhoun, and R. L. Whiting, Science, 184:331 (1974).
4. A. A. Bartlett, Physics Today, p.9 (1976).
5. R. W. Cahn, Nature, 266:106 (1977).
6. See, for example, S. M. Sze, "Physics of Semiconductor Devices," Wiley, New York (1981).
7. A. J. Bard, J.Phys.Chem., 86:172 (1982); A. J. Bard, J.Photochem., 10:50 (1979).
8. A. J. Bard, Science, 207:139 (1980).
9. S. R. Nozik, Annu.Rev.Phys.Chem., 29:189 (1979).
10. R. Memming, in: "Electroanalytical Chemistry," A. J. Bard, ed., Marcel Dekker, New York (1977).
11. M. Wrighton, Acc.Chem.Res., 12:303 (1979).
12. H. Gerischer, in: "Physical Chemistry: An Advanced Treatise," H. Eyring, D. Henderson, and W. Jost, eds., Academic Press, New York (1970).
13. S. R. Morrison, "Electrochemistry at Semiconductor and Oxidized Metal Electrodes," Plenum, New York (1980).
14. A. Fujishima and K. Honda, Bull.Chem.Soc.Japan, 44:1148 (1971); Nature, 238:37 (1972).
15. See, for example, A. Heller, Science, 223:1141 (1984).
16. S. N. Frank and A. J. Bard, J.Phys.Chem., 81:1484 (1977).
17. D. S. Miller, A. J. Bard, G. McLendon, and J. Ferguson, J.Am.Chem.Soc., 103:5336 (1981); M. Spiro, J.Chem.Soc., Faraday Translation, 1, 75:1507 (1979).
18. A. J. Bard, F. -R. F. Fan, G. A. Hope, and R. G. Keil, ACS Symp.Ser., No.211, 93 (1983).
19. H. D. Abruna and A. J. Bard, J.Am.Chem.Soc., 103:6898 (1981); D. C. Bookbinder, J. A. Bruce, R. N. Dominey, N. S. Lewis, and M. S. Wrighton, Proc.Natl.Acad.Sci., 77:6280 (1980).
20. E. L. Johnson, "TI Solar Energy System Development," IEEE Proc.IEOM, p.2, Dec. 1981; "Report to Dept. of Energy," DOE/TI-ER10000, June 19, 1980; J. S. Kilbey, J. W. Lathrop, and W. A. Porter, US Patents 4,021,323, 3 May 1977; 4,100,051, 11 July 1978; 4,136,436, 30 Jan. 1979.
21. J. R. White, F. -R. F. Fan, and A. J. Bard, submitted.
22. M. Krishnan, J. R. White, M. A. Fox, and A. J. Bard, J.Am.Chem.Soc., 105:7002 (1983).
23. A. W. Mau, N. Kakuta, A. J. Bard, A. Campion, M. A. Fox, J. M. White, and S. E. Weber, submitted.

DISCUSSION

Mary D. Archer

Department of Physical Chemistry
University of Cambridge
Cambridge

CHAIRMAN'S SUMMARY

The major themes explored in the discussion following Professor Bard's lecture on solar conversion through photoelectrochemistry were the likely timescale for the widespread utilization of solar energy and the practicability of the necessary accompanying shift in chemical industry from an intensive to an extensive mode of operation. In brief, the discussion provided a frank but at the same time encouraging appraisal of the long-term prospects for utilization of solar energy, in which photoelectrochemical devices would play a role.

Professor Nürnberg agreed with Professor Bard's view that, in the long run, solar energy would play a major role in energy technology. However, he was of the opinion that a pluralistic energy economy, in which solar played only a minor part, would persist throughout this century and well into the next. For solar to play a major role, it would be necessary first to improve the conversion efficiency of solar systems. Even if that were achieved, solar would be economic only in certain geographic location. This implied a complete reorganization in the energy economy of the world, which must act as a brake on the implementation of solar technology. There were many political and economic aspects involved in any major shift in the world's primary energy sources.

Professor Bard, in agreeing with these remarks, commented that the growth of the world's population may mean that the means of energy distribution will become more important. Here the disseminated nature of solar energy has clear advantages, particularly for the less developed areas of the world. For example, chlorine required for water purification might be made in situ, or at least locally, by photoelectrochemical means, rather than centrally by conventional technology.

Professor Dell commented that such application, although valuable, would make only a very minor contribution to global energy demands. Professor Bard responded that he saw solar as the eventual means of world electricity generation, possibly by photovoltaic cells rather than by photoelectrochemical means. There was, moreover, the important market of transportation and the continuing need for liquid and gaseous fuels. Solar photoelectrochemical carbon dioxide fixation and hydrogen and methanol production were important long-term prospects.

49

The importance in conventional chemical industry of maximizing the space-time yield was emphasized by Professor Despic: the diffuse nature of solar irradiance was incompatible with this requirement. It would, for example, be possible to save a great deal of electricity simply by running chloralkali plants at lower current densities. This was not done simply because it was not economic. Professor Bard agreed that the low power density of solar energy implied a major shift from intensive chemical industry to something more like energy farming. Although the areas required for solar to make a significant contribution were large, they could be found within desert areas and by replacement of some conventional construction (roofs etc.) with solar collectors.

There was some discussion of the relative merits of, and the distinction between, coupled photovoltaic-electrochemical systems for the photoelectrolysis of water and other substances and integrated photoelectrochemical systems for the same purposes. The former is a much more highly developed technology and, in Professor Gileadi's opinion, photoelectrochemistry will be unable to displace photovoltaics for electricity production. As regards the relative prospects of coupled photovoltaic-electrochemical systems and integrated photoelectrochemical systems, it was evident that economics would dictate. Professor Parsons remarked that particulate systems were a promising feature of the latter area but that in general the former seemed simpler.

The nature of the Texas Instruments device (Figure 6), on which development work has now ceased, was discussed. It contains buried p-n junctions and may thus be regarded as a set of microphotovoltaic cells coupled to a conventional HBr electrolyzer. Professor Bard preferred to regard all devices in which the light-sensitive semiconductor is immersed in electrolyte as photoelectrochemical in nature, but commented that the issue was partly a semantic one. As regards the location of the photoelectrochemical junction crucial to the separation of light-generated charges, there was a continuum from cases in which it lay immediately at the solution - electrode interface to those in which it was definitely buried. For example, silicon photoelectrode could be effectively stabilized against corrosion by a thin (ca. 5 mm) film of platinum silicide, produced by annealing a deposited layer of platinum. The photovoltage generated by such an electrode is independent of the redox couple in solution, so the effective junction is the buried silicon - platinum silicide interface.

Professor Fleischmann queried the energy payback period for photovoltaic devices. Professor Bard commented that considerations of both the energy and the economic payback period dictated the use of inexpensive thin-film devices containing only small amounts of semiconductors. But even now, crystalline solar cells of conventional design had a practicable payback period in some circumstances. The Solarex Corporation, for example, was able to meet all the energy required by their manufacture of silicon solar cells from roof-mounted panels of such cells.

The question of turn-over numbers of the photocatalyst in particulate systems was raised. Professor Bard commented that these were usually very good, even for difficult processes such as photoinitiated polymerization, which normally cause extensive electrode fouling. It is possible that microparticulate semiconducting photoelectrodes are self-cleansed of surface poisons by strongly oxidizing intermediates, such as hydroxyl radials, produced in the course of the photoelectrochemical process. However, the efficiency (i.e. the quantum yield of photoproduct) of such devices was still very low. Until this was improved, such devices could not be practical and there would be little incentive to improve specificity.

Dr. Archer commented that mass transport to spherical microelectrodes was much more efficient than semi-infinite linear diffusion to planar macro-electrodes, and asked whether this advantage was lost if the microparticles were embedded in a polymer matrix, as described in one part of Professor Bard's talk. Professor Bard replied that this was the case, and added that mass transport to and from, and within, polymer matrix films is now the limiting factor controlling reaction rates in such systems. It will probably be necessary to imitate photobiological systems in the use of ultra-thin membranes to avoid these mass-transport limitations.

Prof. E Gileadi
Tel Aviv University, Israel (communicated)

1. It would seem to me premature to decide that photoelectrochemistry will not be competitive with solid state photovoltaic devices. Although the latter are well ahead at present, it should be remembered that photoelectrochemistry is a relatively new field and the amount of money and man-years spent on research is one or two orders of magnitudes less than that put into the development of solid state photovoltaic devices.
2. The idea of employing photoelectrochemical devices for the industrial production of chemicals by solar energy is very attractive, but probably not very practical. The diffuse nature of solar energy makes it a poor source of power for industrial processes, where the space-time yield and the related cost of investment is a major factor. To illustrate this point consider the world production of chlorine, which amounts to about 30×10^6 tons a year, using 100×10^6 MWh. Decreasing the current density in the electrolytic cells by a factor of two, would lead to a saving of 25 - 30% of the energy. The space-time would decrease by a factor of two and the cost of investment would increase by almost the same factor, making such a saving uneconomical at present and in the foreseeable future.

ELECTROCHEMISTRY AND ENERGY,

ENERGY CONVERSION AND STORAGE

K. Wiesener

Technical University
Dresden
GDR

The rapid development of engineering and the increasing automation
are connected with the full electrification of all technical processes.
This development initiated the demand for high-capacity and mobile power
sources. The penetration of electrical engineering and electronics into
all spheres of economy and of the whole social life is connected with the
use of adequate sources of electric energy. In many cases, independent
carriers of energy are required for mobile applications. As an autonomous
source of electric energy in the range of low and medium power, the electrochemical power source is the most important system for technical and
economic reasons.

Electrochemical power sources are applied in three fields - generation, storage and transport of electric energy.

1. The generation of electric energy can be demonstrated most evidently by the electrochemical fuel cell. After its intensive development
in the sixties, it has been improved technically, but has not yet been used
on a large scale for economical reasons. With respect to the generation of
energy without environmental pollution in connection with the concept of
hydrogen economy it could gain significance in the future.

2. The storage of energy in accumulators is frequently connected with
the requirement that these power sources also have to serve as energy
carrier and supply the energy required for mains-independent operation of
small-size and medium-sized electric systems, esp. for electric vehicles.
They are also important for stationary energy storage in emergency devices
and for load leveling. For these purposes it is necessary to develop
systems with high energy densities, high energy efficiencies and long shelf
life with readily available materials.

3. Lately the importance of the energy carrier function has considerably increased. The point is to develop power sources of a high specific
energy density in the lowest power range for microelectronics as well as in
high power ranges for electric traction.

In the last decade there has been an important development in the
field of electrochemical power sources, solid and liquid aprotic, electrolytes have taken a place in electrochemical power sources. With their use
for heart pacemakers a new and very attractive field in medicine was

created. Recently, a very intensive discussion on organic compounds as electrode materials was initiated among scientists. The property of these systems to store energy, which is closely connected with their electronic conductivity, could be of interest for technical applications in the future. As charge transfer complexes they have already proved to be useful in engineering.

As I have shown by the few examples mentioned, the development of electrochemical power sources is closely connected with the further development of society.

ELECTROCHEMICAL SYSTEMS IN ENERGY

CONVERSION AND STORAGE

V. S. Bagotsky

A. M. Frumkin Institute of Electrochemistry
Academy of Science of the USSR
Moscow, USSR

INTRODUCTION

One of the old definitions of electrochemistry says that electrochemistry is the science of converting chemical energy into electrical energy and electrical energy into chemical energy. To-day this definition covers only a part of the problems of electrochemistry and is not all-embracing. But it correctly reflects the fact that for nearly 200 years electrochemical processes of energy conversion have been playing an exceptional role in science and technology.

When at the very beginning of 1800 the Italian physicist Alessandro Volta constructed his device known as "Volta's pile" (Figure 1) he not only created the first electrochemical device for energy conversion but made mankind realize the existence of such natural phenomenon as the "electric current." Before his experiments the only practical source of electricity was electrostatic generators based on induction of charges and their accumulation. Though these generators set up high voltages on the coatings of Leyden jars (tens of thousands of volts) and produced spark discharges, the electrical charges they guaranteed were virtually infinitesimal – some tens of microcoulombs. No appreciable electric current could be produced by these devices.

Just a few months after the appearance of the Volta pile it was found that the electric current can exert a chemical action. As early as May of 1800, Nicholson and Carlisle carried out water electrolysis. In 1803 the processes of metal electrodeposition were discovered. In 1807 Davy for the first time isolated alkali metals by electrolysis of salt melts. Thus almost simultaneously with the creation of the first electrochemical power source – the "galvanic cell" or "galvanic battery" – many electrochemical processes were discovered and the foundations were laid of the science which to-day we call electrochemistry.

Galvanic cells and batteries also played a decisive role in the development of such branches of physics as electrodynamics and electromagnetism. They were used in experiments resulting in the discovery of the well-known laws of Ampère, Ohm, Joule, Faraday and many others. The existence of galvanic cells favored practical application of electric current. Electrical telegraph and electric motors were invented and galvano-plastics was developed.

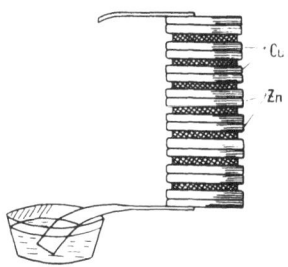

Fig. 1.   Volta's pile

It should always be remembered that during the entire first half of last century electrochemical systems for energy conversion were the sole source of electrical power both for practical purposes and for scientific investigations. For this reason much attention was given to their improvements. A great variety of cells were proposed, such as Daniel's, Bunsen's, Grenet's cells and many others. They were in such great demand that every time a new effective version was suggested its mass production was organized in such a short time that seems amazing even according to modern concepts. Thus, for instance, three years after the French engineer Leclanché created the first manganese-zinc cell their number exceeded 20,000.

No wonder that in the middle of last century the well-known French physicist Arago said that the galvanic cell "is the most wonderful device of all, invented by man, including the telescope and steam engine."

Galvanic cells lost their exclusive position as sole electrical power sources during the second half of the last century when electromagnetic electric current generators were invented. The appearance of these generators allowed heat engines working on fossil fuel (coal, oil, etc.) to be used for production of electrical power. Thus the first thermal power plants were built and at the end of the last century electrical power engineering started to develop rapidly, electric power came to be widely used in homes and industry and large-scale power systems came into being.

In spite of the fact that beginning with the seventies of the last century galvanic cells have ceased to be the only sources of electrical power, it was during this period that their development received a strong impetus and is still processing at a rapid pace. In the initial period, primary cells, and later, storage cells and batteries came to be used for various communication facilities and portable devices. A strong increase of their manufacture in the first quarter of this century was connected with the development of radio engineering and automobile transport. It was in that period that the first serious research in the field was undertaken. Further increase of interest in primary and storage cells is associated with progress in aviation and space engineering and also in microelectronics.

From the energetic point of view all the above-mentioned fields of application can be called "small-scale power systems." For each of the devices mentioned self-contained power sources are needed, not depending on "large-scale power systems" - stationary power plants and power supply lines. In small-scale power systems each individual consumer needs much lower electrical power values and a shorter operation time than one in large-scale power systems (see Figure 2).

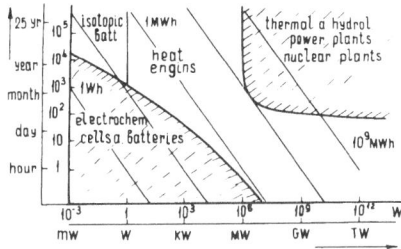

Fig. 2.   Small-scale and large-scale power systems.

The main power supplying devices for small-scale power systems are electrochemical power sources - galvanic cells and batteries, both of primary and storage type, as well as some types of fuel cells.

At present galvanic cells produce only a negligible fraction of electrical power compared to its total amount consumed by mankind. Nevertheless, these cells play a very important role in our life even now. Suffice it to say that the total world production of galvanic cells exceeds 10 billion a year. The overall rated power of all cells together switched on simultaneously for short operation (point B in Figure 2) is comparable with the total power of all electric power stations in the world (point A). It is difficult to visualize what would happen if suddenly as by magic, all galvanic cells should disappear. Not only would all electric torches go out and transistor radios be silent, but all transport facilities, autocars, airplanes, etc. would become stationary, rockets and man-made satellites would be unable to operate, no telephone communication would be possible. The whole tenor of our modern life would be quite different.

DEVELOPMENT OF PRIMARY AND STORAGE CELLS

The history of the development of modern cells can be divided into three stages. The first stage covers the period starting approximately from the sixties of the last century up to the beginning of the Second World War. During that period power sources based on four main electrochemical systems, now called conventional, were developed and found wide use. These are the primary manganese-zinc cells and three types of storage cells - lead-acid, nickel-cadmium and nickel-iron (see Table 1). The manufacture of these power-sources was organized in almost all countries and they came to be used nearly everywhere in small-scale power systems.

Starting about 1940, due to progress in radioelectronics and later in rocket and space engineering, the second stage in cell development began. New systems of power sources were evolved, characterized by higher electrical and operational parameters - higher energy-content per unit mass or volume, possibility of developing higher values of electrical power, longer service life, etc. For the power supply for compact portable devices small-sized leak-proof button cells were evolved. During this period mercury-zinc primary cells, silver-zinc storage cells, water-activated copper-magnesium cells and other cell types were developed and their commercial production was organized. A considerable achievement was the creation of reserve batteries which may be stored for a long time in an inoperative state but can be made serviceable (activated) shortly before being put to use. All these new power sources exhibit much better characteristics that those of power sources based on conventional systems. But at the same time their cost is considerably higher than that of the old versions. More costly and in some cases scarce raw materials are used in their manufacture, such as silver, mercury, etc. For this reason these new

Table 1. Various Electrochemical Systems Used in Primary and Storage Cells

Conventional cells

Manganese-zinc primary cells	Zn /	NH_4Cl, $ZnCl_2$	/ MnO_2
Lead (acid) storage cells	Pb /	H_2SO_4	/ PbO_2
Nickel-cadmium storage cells	Cd /	KOH	/ NiOOH
Nickel-iron storage cells	Fe /	KOH	/ NiOOH

Cells developed in 1940-1960

Manganese zinc alkaline cells	Zn /	KOH	/ MnO_2
Mercury-zinc primary cells	Zn /	KOH	/ HgO
Silver-zinc storage cells	Zn /	KOH	/ AgO, Ag_2O
Copper chloride-magnesium cells	Mg /	$MgCl_2$	/ CuCl
Reserve-type thermal cells	Ca /	$CaCrO_4$	/ (Ni)

New cell types (after 1960)

Nickel-zinc storage cells	Zn /	KOH	/ NiOOH
Nickel-hydrogen storage cells	H_2(Ni) /	KOH	/ NiOOH
Chlorine-zince storage cells	Zn /	$ZnCl_2$	/ Cl_2(C)
Sulfur-sodium storage cells	Na /	β-Al_2O_3	/ S
Iron sulfide-lithium storage cells	Li /	LiCl+ KCl	/ FeS, FeS_2
Zinc-air cells	Zn /	KOH	/ O_2(C)
Iron-air storage cells	Fe /	KOH	/ O_2(C)

Primary lithium cells
Li / aprotic solution / cathode

solvents		active materials for cathodes
Propylene carbonate ⎫		⎧ CuS, CuO
γ-Butyrolactone		MnO_2
Tetrahydrofuran		Ag_2CrO_4
Dimethylsulphoxide ⎭		⎩ $(CF_x)_n$
	Thionylchloride	
	Sulfur dioxide	etc.

kinds of cells were used only in highly specialized fields of application where the conventional cells were not adequate. Thus those new power sources had their own field of application - they did not displace the existing cells from their conventional ones.

The third stage started in 1960. Different versions of cells with high performance for special uses were developed or further improved. During this period the range of rated Wh-capacities and power values of cells was substantially enlarged (Figure 3). Work on miniaturizing cells was continued, specifically in creating cells for electronic wrist watches having the volume of some tens of a cubic centimeter and capacities in the MWh range. On the other side experimental installations in the MWh-capacity and MW-power range were built.

In this third stage of cell development, however, a new tendency is observed - towards creating comparatively cheap power sources that could gradually replace the conventional primary and storage cells in their most extensive applications and whose manufacture would not involve use of comparatively scarce non-ferrous metals or their compounds.

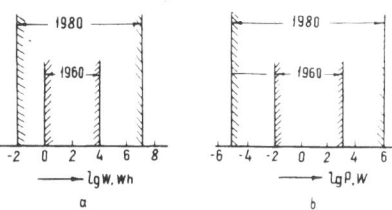

Fig. 3. Range of Wh-capacity (a) and discharge power (b) of electrochemical power sources.

Table 2 lists different substances that are used or in principle could
be used as active materials for positive and negative electrodes of different cell types. These substances are conditionally classified into four
groups placed in the order of decreasing availability. The availability
criteria are; explored reserves in the Earth's crust, ease of mining of
material raw materials and ease of production of the needed substance from
raw materials. It will be seen from the Table that nearly all materials
used in the manufacture of conventional primary and storage cells as well
as in those perfected for special purposes, belong to the third or the
fourth group as regards their scarcity. The reserve of many of these
materials in the Earth's crust are very limited and we are really threatened that in 30-40 years the demand for raw materials of the mass production of cells and batteries in current use will not be satisfied. Suffice
is to say that even now more than a half of the world production of lead is
used for manufacture of lead-acid storage cells. Therefore the problems of
creating primary and storage cells on the basis of materials available in
sufficient quantity for their mass production is a most vital one. Unfortunately, purposeful and sufficiently extensive research in this direction
began comparatively recently - after 1960. As part of this research the
possibility of using almost all less critical materials listed in Table 2
was explored. In developing new versions of power sources investigators
aim of course not only at substituting critical raw materials by less
scarce ones, but also at improving simultaneously the performances of these
power sources as compared to those of the conventional ones.

In the past decades several new types of storage cells have been
developed. The studies on nickel-zinc storage cells are widely known.
In their design these cells have much in common with silver-zinc storage
cells. Though the substitution of costly silver oxides by more readily
available nickel oxides markedly reduces the specific energy of the cells
per unit mass, it is still 1.5 - 2 times higher than in conventional lead
or nickel-cadmium storage cells. Another version of storage cells proposed
in recent years is a cell based on the use of chlorine and zinc. A dis-
tinctive feature of this storage cell is that upon charging, gaseous
chlorine evolves, which at the temperatures below 8°C forms a sold reaction
product with aqueous solutions - chlorine hydrate $Cl_2.6H_2O$. This storage
cell is made up of comparatively easily available materials, its electrical
characteristics are rather favorable.

Most primary or storage cells in use earlier or commercially produced
at present contain aqueous solutions of alkalis, acids or salts as electrolytes. It can be said that the proportion of all possible electrochemical
systems with aqueous electrolyte solutions have now been sufficiently
thoroughly investigated and their performance can be assessed with high
accuracy. On the other side properties of electrochemical systems with
nonaqueous electrolytes - solutions in organic solvents, salt melts or
solid electrolytes - have been studied in much less detail. It is quite
likely that the characteristics of cells with these electrolytes will be

Table 2. Active Materials for Primary, Storage and Fuel Cells Placed in the Order of Decreasing Availability

Group	Positive electrode	Negative electrode
I	O_2, H_2O_2, Cl_2, S, FeS, FeS_2	H_2, NH_3, Fe, Na
II	Br_2, MnO_2, CF_x	Al, Zn, alcohols
III	NiOOH, PbO_2, CuCl, I_2, $CaCrO_4$	Mg, Li, Pb
IV	AgO, HgO	Cd

far superior to those of cells with aqueous electrolyte solutions. Intensive studies as primary and storage cells with nonaqueous electrolytes are characteristic of the third stage.

The efforts to create sulfur-sodium storage cells are widely known. In these cells readily available substances are used as active materials - molten sodium and sulfur working in contact with a solid electrolyte (sodium beta-aluminate). Sulfur-sodium storage cells show rather large values of specific electrical energy. Their working temperature is 350°C, i.e. before use they must be heated up to this temperature. Storage cells with electrodes from iron sulfide and lithium alloys with a melt of chlorides as electrolyte exhibit similar properties. The working temperature of these cells is about 400°C.

The third stage is also characterized by intensive activities connected with the development of fuel cells. Unlike other types of primary or storage cells, fuel cells are capable of operating continuously for long periods of time since in their case the reacting active materials are not introduced beforehand, but are continuously supplied to the electrode surface in the course of operation. Several types of liquid or gaseous reactants can be used in fuel cells - hydrogen, hydraxine, alcohols, carbon monoxide, hydrocarbons, etc. The best results, however, were achieved when hydrogen gas was used as fuel and air oxygen or pure oxygen as oxidant.

The first works on fuel cells date back to the end of last century, but the results obtained then were very modest. The battery of hydrogen-oxygen fuel cells demonstrated by Bacon in 1959 attracted the attention of the scientific community. It was the first time that such a battery showed sufficiently high electrical power values and a satisfactory operation time. This event started the period of tremendous activity directed towards development of fuel cells. Research and development work was initiated in many countries, a number of research centers and universities taking part in this work. For the first time in different countries work on fuel cells was carried out on a national or state scale. The results of this work are a matter of common knowledge. Systems with fuel cells were most successfully used in various space ships, first in "Gemini" and then in "Apollo" and "Shuttle." On these ships they provided not only all power supply necessary during the flight, but the reaction product - water - could be used for the needs of the ship's crew.

If, however, we analyze the result of the great amount of work carried out during the third state - from 1960 up to the present time, we shall realize that there is a great discrepancy between the efforts made and the practical results achieved. Notwithstanding numerous works on new versions of storage cells, none of these is commercially produced and hence has not reached the consumer. In spite of the great efforts exerted in the development of fuel cells and their successful application on space ships, not a single type of fuel cell is produced on a mass scale to find extensive use in "terrestrial" conditions.

Up to now the only real result of this third stage of development of power sources is the appearance of lithium cells - a new group of primary cells with high values of specific energy per unit mass or volume in which lithium - an electrode material with large power capacity - is used in conjunction with electrolyte solutions based on nongaseous (organic or inorganic) solvents. In the past decades various versions of these cells have been evolved and their mass production has been organized in different countries. Now these cells are extensively used as power sources for different electronic devices. The appearance of these cells truly revolutionized the production of primary cells.

However, on the whole the achievements in this field are far from striking. As before, a major portion - more than 90% - of commercial production is formed by primary and storage cells of the four main electrochemical systems. Here we have a case of an amazing longevity of these devices. The foundation of the design and technology of currently used lead acid storage batteries were laid down about 100 years ago and have remained almost unchanged (though, naturally, due to certain particular improvements, the electrical and maintenance characteristics were considerably enhanced). The situation with other conventional systems is similar.

It would be natural to ask why these old versions of power sources are so long-lived and give way to new, more progressive and perfect versions only with great difficulty?

It is not easy to answer this question. One of the possible reasons is related to the following circumstances. The electrical power generated by primary or storage cells is, naturally, more costly than that supplied from the mains. The cost of electrical power produced by power stations is 1-3 cents per 1 kWh, while that generated by primary cells (discarded after a single discharge) is incomparably higher - 30-80 dollars per 1 kWh. For storage cells where multiple repetition of charge-discharge cycles is possible, the cost is less, but still for lead-acid storage batteries it is 15-20 cents per 1 kWh. Of course, this higher cost of electrical power must be compensated for, by convenience in operation. Besides the main advantage of self-contained power sources - independence of the power supply mains - the consumers are primarily interested in such qualities of primary and storage cells as high reliability, freedom from failure, high service life, simplicity of maintenance and operation.

Though new versions of storage batteries have higher specific energy and specific power characteristics, they are as yet inferior to the conventional version just in the operation characteristics mentioned. Nickel-zinc storage cells show as yet an insufficient service life and inadequate reliability. High-temperature sulfur-sodium and sulfide-lithium storage cells are inconvenient in service in that they must be preheated and also their service life and reliability are still low. The necessity for continuous cooling of the electrolyte solution of chlorine-zinc storage batteries limits drastically the possible fields of application of these batteries. In the past years a tendency has begun to appear toward creating storage batteries of a more sophisticated design than in the old version. Thus, for example, in chlorine-zinc batteries and also in certain versions of air-zinc batteries a special system is used for continuous circulation of the electrolyte solution (Figure 4). This complicated design reduces the reliability of the battery considerably and leads to possible premature failure to operate. All these disadvantages preclude wide use of these new storage batteries.

The absence of mass production of fuel cells requires special consideration. This fact has nothing to do with shortcomings of fuel cells themselves. The performance of fuel cells increased considerably during the last two decades (Figure 5). Their successful operation in space ships shows that they can be highly reliable. The present situation with fuel cells seems to be caused by the fact that only such types of cells were brought to a high degree of perfection in which hydrogen or, in some cases, hydrazine are used as fuels. Hydrogen gas is very inconvenient in handling; it is difficult to transport and to store. The successful use of hydrogen-oxygen fuel cells in space ships is partly explained by the fact that liquified rather than gaseous hydrogen can be used in their case. For other applications neither hydrogen stored in steel cylinders or in bound form nor toxic hydrazine complies with the requirements of convenience in operation.

Thus the tasks confronting the designers of electrochemical power sources in the third stage of their development have been solved to a very small degree as yet. This does not mean, however, that the work in this field should be discontinued. Due to raw material problems, the task of creating power sources with the use of more readily available materials remains a most vital one. If this problem is not solved, in 2-3 decades the production of certain kinds of primary and storage cells may become threatened. Undoubtedly, this problem is of major importance for modern electrochemistry.

The new development of new storage batteries with large specific power and high reliability now acquire special importance in connection with the problem of electric vehicles. It has been long recognized that even a partial substitution of electric cars for cars with internal combustion engines would effect a great saving in oil products and also decrease air pollution in large cities. In their electrical characteristics - specific energy and specific power - the existing storage batteries cannot ensure acceptable speeds and mileage of electric vehicles. For this reason the problem of electric vehicles come primarily to that of creating new kinds of storage batteries.

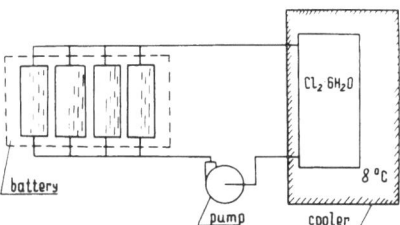

Fig. 4. Schematic of the system with chlorine-zinc storage cells.

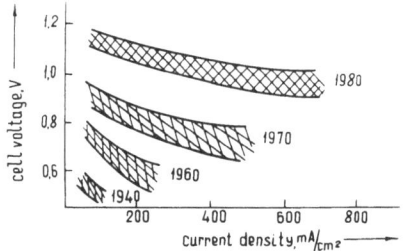

Fig. 5. Performances of fuel cells.

In recent years the possibility of using electrochemical energy conversion principles for solving problems not only of small-scale but also of large-scale power systems has come to the fore in many countries. The reason of this is: the search for new ways of solving energy problems arising in connection with the world energy crisis (approaching depletion of natural fuels) and the advances made in recent years in some areas of theoretical and applied electrochemistry. This problem has three aspects – primary production of electrical power, storing electrical power in systems with irregular power production or consumption and production of convenient energy carriers.

Direct conversion of chemical energy of fossil fuels into electrical energy by means of fuel cells has for a long time been the dream of all electrochemists. It has been long known that the use of heat engines is not the most efficient method of power production based on fossil fuels. This method is based on converting the chemical energy of the fuel primarily into thermal energy, then to mechanical energy and subsequently to electrical energy (Figure 6). The efficiency of converting heat energy is limited by the second law of thermodynamics. Much more efficient would be an electrochemical method without intermediate heat formation similar to that used in galvanic cells. In 1894 the distinguished physical chemist Ostwald clearly expressed this idea in the words "The way to solve the most important of all technological problems – production of cheap electrical power – must now be found by electrochemistry. We have to do with a case for which, just as in any mechanical problem, complete success may be predicted."

The development of modern energetics entailed a paradox – instead of a highly effective electrochemical means of energy conversion a thermodynamically much less effective method of using heat engines came to the fore.

One of the reasons of this paradox lies in the fact that though the thermodynamic efficiency of electrochemical energy conversion is much higher, in its technological and economical characteristics this method is practically inferior to other methods. Another, and a deeper reason is that the principle of operation of heat engines is quite clear – they can be designed in advance and the characteristics of real engines will not differ much from the contemplated ones. On the other hand, in electrochemical devices all processes occur at a molecular level. For the most part, they are not completely understood and cannot be monitored. Therefore, development of electrochemical devices is based on the results of extensive experimental investigations.

Extensive studies of the possibility of increasing the efficiency of hydrogen-oxygen fuel cells and of using other fuels have promoted the formation of a new branch of modern theoretical and applied electrochemistry – electrocatalysis. One of the most interesting achievements in

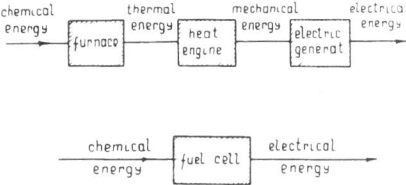

Fig. 6. Two methods of conversion of chemical energy into electrical energy.

this field was the discovery some 20 years ago of the possibility of electrochemical oxidation of hydrocarbons at temperatures 100-150°C, i.e. below the temperatures, at which catalytic oxidation of hydrocarbons by molecular oxygen occurs. Unfortunately, the rate of the electrochemical reaction is far too low for any practical application. As already mentioned, up to now all progress in fuel cells is associated with using hydrogen or at any rate gas mixtures rich in hydrogen. Therefore, the only possible way of electrochemical oxidation of fossil fuels is through their preliminary chemical conversion into hydrogen-containing gas mixtures by steam reforming of natural gas or oil products, coal gasification, etc. These preliminary stages involve energy losses which reduce the overall efficiency of energy conversion.

At present work on fuel cells for large scale power production proceeds in two directions: development of medium-temperature (about 200°C) phosphoric acid fuel cells for oxidation of hydrogen obtained by reforming of hydrocarbon fuel, and development of high-temperature cells with a carbonate melt (about 650°C) or a zirconia based solid electrolyte (900-1000°C) for oxidation of coal gasification products. The first type of cell was used in recent years by United Technologies Corporation in the construction of the well-known 4.5 MW power stations built in New York and Tokyo. The total efficiency of these stations is assumed to be 36-38%, which equals to the efficiency of a modern thermal power station, but less than would be expected for electrochemical energy conversion. Further increase of the efficiency up to 45-50% can be expected when high-temperature fuel cells are used. In this case part of the heat released by the fuel cell battery at 650-1000°C can be used for the endothermic reforming or gasification processes (Figure 7). This way leads to a considerable lowering of the energy losses for the overall process. Furthermore, the higher temperature facilitates solution of two important electrochemical problems: decreases of the hazard of poisoning the hydrogen electrodes by impurities in the fuel gas or by carbon monoxide and increases of the efficiency of the oxygen electrodes.

Medium-temperature cells of the first kind can find application for peak-power needs. Their extensive use as base-load power stations is doubtful since by the time they may become sufficiently developed and wide-spread (in approximately 30 years) hydrocarbon fuel will lose its importance as an energy source. At the same time high-temperature cells of the second kind hold much promise. The practical application of such cells would permit the specific consumption of coal for generation of electrical

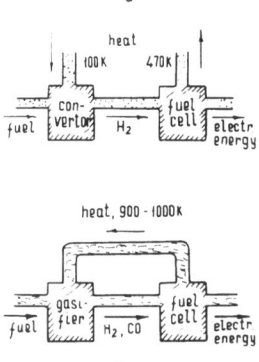

Fig. 7.   Schematic of medium-temperature (a) and high-temperature (b) fuel cells.

power to be decreased by 15-25%. The use of electrochemical devices for
large-scale energy storage in electrical power systems is very promising.
This problem assumes vital importance in connection with the efforts
directed to development of wind, solar and tidal power stations that in
virtue of their particular nature do not generate power continuously. The
use of such energy sources is impossible without intermediate energy
storage. Several methods of energy storage have been proposed, including
water pumping to a higher level, air compression to high pressure, use of
flywheels, etc. Analysis shows that the electrochemical method of energy
storage has undisputed advantages over other methods in that it allows more
compact equipment to be used.

In principle, even existing storage batteries can be used for this
purpose. However such attempts could be hardly described as promising
since storing electrical power in large amounts requires huge batteries
with considerable quantities of nonferrous metals. For example, a plant
for storage of 100 MWh would require a lead-acid battery with about 1000
tons of lead. In this case new versions of storage batteries employing
easily available materials (e.g. sulfur-sodium batteries) assume great
importance.

A different kind of electrochemical energy storage in which an electrolyzer is combined with a fuel cell (Figure 8) will probably prove more
convenient. In the electrolyzer, at the expense of external supply of
electrical power water undergoes electrolysis to evolve hydrogen which is
stored in special gas-holders. To produce electrical power, hydrogen from
the gas-holder is supplied to the fuel cell battery. The principle of this
conversion is rather simple. Unfortunately at the present level the
overall efficiency of such a plant is low – about 50%. The reason for this
is the insufficient efficiency of the available electrolyzers and fuel
cells, resulting mainly from insufficient activity of the oxygen electrodes
in both devices. The reaction rates of cathodic oxygen reduction (in fuel
cells) and anodic oxygen evolution (in electrolyzers) is very low, even on
such active catalysts as platinum or silver. This low activity results in
a considerable potential-shift - polarization - of the electrode even at
low current densities and in corresponding voltage changes (Figure 9) and

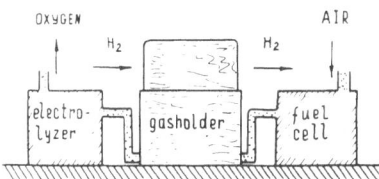

Fig. 8.   Installation for energy storage with electrolyzer and fuel cells.

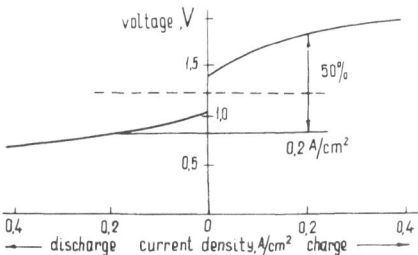

Fig. 9.   Charge-discharge curves for medium-temperature hydrogen–oxygen
storage devices.

energy losses. There are gounds for believing, however, that with further progress in electrocatalysis the problem of increasing the efficiency of oxygen electrodes can be solved. The increase of performances of fuel cells during the last two decades was shown in Figure 5. Figure 10 represents a similar increase for electrolyzers. If, as a result of further studies, voltage values (at current densities 0.2 - 0.4 A/cm^2) of about 1.45-1.50 V for electrolyzers and 1.0-1.05 V for the fuel cells will be achieved, such a combined device will become one of the most suitable means for large scale storage of electrical power. The problem of enhancing oxygen-involving reactions and decreasing polarization of the oxygen electrode is one of the central problems of electrocatalysis and entire electrochemistry.

One of the most important energy problems of the future arising as a result of approaching depletion of liquid fossil fuels and as a result of wider use of nuclear and inexhaustible solar energy in the problem of producing suitable energy carriage for large-scale pipeline transportation and distribution of energy. Hydrogen is one of the promising energy carriers of the future and will find wide application both in energetics and in technology. We are faced with the problem of producing hydrogen not only at the expense of electrical energy but also by partial use of thermal energy of nuclear reactors. One of the ways to achieve this is high-temperature (at 800-1000°C) electrolysis of steam in electrolytic cells with zirconium dioxide based solid electrolyte. It is interesting to note that at high temperatures the oxygen electrode is practically not polarized and therefore such cells are very promising for energy storage devices (Figure 11).

Other ways of hydrogen production by use of thermal energy are based on cyclic combined thermo- and electrochemical processes. One of the best known processes of this type is the cyclic sulfuric acid process shown in Figure 12. Sulfuric acid is thermally decomposed at about 800°C with the formation of sulfur dioxide, oxygen and water. Sulfur dioxide is subsequently reoxidized at the anode of an ambient-temperature electrolyzer with

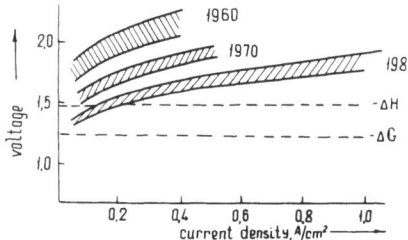

Fig. 10. Performances of electrolyzers.

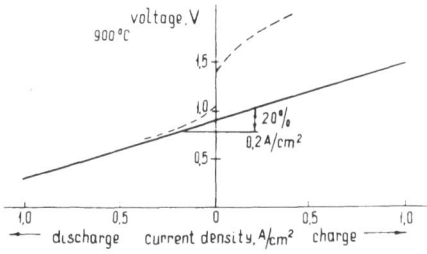

Fig. 11. Charge-discharge curves for high-temperature hydrogen-oxygen storage devices.

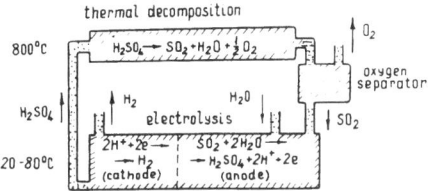

Fig. 12. Sulfuric acid thermo-electrochemical cyclic process.

hydrogen being formed at the cathode. The net result of the cyclic process is splitting of water into hydrogen and oxygen. As a result of the low voltage of the SO_2-depolarized electrolyzer the consumption of electrical power for this process is lower than for conventional water electrolysis and water splitting occurs mainly at the expense of thermal energy. It has been shown that such a coupling of electrochemical and other types of reactions is very beneficial from the energetical and technological points of view.

Another aspect of the problem of producing new energy carriers is synthesis of methanol-like substances by cathodic reduction of carbon monoxide. This is a comparatively new type of electrochemical reaction. The research being done now is of an exploratory character but from the point of view of electrocatalysis it involves many interesting features.

Extensive investigations are being done now on all the problems mentioned. But, unfortunately, in the field of large-scale power systems (unlike the field of small-scale ones) electrochemical devices and processes have not yet found any significant application. Most ideas and conclusions put forward by different scientists and groups are not yet verified on a sufficient large scale. For this reason different opinions are expressed in the literature on the prospects of using electrochemical methods for solving large-scale energy problems. It must be remembered that the use of electrochemical processes has not only advantages in efficiency but also presents many difficulties. The main requirements for large-scale power systems are their adequate economic characteristics. For large-scale electrochemical devices the use not only of platinum catalysis but even of nickel based electrodes can be prohibited from the point of view of costs. To decrease depreciation costs a long service life is required for such devices. A vital problem is also to increase drastically the intensity of different electrochemical reactions, which largely depends on the future progress in electrocatalysis.

A number of questions arise in connection with the foregoing:

1) Can electrochemical processes affect the solution of energy problems of the future?

2) Can they be improved to the point where they will become technologically acceptable and able to compete with other methods?

3) Is it worth while to make considerable efforts and investments for their intensive further development?

The answers to these questions must be based on a thorough analysis of all the results available. A comprehensive analysis should take into consideration both the technological and the economical aspects of this problem. In addition, one should take into account the feasibility of solving a number of related electrochemical problems.

Both an unbiased analysis of the present situation and as intuitive assessment of the possible paths of further development permit the three questions posed above to be answered in the affirmative. Naturally, we shall be faced with many difficulties and disappointments in the future, but yet it seems that the task we have set ourselves can be achieved.

All energy problems are global. As follows from experience, their solution is possible only through the joint efforts of individual scientists and scientific institutions of many countries. The development of electrochemical systems for large-scale energy conversion and storage gives an excellent possibility for further promotion of international scientific cooperation.

DISCUSSION

K. Wiesener

Technical University
Dresden
West Germany

CHAIRMAN'S SUMMARY

The discussion was concentrated on the following topics:

1. New Systems of Power Sources with Non-aqueous Electrolytes

Power sources with liquid organic and inorganic aprotic electrolytes, solid electrolytes and molten salt electrolytes are gaining in significance because by this way it is possible to use very active electrode materials, e.g. lithium with a very negative redox potential. Thus power sources of high energy density are realizable.

Two fields of application are important:

a) small-size power units for electronic and microelectronic devices and
b) larger units for electric traction and for load leveling.

In the first of these fields primary cells with liquid and solid electrolytes are already produced for electronic watches, calculators and pacemakers (e.g. $Li-MnO_2$ or $Li-CF_x$ with $LiClO_4$ dissolved in propylene carbonate or other solvents as electrolyte $Li-I_2$ solid electrolyte cells, where iodine is bound to polyvinylpyridine as a charge transfer complex). At present, intensive investigations on electric recharging of such systems are carried out since the rechargeability of the lithium electrode and the stability of organic solvents in such cells are very serious problems which are not yet solved.

For electric traction and peak load levelling only rechargeable systems are taken into consideration. Because of their higher output storage cells with solid and molten electrolytes working at temperatures 600-800 K are of interest - e.g. sodium-sulphur storage cells with a sodium ion conducting solid electrolyte or lithium-iron sulphide storage cells with molten salt electrolytes. Therefore, in future fundamental investigations on the electrolytes' structure and on the transport mechanisms must be intensified and such electrolyte systems must be developed which have high ionic conductivity, high corrosion resistance in contact with the active components, high resistance to thermal shocks and which can be easily and reproducibly produced from available materials.

In addition to the oxygen conducting solid electrolytes for fuel cells (and also for oxygen measuring devices) β-Al_2O_3 and Nasicon as selective sodium ion conductors are of great importance for medium and high temperature systems.

If the materials and corrosion problems are solved high temperature fuel cells could be used in the framework of a hydrogen economy for the conversion of hydrogen into electric energy. Cells with molten salt electrolytes immobilized in a ceramic matrix could be applied at 850-950 K and with solid oxygen ion conducting electrolytes at 1200-1300 K.

2. Organic Substances as Solid Electrolytes and Electrode Materials for Electrochemical Power Sources

When ion exchange membranes are not taken into account, this subject was dealt with fundamentally only during the last 5-10 years. The point is that organic substances have a poor electronic conductivity, and when they are used as electrolytes they also show a poor ionic conductivity. Recently it was shown that certain compounds with cyclically conjugated bounds of special structure (e.g. polyacetylenes doped with lithium ions) have a conductivity which nearly corresponds to that of graphite. However when these materials are used as rechargeable electrode materials the load densities obtained so far are low because their equivalent weight is not below 100 (it is even 200-500 for most of the compounds including the dopant), that means that the charge densities are lower than for the well known electrode materials such as zinc, nickel, cadmium or lead, to say nothing of lithium. Nevertheless, this field of application is of interest especially as to the charge-transfer complexes and to the selective separation affects obtained in solid polymeric electrolytes. Therefore research in this field should be promoted. Perhaps in this way rechargeable small-size energy sources for use in microelectrodes could be obtained, e.g. on a sandwich base containing a solid organic electrode material, a solid organic electrolyte and a lithium foil which are directly enclosed in a printed circuit board.

3. New Power Sources for Electric Traction and Load Leveling

At present, apart form the mentioned medium temperature storage cells with solid or molten electrolytes, new types of metal-halogen storage cells are developed, which could be used in electric traction. On the whole, the Zn-Cl_2 and Zn-Br_2 systems are most promising. The problem is to store halogens and to find materials with a sufficient corrosion resistance. There are some problems caused by liquid junctions in batteries with a great number of cells connected in series. Metal (Al,Zn)-air cells have a high energy density but are recharged only with a high expenditure energy.

To sum up, electrochemical power sources as energy carriers for electric traction are only competitive with respect to liquid fuels when from the economic point of view electric energy is much cheaper than that of liquid fuels and when the disadvantages of mains-independent traction with storage cells (lower range, time consuming recharging compared with refuelling of engines) are accepted.

Concerning the problem of load leveling, a combined system of hydrochloric acid electrolysis and hydrogen-chlorine fuel cells must also be taken into consideration. The chlorine electrode's overvoltage is considerably lower than that of the oxygen electrode in an analogous hydrogen-oxygen system. But also here the problem of chlorine storage continues to exist.

Finally, the necessity of a complete working up of exhausted power sources to recover secondary raw materials and to reduce environmental pollution was discussed.

The further development in the field of electrochemical power sources will be considerably influenced by the interaction of fundamental research (e.g. on electrocatalysis, solid state structure, transport phenomena, electrochemical kinetics and technology) and applied research, where progress will depend decisively on a successful cooperation in science between electrochemists, physicists, experts in materials, engineering, chemical engineering, electrical engineers a.o. Research work will probably concentrate on special power sources which are advantageously applied only in certain fields, as illustrated by the lithium solid electrolyte cell for pacemakers.

Prof. E. Gileadi
Tel Aviv University, Israel (communicated)

1. For the feature widescale use of high energy-density primary and secondary batteries it is important to develop methods for the determination of the state of charge of the battery. Continuous and reliable monitoring of the state of charge of the batteries in an electric car, for example, would seem to be an essential requirement for their large scale acceptance.

2. Energy storage by electrolysis of water and recombination of the products in fuel cells has a low efficiency, because of the poor performance of the oxygen electrode. It was shown [E. Gileadi, S. Srinivasan, F. J. Salzano, C. Broun, A. Beaufrer, S. Gottesfeld, L. J. Nuttal and A. B. LaConti, J.Power Sources, 2:191 (1977/78)] that the "electric to electric" (ETE) efficiency is less than 40%. Electrolysis of HCl could lead to an ETE efficiency of about 70%. Storage of chlorine should not present new technological problems in view of the wide production, transportation and use of chlorine. Other systems such as Zn/Br_2 or Zn/Cl_2 will probably also be more efficient than water electrolysis for energy storage.

3. Electrochemists should consider the poor performance of stacks of batteries and fuel cells connected in series, compared to the performance of single cells. The solution to this problem is probably by microprocessor monitoring and control of each of the cells in the stack, making it possible to disconnect a malfunctioning cell for later replacement. Although not an electrochemical problem per se, its solution could increase the viability of electrochemical energy conversion devices considerably.

HYDROGEN AS AN ENERGY VECTOR IN THE 21ST CENTURY

R. M. Dell

Materials and Developments Division
Atomic Energy Research Establishment
Harwell, England

INTRODUCTION

In recent years the suggestion has arisen that hydrogen will become the universal energy vector and fuel of the post petroleum era. This paper reviews in outline the Hydrogen Energy concept, with particular reference to the role of electrochemistry in the generation of hydrogen by water electrolysis and its utilization by combustion in a fuel cell.

Some likely developments in the world energy scene over the next century are discussed and the complex interplay of technical, environmental, economic and political factors is emphasized. In this timescale hydrogen is not seen as a universal fuel, rather as an intermediate in the manufacture of synthetic liquid and gaseous fuels from low grade fossil sources. This hydrogen will derive principally from steam-reforming reactions. Electrolytic hydrogen is more likely to be manufactured and used by the chemicals industry, utilizing electricity derived from nuclear power or renewable energy sources. This will happen only when electrolytic hydrogen is competitive with hydrogen made by steam reforming.

Hydrogen has other possible future roles – as an energy vector for transmitting and storing surplus electricity and as a transport fuel. As an energy vector the prospects are dependent upon developments in fuel cells for reconverting the hydrogen back to electricity. Hydrogen-fuelled road vehicles also depend upon fuel cell technology, while in aviation the attractions of liquid hydrogen as an aircraft fuel are mentioned.

It is concluded that the 21st century is likely to be a transitional period in the world energy scene, with hydrogen playing an increasingly important role as an intermediary in fuel synthesis and as an energy reactor, possibly leading to a full Hydrogen Energy scenario in later centuries as fossil fuels become progressively depleted. Within this overall framework, electrochemistry has an important role to play in the development of optimized electrolyzers and fuel cells.

HYDROGEN FUEL CELLS

The decomposition of water by the passage of an electric current was known in the eighteenth century, but it was not until 1834, when Faraday enunciated his well known laws of electrolysis, that the phenomenon was put on to a quantitative basis. Shortly afterwards, in 1839, the first hydrogen-oxygen fuel cell was constructed by Sir William Grove[1]. To these two scientists belongs the distinction of demonstrating the basic electrochemical principles on which the modern "Hydrogen Energy" concept is based. It was left to Jules Verne in 1870 to foresee that water could, in principle, provide a cheap, inexhaustible supply of hydrogen fuel[2].

Industrial scale electrolyzers were developed early in the 20th century for the manufacture of chlorine and caustic soda from brine, and for the commercial production of hydrogen used in ammonia synthesis. Large water-electrolysis plants were constructed in Norway and Canada in the 1930's, based on cheap hydroelectric power, and the hydrogen so produced was used in fertilizer manufacture. With the advent of natural gas and low cost petroleum, hydrogen production moved toward catalytic steam-reforming of hydrocarbons, and water electrolysis became less significant.

Fuel cells remained essentially dormant until the pioneering work of Bacon in the 1950's at Cambridge on alkaline electrolyte, high pressure, hydrogen-oxygen cells[3]. He developed porous nickel electrodes of a sophisticated design and built fuel batteries of 6 kW output operating at 3-4 MPa gas pressure. This system was chosen in the United states as the power unit for the NASA-Apollo manned space missions and was further developed by Pratt and Witney into a highly successful, light-weight fuel cell. Subsequently, 12.5 kW alkaline cells have been employed as power sources for the Space Shuttle Orbiter.

In parallel with these development, work has been carried out on phosphoric acid electrolyte fuel cells which operate at 200°C and are able to burn hydrogen containing CO_2 impurity, using air rather than pure oxygen as the oxidant. They find application as combined heat and power generation in 40 kW sizes, while much larger units (4.8 MW) have been installed in Manhattan and in Tokyo for peak load generation in the electricity supply system[4].

While these electrolyzer and fuel cells developments were undertaken for strictly commercial reasons, there has sprung up in the last two years a quite different interest in hydrogen as a future universal energy vector and fuel for the post-petroleum era. Here the motivation is much less commercially based and stems from considerations of economic growth, resource depletions and environmental pollution. Hydrogen is seen as the ultimate fuel, available in limitless quantity from water by means of electricity derived from non-hydrocarbon sources, and capable of being combusted back to water in a closed cycle with no pollutants being released[5]. While some scientists see this as a vision of the future, others dismiss it as quite unrealistic. Even the sceptics, though, will mostly agree that hydrogen will play an increasingly important role in future in the manufacture of synthetic gaseous and liquid hydrocarbon fuels from sources such as shale oil, tar sands, bitumens and coal. As we shall indicate below, the true situation is highly complex and it is by no means obvious which school of thought is the more nearly current.

ENERGY SCENE IN THE 21ST CENTURY

Although a few, far-sighted individuals gave warning in the 1960's of the eventual depletion of petroleum resources, it was not until the sharp rise in oil prices in 1973 and the formation of the OPEC cartel that the general public became aware of the situation. This was reinforced by a further major escalation in the price of liquid fuels in 1979. For the first time, the price began to reflect the market value of the commodity rather than its production cost. This stimulated oil exploration and allowed the development of new oil fields (e.g. Alaska, North Sea) which would not formerly have been economic. At the same time, the high price of oil led to conservation in its use (e.g. smaller, more economical cars), and the substitution of alternative fuels where possible (natural gas for space heating, coal and nuclear power for electricity generation). These measures led to a restoration of equilibrium in the oil market which still persists today.

What will be the situation in the future? During the remainder of this century the prognosis is for an approximately steady consumption of oil and natural gas worldwide, with economic growth being accommodated by increased utilization of coal, although with some growth also in nuclear energy, synthetic fuels, hydroelectricity and other forms of "renewable" energy. Within this overall framework the situation will vary widely from country to country. Looking to the 21st century, supplies of readily accessible liquid hydrocarbons will progressively become depleted. This will also be true for natural gas. Liquid and gaseous hydrocarbons are regarded as premium fuels, because of their cleanliness (after appropriate refining or pretreatment) and because of their unique value as transport fuels and for space heating, respectively. It follows that fluid fuels will always command a premium price. When supplies of the natural commodity are no longer available, synthetic liquid and gaseous fuels will be manufactured from lower grade sources such as tar sands, shales, bitumens and coal. These synthetic processes, carried out on a large scale, pose massive engineering and environmental challenges and will require vastly increased quantities of hydrogen. Just when these processes will be taken up is a matter of economics and politics, and the timing will vary from country to country. In any event, on a century-long timescale, although we face formidable technical and financial challenges in setting up a new synthetic fuels industry, we may feel reassured that there is no shortage of potential feedstocks for that industry.

In the aftermath of 1973 there was a surge of interest worldwide in the so-called "renewable" energy sources - solar energy, geothermal energy, wind energy, wave energy, biomass etc. Enthusiasts for these energy forms saw them as supplanting fossil fuels and nuclear power. Unfortunately, there are two major problems. The first is economic; many of the renewable energy sources simply do not stand up to detailed economic analysis at today's fuel prices. There are some exceptions to this general statement, and in certain counties renewable forms of energy will play a useful role in the foreseeable future, as hydroelectric power has done in the past. An example would be solar water heating in countries with appropriate climates. The incentive to exploit a particular form of renewable energy will be especially strong in countries which have that resource, but lack indigenous fossil fuels (e.g. wind energy in Denmark, wood burning in Sweden). Globally, though, the overall contribution of the renewables to the world's energy supply is likely to remain small for many decades yet.

The second problem with renewables is that, unlike fossil fuels, they do not usually have a storage component associated with them. Hydrocarbon fuels are not just a convenient, transportable form of energy; they are also an energy store - quite literally, stored solar energy from bygone millenia. Renewables mostly lack this essential storage element. This is particularly unfortunate as they tend to be erratic or irregular in intensity, often with the peaks in supply not coinciding with the peaks in energy demand. For example, solar heating is most available in summer and most needed in winter. Other renewables (wind, wave energy) are best converted into electricity, but because of their erratic natures, electricity utilities will give credit only for the fuel saved and not for savings in capital costs. This factor militates against the economics of renewables as a source of electricity.

One way around this problem is to build a storage component into the electricity supply network. This is already done to a degree with pumped hydro-storage. Other possible storage media are heat (in the UK night storage heaters,) compressed gas or chemical storage. The latter may take the form of secondary batteries, from which electricity may be recovered directly, or electrolytic hydrogen which may be sold as a chemical or reconverted to electricity in a fuel cell.

After the initial euphoria with renewable energy in the mid 1970's, a sense of realism/pessimism set in and by 1980 a general feeling emerged among energy analysts in major developed countries that a future national energy strategy must be built around some combination of conservation, coal and nuclear technology (the so-called "COCONUT" strategy). This strategy does not deny that renewable energy sources have a contribution to make, but sees it as small in overall national terms. While all countries accept the need for conservation as a key component of strategy, they have very different views on the contributions of coal and nuclear to their overall future energy strategies.

From a technical and economic viewpoint, COCONUT is still, in 1984, the only viable medium term strategy for most industrial nations. However, there are two other factors which have arisen to complicate the issue. One is public concern over the safety of nuclear power and the disposal of radioactive waste. The other is similar public concern over acid rain and the so-called "greenhouse effect" caused by the build-up of CO_2 in the stratosphere; both of these undesirable effects have been ascribed to the burning of fossil fuels. These concerns are the subject of extensive public debate and professional evaluation at present and it is not profitable to speculate on the outcome[6]. Clearly, conservation alone is an inadequate strategy and the renewables will be continuously monitored and reassessed in the light of additional financial burdens which may fall on nuclear power and coal-burning to produce technical solutions which satisfy the public as regards safety and environmental issues.

THE FUTURE ROLE FOR HYDROGEN

The proponents of the Hydrogen Energy concept (or "Hydrogen Economy" as it is sometimes known) point to the undoubted attractions of hydrogen as an energy vector for the future. These may be summarized briefly:

(1) Hydrogen is universally available in the form of water from which it may be extracted electrochemically using electricity derived from nuclear, hydro or other renewable energy sources.
(2) Hydrogen may be transmitted over long distances in buried pipelines which are cheaper to construct and operate than electricity grids, and are more environmentally acceptable.
(3) The gas in the pipeline, and also in underground storage caverns, provides a storage component within the electricity supply system.
(4) Hydrogen is the ideal fuel for use in fuel cells to regenerate electricity.
(5) Hydrogen combusts cleanly into water, thereby closing the environmental cycle with no significant pollutants being formed.

It will be seen that hydrogen is the vector which conveys primary energy from the place and time where it is available to the place and time where it is needed. Just as electricity is a common and convenient form of secondary energy derived from a number of primary sources, so hydrogen may be seen as an "add-on-extra" to electricity giving the added dimensions of (1) storage, (2) cheap transmission and (3) an environmentally acceptable fuel for combustion. Attractive though this concept may seem at first sight, it is necessary to evaluate any proposed application quantitatively, having due regard to technical and engineering factors, efficiencies and costs. Only then will a true picture be obtained.

To many purists, Hydrogen Energy is a concept for the future, inspired by environmental and resource consideration, and has no connection with fossil fuels which are seen as finite and limited. While this may be true after, say, the 24th or 25th century (not so far away on an historical perspective), for the present discussion of the 21st century, I believe that we must start from where we find ourselves rather than where ultimately we wish to be. This involves taking three other important facts into consideration:

(1) Hydrogen is today manufactured predominantly for use as a chemical rather than as a fuel.
(2) Most hydrogen is derived from water by reaction with fossil fuels (steam reforming) rather than by electrolysis.
(3) A large, expanding market for hydrogen in the manufacture of synthetic liquid and gaseous fuels is foreseen for the 21st century based on low grade fossil fuels as feedstock.

For these reasons, I should like to expand the Hydrogen Energy concept to include hydrogen derived from fossil sources and/or used in the manufacture of synthetic fuels rather than restrict the discussion to the post-fossil fuel era. The latter is too far away for sensible technical and economic evaluation.

At present hydrogen finds three major chemical uses:

- ammonia (fertilizer) manufacture,
- in petroleum refining for reforming and upgrading products,
- in methanol synthesis.

Table 1 shows the quantities of hydrogen used for these purposes in the UK in 1980 and some predictions for 2025. The projections are based upon rational, semiquantitative considerations and lead to a modest growth rate in hydrogen use (80% over 45 years). Of course, this represents an extrapolation of current uses and takes no account of possible requirements for manufacturing synthetic fuels. Premium gases and liquid fuels have a hydrogen/carbon atom ratio from 4 (methane) to 2.25 (octane) compared to < 1 for low grade fossil fuels. Clearly, much hydrogen will be required for upgrading these feedstocks to synthetic fuels, as well as for the removal of sulphur by catalytic hydrodesulphurization.

Most hydrogen is manufactured today by the steam reforming of natural gas or light naphtha as these are the most economic routes. Approximately one quarter of the world's hydrogen is made by the partial oxidation of residual oils and only a tiny faction (< 5%) by electrolysis, generally in

Table 1. Chemical Uses of Hydrogen in the UK 1980-2025[7]

Use		Hydrogen Mt/year	
		1980	2025 (Projected)
Ammonia production		0.36	0.5
Methanol production		0.08	0.2
Refineries		0.12	0.2
	Totals	0.56	0.9

Table 2. Hydrogen Production Processes - Worldwide

Hydrogen Production Process	%
Catalytic steam reforming	65
Partial Oxidation	25
Coal/Coke Based	7
Electrolytic	3

regions of low cost hydroelectricity. This situation may well change and the pendulum swing back to electrolysis, as in the 1920's if the price of natural gas rises with respect to that of electricity. For this to happen it is essential to have cheap electricity derived from large, efficient nuclear or coal fired power stations.

Another possibility is to site ammonia production plants in regions where cheap hydroelectric power is available. This has already happened, long ago, in Norway, Canada and Egypt (Aswan Dam.) Other remote areas with untapped hydropower include Alaska, Iceland and Brazil. As one moves to more remote sites of hydropower, it may be advantageous to pipe the hydrogen to a more convenient location to build a chemical plant rather than to process it in-situ.

Whatever the source of electricity, the salient point to note is that ammonia synthesis is a high value end-use for hydrogen and, unless the electrolytic route can compete economically with steam reforming for this premium market, the prospects for electrolytic hydrogen entering the general fuels market are not good.

Hydrogen made by the steam reforming of low grade sources is likely to be used first for the manufacture of synthetic fuels. This is not simply a matter of the relative costs of fossil and electrolytic hydrogen, but is also for institutional reasons, as the synthetic fuels will be made by the traditional fuels industries using well established refinery procedure. Electrolytic hydrogen, where it can compete, will be used first in chemicals manufacture, principally ammonia and methanol, by the chemical companies.

The transmission of bulk storage of hydrogen, as required for an energy vector, are not believed to pose especially difficult problems. Already, underground pipeline installations exist in Germany (Ruhr Valley) and in Britain (Teeside) for conveying high pressure hydrogen over many kilometers from one chemical plant to another. Considerable experience of operating these installations has been accumulated. Bulk storage of high pressure gases in underground aquifers, salt caverns or exhausted natural gas fields is established technology. In the United States the question of bulk storage and shipment of liquid hydrogen has been addressed as part of the space program. Rail tank cars holding 18,000 gallons of liquid hydrogen have been transported safely around the country and a 900,000 gallon liquid hydrogen storage tank has been used at the Kennedy Space Center, Florida. The bulk handling of liquid hydrogen may be regarded as essentially solved, thanks to the space program.

Most of the technology appropriate to hydrogen as an energy vector seems to be established, at least in outline. Much remains to be done by way of optimization and scale-up, as we shall discuss for the particular cases of electrolyzers and fuel cells. Such work is needed to define the economics of the process; a full knowledge of the economics is vital when one has to choose between options for large-scale energy projects of national importance.

HYDROGEN AS A TRANSPORT FUEL

Today the internal combustion engine reigns supreme. Virtually all road transport, sea transport and aviation are dependent upon it. Only in the field of railways is there a viable alternative (electric traction) and even there electrification is by no means complete because of capital cost considerations. The overwhelming predominance of the internal combustion engine (including gas turbine in aviation) is associated with a total dependence on petroleum-derived fuels for our transport requirements.

The question to be asked is "will it ever be thus?" The easy and unimaginative answer is: "Yes, by virtue of identical fuels synthesized from coal and other low grade fossil sources". However, this begs a lot of questions - technical, economic, social, environmental and political. Without exploring these in depth right now, suffice it to say that the total replacement of the existing petroleum industry by a new synthetic fuels industry based on tar sands, shale oil or coal will pose major headaches for the engineer, the financier and the politician alike. Again the outline technology exists; it is its optimization, scale-up and implementation which will prove so difficult and so expensive.

So what are the alternatives? There are no easy answers, except for electrification of railways (and even this is expensive). Rather there are numerous speculative options to be assessed and evaluated in the years ahead.

Aviation

Although it is generally held that the jet engine burning kerosene is unlikely to be supplanted for aviation, there is one other possibility on the horizon - the hydrogen engine. For aircraft designers, hydrogen has a unique and most attractive property - its high gravimetric energy content compared to conventional aircraft fuel. For an equivalent energy content its mass is only 38% of that of kerosene, (Table 3). This permits a substantial increase in aircraft payload and/or range when using liquid hydrogen on the fuel. On the other hand, the volumetric energy of LH_2 is only one quarter that of kerosene so it is necessary to provide four times the volume to accommodate the equivalent amount of fuel. This poses a serious problem to which aircraft designers in USA are addressing themselves. Possible solutions include wing tip pods as additional fuel tanks, storage of LH_2 in the fusilage above the passengers' heads, or amidships in a stretched aircraft. Preliminary economic assessments which have been made suggest that aviation is the one application (other than rockets) where liquid hydrogen is likely to be cost-effective as a fuel. Much of the basic LH_2 technology for this application already exists, thanks to the space program, but considerable development work would be needed to apply it to the aircraft industry. The cost of building and demonstrating a full size, prototype, LH_2 fuelled airliner will be formidable and for the present people are contenting themselves with design studies[8,9].

Table 3. Comparison of Kerosene and Liquid Hydrogen as Fuels

Fuel	Mass (kg/GJ)	Mass Index	Volume (m^3/GJ)	Volume Index
LH_2	8.33	0.38	0.112	4.00
Kerosene	22.0	1.0	0.028	1.0

Liquid hydrogen is not suitable as a fuel for road vehicles because of the comparatively high cost of liquefaction. The alternatives are to store hydrogen as compressed gas in cylinders, or in the form of a metal hydride which may be decomposed to provide fuel. The metal hydride bed can be recharged with hydrogen when exhausted.

Some research is in progress on light-weight alloy gas cylinders strengthened by overwrapping with a fiber-reinforced plastic composite, but it is not yet clear whether this will be a cost-effective solution.

Several groups of research workers in USA and Europe have investigated metal hydrides as a means for on-vehicle storage of hydrogen fuel, either solely or in conjunction with gas cylinders. The hydrides of the alloys LaNi$_5$ and TiFe have suitable dissociation pressures at ambient temperature, but both are heavy and LaNi$_5$, in particular, is expensive. Magnesium nickel hydride (Mg$_{0.9}$Ni$_{0.1}$H$_2$) is better on mass and cost grounds, but has to be heated to 300°C before it has a suitable dissociation pressure. Despite these problems, the Daimler Benz company in Germany has converted a number of Mercedes-Benz vehicles (cars, vans and a mini-bus) to hydrogen operation and equipped them with hydride storage beds[10]. The problem of adapting the internal combustion engine to burn hydrogen was readily solved. In the case of the mini-bus there was a Ti-Fe hydride bed (ambient temperature) for start-up and a Mg-Ni hydride bed (300°C) to supply some of the hydrogen when hot.

Despite the success of these German demonstration vehicles, the future of hydrogen-fuelled road transport must be regarded as highly speculative. If the hydrogen is generated electrolytically, the overall efficiency from primary fuel source to tractive effort at road wheels is extremely low (<5%) as one is going through two Carnot heat engine cycles, one at the electricity power station and the other in the vehicle's combustion engine. The efficiency is, of course, rather higher if hydrogen is prepared directly from fossil fuels by steam reforming. In addition to the efficiency argument, there are many other problems to be overcome in the introduction of hydrogen-fuelled road transport (Table 4).

Table 4. Problems of Introducing Hydrogen-fuelled Transport

1. Inertia required to change an existing technology, industry and way of life.

2. Long timescale for introducing new energy industry and the large capital investment needed.

3. Technical problems of manufacturing, distributing and storing hydrogen in a form suitable for transport applications on the megaton scale.

4. Technical problems of carrying hydrogen on the vehicle.

5. Manufacture of hydrogen at a price which competes with natural or synthetic liquid fuels.

6. Safety of hydrogen as a transport fuel.

7. Problems of the progressive introduction of hydrogen-fuelled vehicles into the transport system.

8. Competition from electric vehicles.

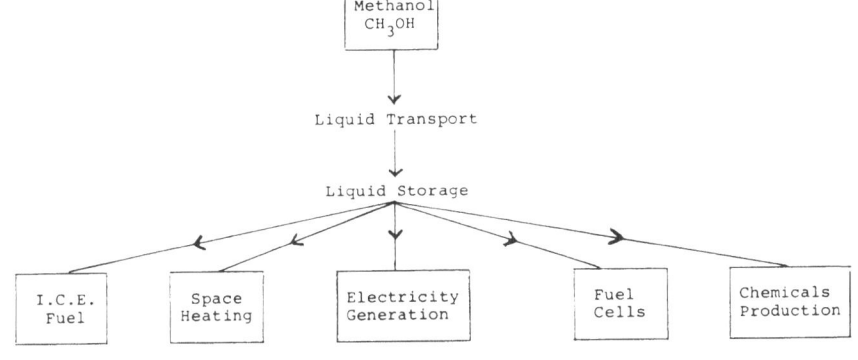

FOSSIL FUELS   RENEWABLE FUELS

Fig. 1. Methanol as an energy vector and fuel.

One way to improve the overall energy efficiency of the hydrogen fuelled vehicle is to burn the hydrogen in a fuel cell to drive an electric vehicle. This approach is discussed in more detail later.

A possible way to overcome the hydrogen storage problem is to use hydrogen to synthesize methanol (CH_3OH) which has all the convenience of a liquid fuel. Methanol has been described, with scant regard for structural chemistry, as "two molecules of hydrogen made liquid by one of carbon monoxide". As such, it may be thought of as a transitional stage between petroleum and hydrogen energy. Its energy content is inferior to that of petroleum, but it may be used as a fuel for internal combustion engines, either alone or as an additive to petrol. Methanol is an extremely versatile fuel, both in terms of the sources from which it can be derived and the uses to which it can be put. We may consider the Methanol Economy (Figure 1) as an alternative to, or aberrant of, the Hydrogen Economy. The utilization of methanol directly in a fuel cell to power an electric vehicle is a particularly attractive long term goal. While devotees of hydrogen energy would not necessarily accept methanol as a substitute fuel, it must be admitted that from a resource and pollution aspect it is more desirable than petroleum, while as a road transport fuel it is much easier to visualize than hydrogen. It may well have a role to play, along with other alcohols (ethanol and isopropyl alcohol) in diversifying the fuel base of road transport in the next century.

HYDROGEN PRODUCTION BY ELECTROLYSIS

Economic and Strategic Aspects of Electrolytic Hydrogen

From any energy strategy viewpoint, there would seem to be three future situations in which it might prove economic to generate hydrogen electrolytically:-

(1) In remote or lightly populated regions of the world where the supply

of renewable energy, particularly hydropower or wind energy, exceeds the local demand for electricity.

(2) As a means of supply/load levelling in the electrical utility network, to store surplus nuclear electricity generated at off-peak periods, or excess electricity available from reasonable energy sources at periods of peak supply (e.g. in gales for wind energy or at spring tides for tidal energy).

(3) Using nuclear power plant dedicated to hydrogen production.

Although the electrolytic hydrogen so produced will be used first in the premium chemicals market, this need not continue to be the case. For example, a country which has a high installed capacity of nuclear power, but is short of hydrocarbon fuels (e.g. France), may find it economic to use electrolytic hydrogen to manufacture synthetic fuels. This raises the interesting concept of a hydrocarbon or methanol molecule which is part fossil, part nuclear in its origins.

At the moment we are far from this situation. Not only, as we have seen, is most hydrogen manufactured by steam reforming or partial oxidation, but where by-product hydrogen is produced electrolytically on a fair scale, as in the chlor-alkali industry, it is frequently burnt to raise process steam for the factory. This cannot be an economic use of pure hydrogen, but it illustrates the logistic problems and capital costs involved in putting a byproduct gas to better use. If a suitable fuel cell were available, the by-product hydrogen would be used to generate D.C. electricity to feed back to the chlor-alkali electrolyzer.

With water electrolyzers designed specifically for hydrogen production, oxygen is an inevitable byproduct and the economics of the overall process will depend very much upon whether this can be put to good use. Possibilities include steel making, the manufacture of high calorific value gas and the treatment of sewage. The combination of a dedicated nuclear reactor and electrolyzer to produce hydrogen and oxygen, with an ammonia synthesis and fertilizer plant to consume the hydrogen and a steelworks to utilize the oxygen, would be a powerful industrial complex, particularly if located close to a source of good quality coking coal and a seaport capable of berthing bulk iron ore carriers. Because of progressive changes in the pattern of energy supply and use in the 21st century, it is likely that there will be corresponding developments in the composition and siting of large-scale industrial complexes. Inevitably, such changes are slow to take place as the capital investments involved are enormous.

Electrolyzer Development

The change in Gibbs energy associated with the electrolysis of one molecule of water may be written

$$\Delta \bar{G}^O = \bar{G}^O_{H_2} + 0.5 \; \bar{G}^O_{O_2} - \bar{G}^O_{H_2O}$$

where $\bar{G}^O$ represent the molar Gibbs energy of the indicated species at a reference pressure of 1 atm. (0.101 MPa).

since $\Delta \bar{G}^O = 2 \; E^O_{rev} \; F,$

where F = the Faraday constant (96,500 coulombs mole^{-1})
and E^O_{rev} is the thermodynamically reversible potential, it follows that

$$E^O_{rev} = \frac{1}{2F} \; (G^{-O}_{H_2} + 0.5 \; G^{-O}_{O_2} - G^{-O}_{H_2O}) \; .$$

At 298K E_{rev}^o = 1.299 V, decreasing almost linearly to 1.088 V at 473 K (Figure 2). Hence there is, theoretically, a significant saving in electrical energy to be made if electrolysis is carried out at 200°C compared to ambient temperature.

In practice the situation is more complicated than shown as there are three further, enthalpic factors to take into account:

i) the entropy term $T\Delta S$ which increases with temperature; this almost cancels out the decrease in ΔG so that the total enthalpy of reaction (ΔH) is almost independent of temperature (1.47 V at 25°C, 1.49 V at 200°C); this is the so-called "thermoneutral" voltage (Figure 2.)

ii) the latest heat of vaporization of the water vapor contained in product gases.

iii) the energy required to pre-heat the feed water to the electrolyzer temperature.

In principle, if the electrolyzer were operated isothermally, the heat corresponding to (i) and (ii) could be absorbed from the surroundings (e.g. from a heating coil or heated jacket). In practice, it is found that due to IR losses in the cell (see below) the heat corresponding to (i) and (ii) is effectively supplied electrically and only the feed water preheating (iii) is of thermal origin. These factors, taken together, make operation at high temperature less favorable than suggested by a consideration of the reversible voltage alone. Moreover, at higher temperatures there are materials compatibility problems and, above 100°C, sealing problems associated with the operation of cells under pressure.

Increasing the operating pressure from 0.1 MPa to 2.5 MPa leads to an increase in the reversible voltage by about 0.7 V (independent of temperature). This represents the increase in ΔG corresponding to the Gibbs energy of compression of the product gases at unit efficiency. For many applications gas is required at high pressure and it is more efficient to operate a pressure electrolyzer than to compress gas liberated at 1 bar pressure. Moreover, high pressure operation reduces the water vapor content of the product gases (on a mass basis) and so reduces enthalpic factor (ii).

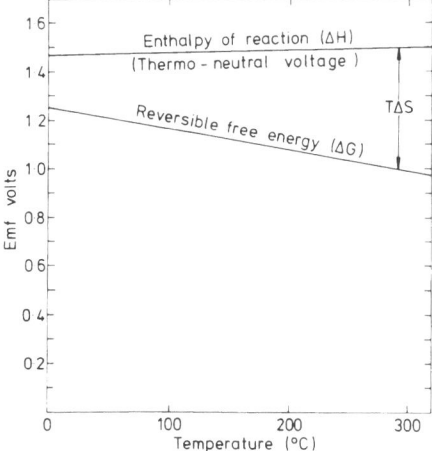

Fig. 2. Theoretical voltage vs. temperature relationship for electrolysis of water.

There are two basic types of electrolyzer, the monopolar (or "tank") type and the bipolar (or "filter press") type (Figure 3). In the tank type the anodes and cathodes are each connected in parallel so that, in effect, the entire electrolysis module is a unit cell. Individual anodes and cathodes are separated by a porous diaphragm, e.g. of asbestos. In the bipolar design there is a metal bipole electrode which has the anode electrocatalyst on one face and the cathode catalyst of the adjacent cell on its other face. The cells in one electrolysis module are thus series connected and the module operates at a higher voltage and lower current than the tank-type design. In building up an electrolysis plant from unit modules, the tank-type modules are series connected while the bipolar modules are parallel connected, thereby leading to reasonable values for the overall voltage and current requirements.

As mentioned in the Introduction, large industrial-scale water electrolyzers were developed in the 1920's and 1930's for synthetic ammonia production. Among the major companies who manufacture this plant are Brown Boveri (Switzerland), Lurgi (Germany), Norsk Hydro (Norway), De Nora (Italy) and Electrolyzer Inc. (Canada.) A recent review of industrial water electrolysis, present and future, has been written by Leroy[11]. The Canadian company has specialized in monopolar (tank-type) electrolyzers, while most of the other manufacturers have concentrated on bipolar design. In each case the electricity consumption lies in the range 4.3-4.6 kWh $(Nm)^{-3}$* of hydrogen gas produced. A typical, large, bipolar electrolyzer module, operating at 60-80°C and ambient pressure, has a hydrogen output of 250-750 Nm^3/h for a power consumption of 1.15-3.5 MW. These modules will be connected in parallel in a fertilizer plant which may consume 20-60 MW of hydro power. The largest industrial water electrolysis plants built to date consume $\sim$ 100 MW; this is still only one tenth of the power output of a commercial nuclear reactor (e.g. a Pressurized Water Reactor) which is indicative of the scale-up problems to be faced in tailoring a water electrolyzer to a nuclear power reactor.

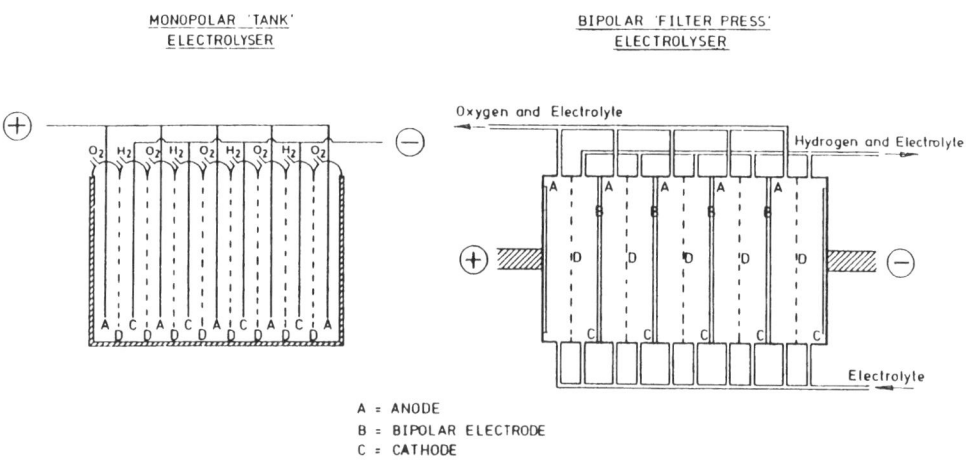

A = ANODE
B = BIPOLAR ELECTRODE
C = CATHODE
D = DIAPHRAGM

Fig. 3. General arrangement of monopolar and bipolar electrolyzer batteries.

---

*$(Nm)^{-3}$ normal cubic metre; i.e. gas at standard temperature and pressure 273 K and 101.3 kPa.

During the past few years, thanks largely to the initiative and foresight of the Commission of the European Committee in mounting a research and development program on hydrogen energy, there has been increased industrial and academic interest in improving the performance of water electrolyzers. Most of this research has been directed towards reducing the cell operating voltage and increasing the current density, so as to minimize both the energy consumption and the capital cost of the plant. Increasing the scale of operations is an important secondary objective.

In addition to the enthalpic factors discussed earlier, the operating cell voltage exceeds the reversible potential by an amount which represents electrical losses in the cell. These arise from four different sources (Figure 4):

  i) Resistive losses in the electrolyte.
 ii) Overvoltage at the anode.
iii) Overvoltage at the cathode.
 iv) Resistive loss in the electrodes.

The cell voltage increases sharply with current density, predominantly because of (i). It is therefore important to choose an electrolyte of maximum conductivity and for this reason aqueous KOH is usually employed. Increasing the concentration of the KOH solution (e.g. from 30-40 wt%) increases its conductivity. There is also a second effect which becomes increasingly significant at higher temperatures; the reduced vapor pressure associated with higher concentration results in a lower water vapor tension in the product gases and so a reduction in the enthalpic effect.

Research on reducing the overvoltages at the electrode is directed towards producing better electrocatalysts. Cells with base metal electrodes (e.g. mild steel cathodes, nickel anodes) operate at, typically, 2.1 V and 0.2 A/cm² at 70°C and atmospheric pressure. By using cathodes electroplated with e.g. nickel boride or mixed sulphides, and anodes activated with transition metal oxides such as spinels or perovskites it is possible to reduce the operating voltage by 400 mV to around 1.7 V at 0.2 A/cm². This represents a significant improvement in electrical efficiency. In order to reduce the capital cost of electrolyzers it is necessary to increase the working current density and bipolar cells operating at up to 1 A/cm² have been designed. High current densities necessarily result in reduced electrical efficiency and a trade-off must be made between capital costs and operating (electricity) costs.

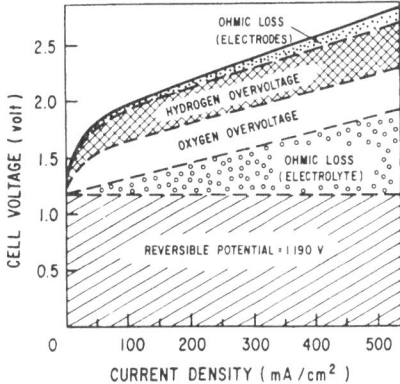

Fig. 4.   Reconstruction of the voltage/current density performance of an unactivated unipolar electrolyzer [taken from Ref.11 by permission of Pergamon Press].

Other research has been directed towards new materials of construction for pressurized electrolyzers using concentrated KOH electrolyte at temperature exceeding 100°C. This poses some difficult materials science problems steaming from stress corrosion cracking of steels, degradation of gaskets, etc. A particular problem has been the replacement of asbestos cloth separators by more esoteric materials, e.g. sheets of porous "Teflon" impregnated with potassium titanate.

A particular approach adopted by General Electric in U.S.A. is the solid polymer electrolyte (SPE) cell in which the porous cloth-type separator is replaced by a polymeric ion exchange membrane which is conductive to cations (Figure 5). The particular membrane employed, NAFION, is a perfluorsulphonic acid polymer which is extremely stable in both acid alkaline solution. Appropriate electrocatalysts are coated on each face of the polymer sheet and these are contacted by a metal mesh current collector. Further research is aimed at reducing the cost and improving the electrical efficiency of the system to make it competitive with conventional electrolyzers.

Fuel Cells

A fuel cell is, in essence, an electrolyzer working in reverse (Figure 6.) It follows that considerations are reversed also and high equilibrium conversions are favored by low temperatures and high pressures. However kinetic factors (electro-catalysis, polarization) cell for elevated temperatures and a compromise must be made. At higher temperature the overall efficiency can be maintained through waste heat recovery.

As mentioned earlier, two types of fuel cell have been developed beyond the prototype stage. These are the alkaline (KOH) electrolyte hydrogen-oxygen cell and the phosphoric acid and electrolyte hydrogen-air fuel cell. Both of these developments took place in the United States. The stimulus for the alkaline fuel cells was provided by the NASA space program, while that for the phosphoric acid fuel cell (PAFC) came from the public service gas and electricity utilities. The gas utilities, who were

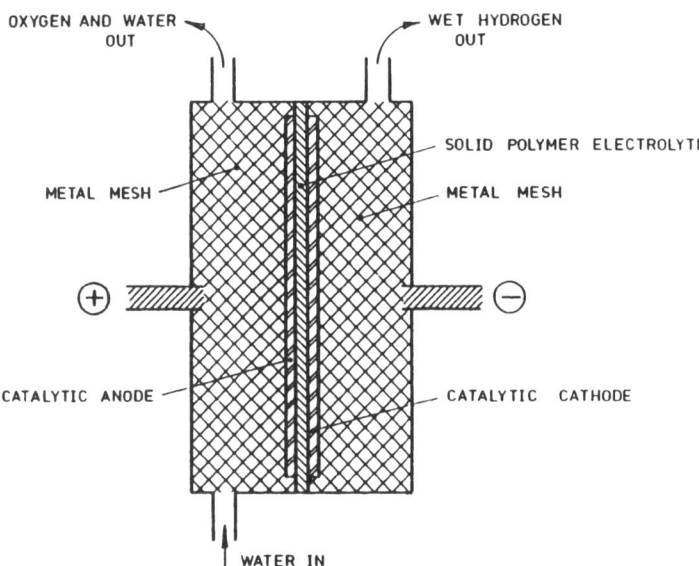

Fig. 5. Schematic of General Electric solid polymer electrolyte cell.

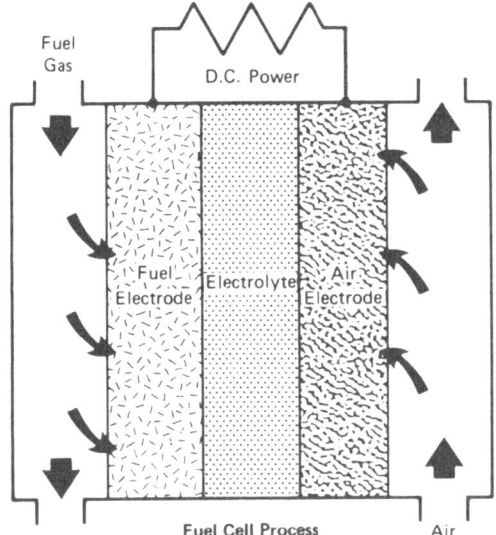

Fig. 6.   Schematic of fuel cell.

first in the field in the 1960's, saw new markets for their ample supplies of natural gas, while the electricity utilities became interested in fuel cells for peak load generation.

These are not the only possible applications for fuel cells as shown in Table 5. Among the static uses, the co-generation of combined heat and power is a potentially important integrated energy system for installation in factories, hotels, apartment blocks, hospitals, department stores, or wherever there is a requirement for both heat and electricity. The overall efficiency (useful heat and electricity) can be as high as 80%, compared to ∿ 40% for electricity alone. PAFC's of 40 kW output have been developed by United Technologies Corporation for this application and these are currently being evaluated in field trials. Within certain limits it is possible to adjust the ratio of electricity/heat produced to suit the application and the season, but broadly speaking fuel cells generate in the ratio 1:1 whereas diesel generators provide a greater proportion of heat (ratio 1:2.5.) These two types of CHP systems therefore serve different markets depending upon the ratio electricity/heat desired.

The fuel for the PAFC is either naphtha or natural gas. These fuels must first be reformed to produce hydrogen and CO_2, and treated to remove any CO or H_2S which would poison the platinum catalyst. Thus it is necessary to have a fuel processor unit proceeding the fuel cell stack. The electricity produced is, of course, D.C. and so for many applications it is necessary to invert the output to A.C. Thus a complete fuel cell plant consists of three units – a fuel reformer, a cell stack assembly and an electrical invertor. The cost of the two auxiliary units is by no means insignificant.

United Technologies Corporation has also built the very large (4.8 MW) demonstration PAFC units in Manhattan and Tokyo for the electricity utilities. These are designed for peak generation near the load centers so as to smooth the demand on the transmission and distribution systems. Base load generation by means of fuel cells appears to be well in the future as the cost targets are such that only advanced, high temperature fuel cells are likely to be economic and these have severe developmental problems of a materials science nature yet to be solved.

Table 5. Applications for Fuel Cells

Static		Mobile	
1.	Combined heat and power (factories, hotels, apartments, etc.)	1.	Heavy goods vehicles
2.	Peak generation in cities.	2.	Buses
3.	Peak generation at power stations (to replace gas turbines).	3.	Railways
		4.	Military
4.	Base load generation	5.	Spacecraft

The situation as regards mobile applications (Table 5) is, with the exception of power supplies for spacecraft, more problematical than static applications. The potential attractions of the fuel cell electric vehicle compared to the conventional internal combustion engined vehicle include:

- Higher fuel efficiency, especially at low power.
- Silent operation.
- No pollutants.
- Modular construction.

Of course, secondary (rechargeable) batteries have the same desirable characteristics, but their limitations include excessive mass (in the case of lead/acid) and limited range between recharges. Fuel cells do not suffer so much from these problems, but have limitations of their own including excessive bulk, low peak power output, high cost, problems of fuel storage and air purification. Phosphoric acid fuel cells can run on liquid fuels but need an on-board reformer which is both expensive and bulky. Alkaline fuel cells, operating on hydrogen, pose problems of hydrogen storage and the need for scrubbers to remove CO_2 from the air. Suggestions have been made for fuel cell/battery hybrid electric vehicles, with the fuel cell providing the base power output and the desired vehicle range, while peak power requirements for acceleration and hill climbing are met by the storage battery.

Altogether, the development of a commercially viable fuel cell powered vehicle poses a formidable challenge to technology and engineering. There is general agreement that the combined characteristics of long operating range, bulky power unit and limited peak power output point in the direction of large vehicles (buses, trucks and railway trains) rather than private cars. In Europe the company ELENCO N.V. (Belgium) is making a determined effort to develop an alkaline electrolyte H_2/air fuel cell for powering an electric bus.

In addition to the two principal types of fuel cell discussed so far, there are also three other types in various stages of development. The outline properties of all five types are summarized in Table 6. The sulphuric acid electrolyte cell is designed for the direct conversion of methanol without the need for a reformer. A great deal of work has been done by Shell Research in the U.K. and the Netherlands on this system. Although the methanol direct conversion fuel cell has been demonstrated in principle, there remain considerable technico/economic problems to be solved in reducing the inventory of platinum catalyst on the electrodes while still maintaining an adequate conversion rate. For the present, work on this fuel cell appears to have stopped and there is a feeling in the

Table 6. Principal Types of Fuel Cell

Electrolyte	Electrode Catalyst	Operating Temperature	Fuel
Alkaline (KOH) (circulating)	Nickel cathode Steel anode or Pt on C electrodes	50-150°C	H_2 (pure)
Phosphoric Acid in matrix	Platinum on Carbon	200°C	H_2 (free of H_2S, low CO)
Sulphuric Acid	Platinum on Carbon	60-90°C	Methanol
Molten Carbonate in Ceramic Matrix	Porous Nickel	650°C	H_2 (free of H_2S)
Solid Oxide	Conducting Oxide cathodes Ni or Co anodes	850-1000°C	H_2 (low purity)

U.K. that further progress is dependent upon a fundamental advance in electro-catalysis.

Interest in high temperature fuel cells stems from:

- Avoiding electro-catalysis problems, and the opportunity to eliminate platinum from the electrodes.
- Being able to accept a less pure fuel gas.
- Producing by-product heat at a temperature high enough to be useful for industrial processing.
- Being more economical for base-load generation.

To balance these attractions are two disadvantages:

- The severe materials science problems which these cells pose.
- The fact that, thermodynamically, exothermic reactions are favored at low temperatures rather than high temperatures; the conversion of fuel at equilibrium will therefore be less than at low temperatures. In practice, the conversion achieved in low temperature cells is kinetically determined and is less than in high temperature cells.

The two types of high temperature fuel cell are quite different from each other (Table 6). The molten carbonate fuel cell, which operates at ∿ 650°C, has a metal anode (nickel), a conducting oxide cathode (e.g. lithiated NiO) and a mixed Li_2CO_3/K_2CO_3 fused salt electrolyte. Sulphur attack of the anode, to form liquid nickel sulphide, is a severe problem and it is necessary to remove H_2S from the fuel gas to <1 ppm or better. However, CO is not a poison. Other materials science problems include anode sintering and degradation, corrosion of cell components and evaporation of the electrolyte. Work continues on this fuel cell in U.S.A. and there is some optimism that the problem will be solved within 10 years.

The second type of high temperature fuel cell, the solid oxide electrolyte fuel cell, is even more doubtful and futuristic. This operates at 850-1000°C and depends upon conduction of $O^=$ ions through a solid oxide electrolyte. In the past ZrO_2, stabilized in the cubic phase by dissolved

CaO or Y_2O_3, has been used, although new conducting oxides capable of operating at rather lower temperatures are being discovered. There are many other materials problems to be solved as well as design and safety problems which put this type of fuel cell well into the 21st century. Both types of high temperature fuel cell are suitable only for static installations in comparatively large units and, if they ever come to pass, will be used by the electricity utilities or in industry.

Reverting to the more mature acid and alkaline electrolyte cells, a comparison of their principal characteristics is made in Table 7. On balance, the PAFC is favored for medium-large scale static units and the alkaline fuel cell for small-medium scale portable or mobile units. For vehicle traction the key factors favoring the alkaline cell are cold start ability, higher power output and a recirculating electrolyte to remove heat. There remains the problem of storing hydrogen on board the vehicle or reforming a suitable, clean fuel such as methanol.

The electrode reactions for the two types of aqueous electrolyte fuel cell are shown in Table 8. At each electrode it is necessary to have a three-phase contact between the gas, the electrolytes and the solid electro-catalyst and current collector. The art of designing a fuel cell electrode lies in ensuring that the gas/electrolyte interface is stable within the electrode structure. Clearly, if the electrode dries out it will be imperative; conversely if it floods, reactant gas will be excluded from the electro-catalyst. Most designs of cell stack incorporate a diffusion grid to admit the gas and a composite electrode structure which admits electrolyte on one face but prevents its passage through to the gas side. This may take the form either of a dual pore structure, with the fine pore layer adjacent to the separator, or a "wet proof" type electrode which is hydrophilic on the separator side and hydrophobic on the gas side (Figure 7). Considerable ingenuity and physico-chemical skill has gone into design and construction of these electrodes so as to ensure their stability and long life, together with the maximum power output from the minimum content of platinum.

Table 7. Comparison of Acid and Alkaline Fuel Cells

Property	Phosphoric Acid Fuel Cell	Alkaline (KOH) Fuel Cell
Fuel gas (H_2) purity	Tolerant to CO, CO_2 impurities	Only pure H_2 may be used
Air scrubbing	Not needed	Essential
Temperature of operation	200°C	60–120°C
Suitability for CHP	Suitable	Operating temperature is generally too low
Cold start-up	No; must be kept hot	Yes; fast start-up
Power density	Low	High, when run with pure O_2
Circulating electrolyte	No	Yes; aids cooling
Electrical conversion efficiency	40–50%	50–60%

Table 8. Electrode Reactions for H_2/O_2 Fuel Cell

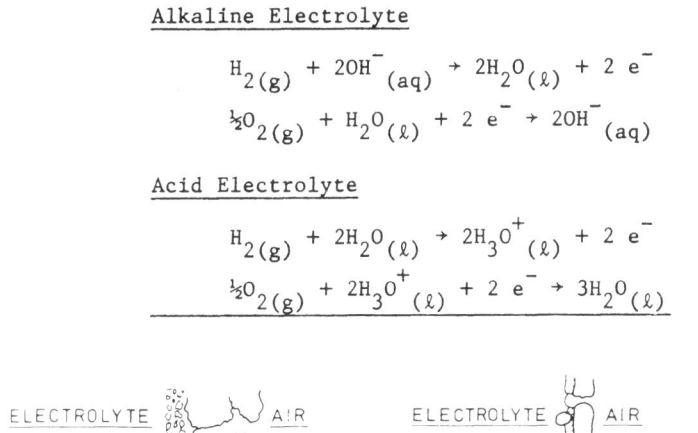

Alkaline Electrolyte

$$H_{2(g)} + 2OH^-_{(aq)} \rightarrow 2H_2O_{(\ell)} + 2\ e^-$$

$$\tfrac{1}{2}O_{2(g)} + H_2O_{(\ell)} + 2\ e^- \rightarrow 2OH^-_{(aq)}$$

Acid Electrolyte

$$H_{2(g)} + 2H_2O_{(\ell)} \rightarrow 2H_3O^+_{(\ell)} + 2\ e^-$$

$$\tfrac{1}{2}O_{2(g)} + 2H_3O^+_{(\ell)} + 2\ e^- \rightarrow 3H_2O_{(\ell)}$$

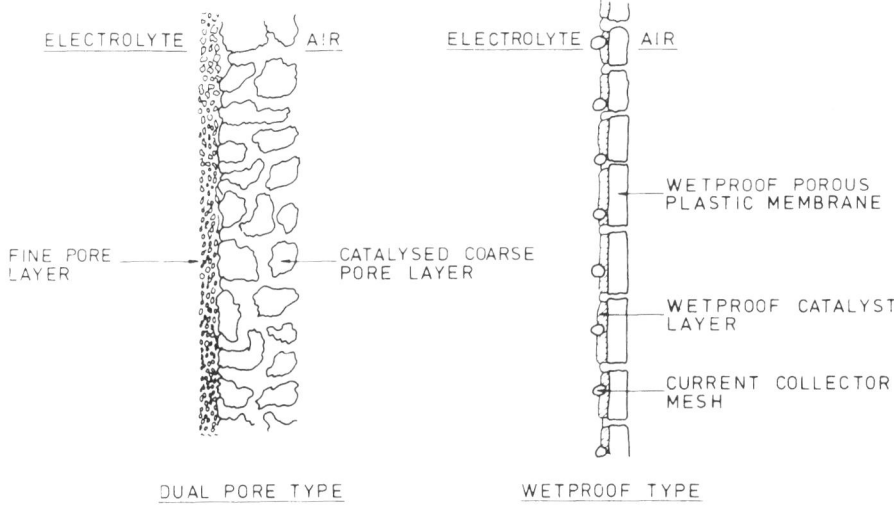

Fig. 7. Schematic of two types of air electrode (cross sections).

Generally, the platinum is dispersed in the form of microcrystallites supported on carbon particles which are formed into an electrode with an appropriate plastic binder. Control of porosity and pore size distribution are critical factors in electrode optimization. The individual cells are joined together in a bipolar array. This necessitates careful sealing around the bipole to prevent mixing of the reactant gases and/or electrical shorting though the electrolyte. Concentrated KOH solution is notoriously difficult to contain, particularly at elevated temperatures. The water formed by the discharge reaction must be removed as steam to prevent dilution of the electrolyte. All in all, the development of fuel cells poses some difficult, but challenging, scientific and engineering problems. Steady progress is being made and technical solutions to certain of the problems are now in sight.

CONCLUSIONS

In this paper an attempt has been made to indicate the important role that electrochemistry will play in the future world energy scene, particularly in the context of hydrogen as an energy vector of the next century. The author does not subscribe to the view that, on this timescale, hydrogen will become a universal fuel. Rather, the 21st century is seen as being a period of transition and change in primary energy sources and manufactured fuels, possibly leading ultimately to a Hydrogen Economy in later centuries.

During the transitional period, which may last more than a century, the emphasis will be towards:

- Conservation of petroleum for premium uses.
- Manufacture of synthetic fuels, both liquid and gaseous from lower grade hydrocarbon fuel stocks and from coal.
- Development of nuclear power for electricity generation.
- The adoption of renewable energy sources for electricity generation in locations where these prove to be economic.
- Diversifying the fuel base of transport.

On this scenario, hydrogen will have a number of possible major roles to play:

(1) In the up-grading of solid or low grade hydrocarbon feedstocks to liquid and gaseous fuels.
(2) In the manufacture of methanol.
(3) As a fuel for use in fuel cells for combined heat and power schemes and for local electricity generation in cities (urban load leveling).
(4) As an energy storage medium for nuclear power and renewables.
(5) As a vector for transmitting renewable energy to the point of use.
(6) As an aircraft fuel (LH_2).
(7) As a fuel for fuel-cell powered electric vehicles.

It is difficult to be at all quantitative as to when and to what degree these various possible applications will come to pass. Among the many factors which will determine the future energy scene are technical factors (advances in fuel cells, electric vehicles, electrolyzers, LH_2-fuelled aircraft, etc.), environment factors (SO_2 emissions, mining of fossil fuels, etc.) and, of course, the ubiquitous economics and politics which control all major human activities. What does seem clear is that, in the early years, synthetic fluid fuels will be manufactured by steam reforming, both for economic reasons and for institutional reasons associated with the expertise of the petroleum and gas industries. Electrolytic hydrogen will enter upon the scene more slowly, as it will be dependent upon the availability of cheap or surplus electricity and will tend to be produced by the chemical industry or electricity utilities rather than by the fuel industries. Moreover, its first use is likely to be for chemical synthesis, rather than as a fuel.

The role of hydrogen as a transport fuel is questionable. Perhaps the best prospect is for LH_2 as an aircraft fuel. For road transport one is very dependent upon future technical advances in fuel cells and in hydrogen storage, as well as development in competing alternatives, such as advanced batteries for electric vehicles and improved internal combustion engines.

Within this overall scenario, it is evident that future developments in electrochemistry have the potential to make a major impact on society in the 21st century in terms of energy production and use, public and private transport and environmental impact. As research scientists, it is not our role to decide, or even to predict, what society will choose to do in the future. Rather, our role is to identify technical opportunities and to develop engineered options from which society may select the preferred choice when the timing is correct and in the light of prevailing political, economic and other factors. In this context, electrochemists face the challenge of clarifying the contributions that their science can make, within a realistic engineering and economic framework, to the future world energy scene.

REFERENCES

1. W. R. Grove, Phil.Mag., 14:127 (1839); 21:417 (1842).
2. Jules Verne, "The Mysterious Island," Part 2, "Abandoned," Ch.11 (1870).
3. F. T. Bacon, Brit.Pat.725661 (1955). Review in J.Electrochem.Soc., 126:1, 7C-17C (1979).
4. D. P. Gregory, Fuel cell research and development in the United States, in: "U.K. Report/Science and Engineering Research Council, RL-82-055," ed., G. E. Gallaher-Daggitt (1982).
5. vd. Publications of the International Association for Hydrogen Energy and the International Journal of Hydrogen Energy.
6. G. D. Howells and A. S. Kallend, Acid rain - the CEGB view, Chem.Brit., 20:407 (1984).
7. K. F. Langley, "The Future Role of Hydrogen in the U.K. Energy Economy," U.K. Dept. of Energy Report ETSU R15 (1983).
8. G. D. Brewer, A plan for active development of LH_2 for use in aircraft, Internat.J.Hydrogen Energy, 4:169 (1979).
9. K. G. Wilkinson, An airline view of LH_2 as a fuel for commercial aircraft, Internat.J.Hydrogen Energy, 8:793 (1983).
10. H. Büchner, The hydrogen/hydride energy concept, Internat.J.Hydrogen Energy, 3:385 (1978).
11. R. L. LeRoy, Industrial water electrolysis: present and future, Internat.J.Hydrogen Energy, 8:401 (1983).

DISCUSSION

G. Hills

University of Strathclyde
Glasgow
Scotland

CHAIRMAN'S SUMMARY

Professor H. W. Nürnberg (Jülich) referred to several aspects of Dr Dell's
paper. To begin with, he expressed his support for the general strategy of
the hydrogen economy as set out in this paper. He agreed that the
establishment of a comprehensive hydrogen economy was not an immediate
prospect but that hydrogen would find an important preliminary role in the
manufacture of other liquid fuels.

Professor Nürnberg then referred to the speaker's remarks concerning the
inefficiency of involving two Carnot cycles in the production and consumption of hydrogen. He remarked that this was a serious matter only if
fossil fuels remained the primary power source. Where the primary source
was solar power or nuclear power, then this consideration was less important. Apart from the attendant thermodynamic inefficiency, the involvement
of fossil fuels perpetuates the problems of atmospheric pollution including
that of carbon dioxide. These problems will only be solved by using the
correct primary energy sources.

Secondly, reference was made to the effects of increasing the temperature of the electrolysis of water. Whilst the thermodynamic gains might
be modest, there were significant improvements in the kinetic performance.
Finally Professor Nürnberg referred to the current-voltage curves cited in
the paper. He stressed that the performance of the ambient temperature
alkaline cells developed at Jülich was already considerably imposed.
Current densities of 4.5 kA m^{-2} can be sustained at an operating voltage of
1.5 V, i.e. about 1 V less than that cited in the paper. These new cells
therefore offer an immediate increase in efficiency of 30 per cent over
those at present on the market. The new Lurgi cells are essentially of
Julich design and no longer based on the older, pressure system. The
performance of the new cells is at least as good as those based on the new
solid polymer electrolytes. The polymer systems also suffer the disadvantages of high capital investment. The Julich cells use nickel electrodes,
new nickel oxide diaphragms and new cell configurations. The cost of
producing hydrogen from them is now only 10-15 per cent higher than that of
the steam reforming process which, it is worth noting, is also a polluting
process. Electrolytic hydrogen is clean hydrogen and therefore environmentally desirable. Several studies have concluded that by the year 2030,
as fossil fuels became more scarce or more expensive, there will be an

established need for electrolytic hydrogen, i.e. hydrogen from the splitting of water.

In reply, <u>Dr Dell</u> remarked that in the written text there is an acknowledgement of the extensive work done by the European Community or hydrogen production. It is an excellent example of international co-operation. He did not support the concept of thermochemical cycles and doubted whether the investment in R and D in that field had been worth while. In contrast, the researches into electrolyzer design seemed to have been a good investment.

He agreed that the problem of a double Carnot cycle disappears when the primary energy source is no longer a fossil fuel. It remains true that a successful hydrogen economy depends on cheap electricity and consideration must be given as to where it will come from. By the 1940's, Norway had developed abundant hydropower. Electricity there is still cheap. More recently, Iceland has found itself with excess electric power and has considered storing and using that in the form of hydrogen. More recently still, France has made a massive investment in nuclear power and this seems an obvious basis for the production of hydrogen. Indeed, since most power stations, and certainly nuclear power stations, function best at constant load, then stand-by electrolyzers become a necessary component of an integrated energy generation - energy storage program. Under these circumstances, the excess electricity is virtually free and it would not be surprising if France became a leading proponent of water electrolysis and the hydrogen economy.

<u>Dr Hills</u> asked why neither Dr Dell nor Professor Nürnberg had referred to the room-temperature electrolytic reforming of carbon or coal by using it to depolarize the anodic production of (waste) oxygen. In reply, <u>Dr Dell</u> acknowledged that the development had been noted but that it had not yet been scaled up or shown to work in an industrial electrolyzer.

<u>Professor H. Wendt</u>, (Technical University, Darmstadt) then referred to the considerable program of work on the electrolytic basis of the hydrogen economy that had been launched seven years ago under the auspices of the European Community. It was noted that France and Belgium had taken steps to increase their output of nuclear power greatly, which would leave them free to dispose of considerable quantities of excess electricity. As a result, attention was given to the range of uses and methods of production of hydrogen. In the end, however, effort was concentrated on the electrolysis of water, especially under alkaline conditions where it was recognized that considerable progress would be made.

It has always been evident from the current-voltage curves of conventional electrolysis that there remained a substantial ohmic drop resulting from the relatively high internal resistance of the cell. The consequential power loss made these systems uneconomic and the principal target of the studies described here was to improve the technology of cell construction and cell operation such that these ohmic losses were reduced to a minimum.

For example, an increase in operating temperature significantly reduces the resistance of the electrolyte. A two-fold or three-fold increase in conductivity is readily achieved in this way, i.e. by increasing the working temperature from 100 to 150°C. Under these circumstances it is no longer possible to use asbestos as a cell material and new diaphragms are needed. In designing such diaphragms, every effort needs to be made to reduce electrode separation. So-called zero separation cells were therefore developed on the basis of extremely thin electrode separators. These were composite structures, the strength and rigidity of which allowed the previous separation thickness, for example, 2 mm in the case of asbestos, to be reduced to 0.2 mm or less. The four kinds of diaphragms studied were as follows:

(1) Polysulphone impregnated with hydrophilic dispersion of antimony pentoxide,

(2) polymer coated asbestos,

(3) teflon tissue coated with a hydrophilic material,

(4) metal mesh coated with a porous oxide.

The use of such separators reduced the ohmic resistance from 1Ω cm^{-2} to 0.02Ω cm^{-2}.

Pressed between appropriate electrodes, these diaphragms form the basis of highly efficient electrolyzers. The most effective anodes (stable as well as catalytically active) were made of a cobalt spinel. The cathodes were of finely divided nickel doped with molybdenum (or other transition metals). The resultant hydrogen evolution overvoltage was reduced to 150 mV at 1 Acm^{-2}. The anode overvoltage was not reduced to the same degree and remained at 250 mV at 1 Acm^{-2}. The resultant current-voltage characteristics were nevertheless a considerable improvement on those of previous cells. The new design is being incorporated into large scale electrolyzers of the Lurgi type and in the manner and with the results as described by Professor Nürnberg.

If one now examines the techno-economic considerations, it is evident that the new electrolyzers come close to being economically viable. The hydrogen energy price is good but remains wholly dependent on the price of electricity: At present commercial electricity prices the cost of electrolytic hydrogen is still too high compared with that of the chemical process. Only when the new technology can be fed with cheaper electric power is it likely to be widely adapted. What can be said is that the enabling technology has been developed considerably. A door to the hydrogen economy has been opened wide.

Professor K, Wiesener, (Technical University, Dresden) returned to the subject of electrode depolarization and drew attention to the use of hydrogen itself as a depolarizer. In many metal electrowinning processes, considerable inefficiencies arise at the anode. These can be reduced substantially by introducing hydrogen and changing the anode reaction to the oxidation of hydrogen. Examples were given of the effect of using tungsten carbide anodes depolarized by hydrogen. The consequential reduction in cell voltage was 1.6 V.

It is therefore proposed that the waste hydrogen from say, the chloralkali electrolysis process might be used (in tandem) to depolarize ready metal winning cells. The resulted overall reduction of overvoltage energy ions could be as much as 0.8 V.

Prof. H. W. Nürnberg and J. Divisek
Nuclear Research Center, Jülich (communicated):

At present the hydrogen production is still dominated by steam reforming of fossil fuels. Only 3% of the total hydrogen production are obtained by electrolysis. This electrolytically produced hydrogen is at present only a special chemical commodity for applications where hydrogen of high purity degree is definitely mandatory. The reason for this situation is that conventional alkaline water electrolysis (AWE) is technically imperfect and requires much too large amounts of electricity.

Meanwhile the situation has changed substantially, particularly as consequence of the achievements and results obtained since 1977 in the "Hydrogen Technology Program" of the International Energy Agency (IEA) and in the "R & D Hydrogen Program" of the Commission of the European Communities. Both programs demonstrated the new potentialities for enormous improvements in electrochemical hydrogen production, a basic prerequisite for the use of hydrogen as an energy vector either indirectly via the improvement of cheap fossil fuel or directly as hydrogen fuel.

The improvements achieved by our Institute in Jülich have been particularly successful and fundamental with the development of the new "Jülich Alkaline Water Electrolysis (JAWE)" which has been a project in Annex IV of the IEA-program "Hydrogen Technology - Electrolytic Hydrogen Production"[1-9].

By the JAWE-process, now being introduced into the industrial production of future electrolysis plants, three fundamental major improvements have been achieved.

The operation temperature of the cell is increased to 100°-120°C. This improves the electrode kinetics and causes a decrease of the iR-drop in the electrolyte without causing corrosion problems at the same time.

The decisive prerequisite for the increased operation temperature was the development of a new ceramic diaphragm based on NiO[1-4]. This ceramic diaphragm is absolutely stable in hot potassium hydroxide solution, has excellent separation properties for both product gases, hydrogen and oxygen, and a very low ohmic resistance. Thus, iR-losses amount to only 50-100 mV at 5000 A m^{-2} and 100°C. The costs of this new diaphragm are low and do not exceed that of the hitherto conventional asbestos diaphragm. In contrast to this new NiO-diaphragm other new diaphragms based on teflon/potassium titanate are not reliable, due to the solubility of potassium titanate in hot KOH[5].

Also highly catalytically active Raney nickel electrodes have been developed. Their production is possible at remarkably low cost by cathodic deposition of a Ni/Zn alloy and subsequent activation by a treatment with hot KOH[6]. These electrodes are used as cathode and anode. Their oxygen overvoltage is below 200 mV and their hydrogen overvoltage less than 100 mV at current densities of 4000-6000 A m^{-2} and 100°-120°C[7].

Thus, at current densities providing a hydrogen production rate at least twice as high as that in the conventional electrolysis plants operated hitherto a cell voltage of only 1.5 to 1.6 V is required. The low cell voltages were confirmed experimentally in long term studies (over 6000 h) of the new JAWE electrolysis concept using a sandwich construction with "zero gap" between the electrodes and the new hydrophilic ceramic diaphragm.

The new situation with respect to cell voltage improvement by the JAWE concept is demonstrated in Figure 1. It is to be expected that the stable total cell voltage of only 1.5 V will be required under technical conditions at 120°C for an economically optimal current density of 4000 - 5000 A m^{-2}. This corresponds to a saving of 600 mV or more compared with present conventional alkaline water electrolysis (AWE) plants operated, due to the limitations caused by the asbestos diaphragm, at 80°C. The construction element of the new JAWE electrolyzer are scarcely more expensive than those of the present conventional electrolyzers, although the specific current load and consequently hydrogen production rate is for the JAWE at least twice as high. As a consequence the specific capital costs for the JAWE electrolyzer are evidently lower than for conventional devices. The most important factor is, however, that compared to the conventional electrolyzers the electricity saving[9] and consequently the decrease of the energy cost amounts to 20-30% for the JAWE (Figure 2). The Solid Polymer Electrolysis (SPE) also provides a low cell voltage and an electricity demand comparable to that of JAWE (Figure 1); yet the capital costs are significantly higher for the SPE[9], see Figure 2.

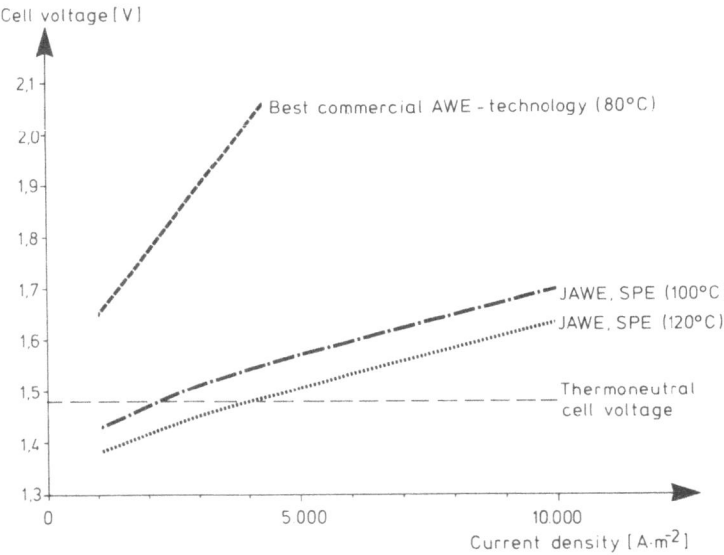

Fig. 1.   Comparison of required cell voltage between conventional alkaline water electrolysis (AWE) and advanced JAWE.

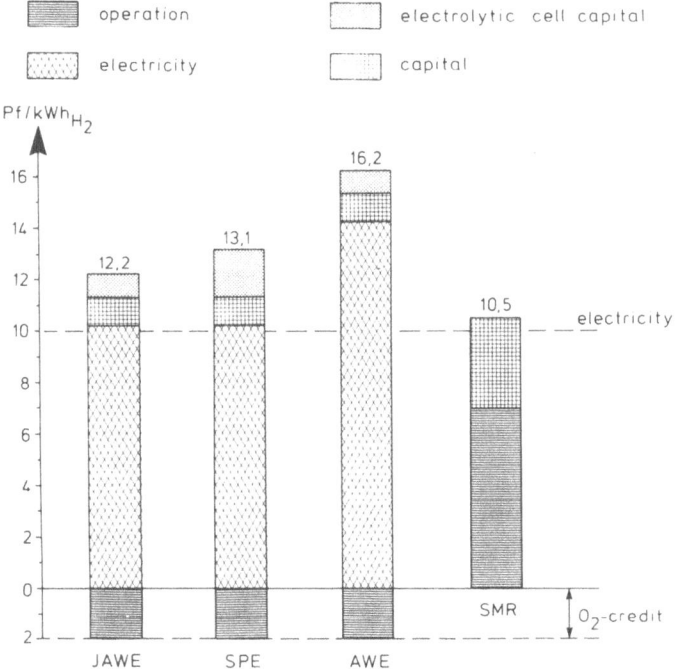

Fig. 2.   Costs (in German Pfennigs) of various hydrogen production processes.

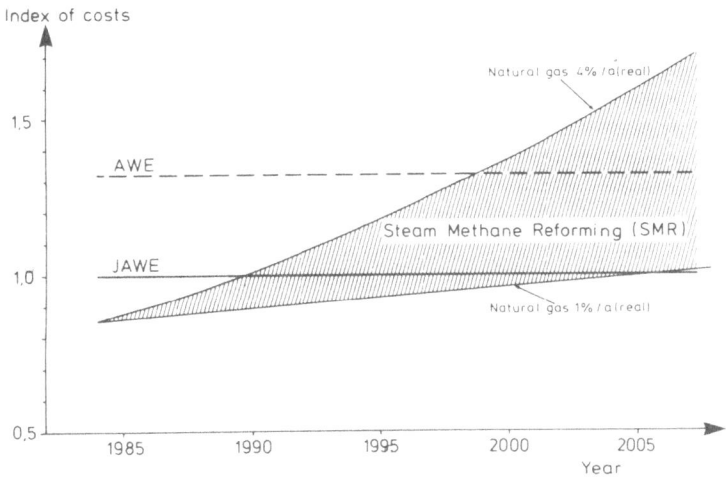

Fig. 3. Development of inflation corrected costs for hydrogen production by SMR in comparison to electrolytic hydrogen production by present conventional alkaline water electrolysis (AWE) and advanced JAWE.

Recent economic assessment[8,9] shows that the total costs of electro-chemical hydrogen production with the advanced JAWE technology in 1984 already exceeds the costs of Steam Methane Reforming (SMR) hydrogen only by 20% (Figures 2, 3). This cost consideration includes an oxygen bonus corresponding to 15% of the total costs (Figure 2). Without this bonus the hydrogen costs for JAWE in 1984 exceed those for SMR hydrogen production by 35%. However, the outlook in the near future for electrolytic hydrogen is quite bright[9]. Under the quite conservative assumption that the real price increase, i.e. corrected for inflation for natural gas, amounts to only 1-4% per year, it becomes evident that under the proviso of constant, inflation corrected, costs for JAWE electrolytical hydrogen production could become competitive to SMR hydrogen production still in this century (Figure 3) regardless of the existence of an oxygen bonus[9]. The present cost situation[9] for JAWE, SPE, conventional alkaline water electrolysis (AWE) and SMR is compared in Figure 2.

REFERENCES

1. J. Divisek, H. Schmitz and J. Mergel, Chem.-Ing.-Technik., 52:465 (1980).
2. J. Divisek, H. Schmitz and R. Hesse, Poröses diaphragma für alkalische elektrolysen, dessen herstellung und verwendung, D. Patent 3031064.
3. P. Malinowski and J. Divisek, Diaphragma auf NiO-basis und verfahren zur herstellung desselben, D. Patent 3318758.4.
4. J. Divisek and H. Schmitz, Int.J.Hydrogen Energy, 7:703 (1982).
5. G. Imarisio, Present Status and Prospects of the CEC Hydrogen Programme, in: Proc. IEA Workshop "Electrolytic Hydrogen Production," A. Mezzina, ed., Annex IV, Brookhaven Nat. Lab. (1983).
6. J. Divisek and H. Schmitz, Porous electrodes of a bipolar electrolytic cell for water electrolysis, in: "Extended Abstracts of 32 ISE Meeting, Dubrovnik," B. Lovrecek, ed., 2:1101-1104 (1981).
7. J. Divisek, P. Malinowski, J. Mergel and H. Schmitz, Improved con-struction of an electrolytic cell for advanced alkaline water electrolysis, in: "Hydrogen Energy Progress V," T. N. Veziroglu and

J. B. Taylor, eds., Pergamon Press, Oxford, pp.655-664 (1984).

8.   J. Divisek, New developments in alkaline water electrolysis for pro-
     duction of hydrogen as an energy vector, in: "Proc. 3rd Int. Symp.
     on Advances in Electrochemical Science and Technology," Transaction
     of the SAEST, Karaikudi, India (1984).

9.   M. W. Nürnberg, Some general aspects of the significance and potenti-
     alities of hydrogen in future energy technologies, in: Proc. 3rd
     Int. Seminar "Hydrogen as an Energy Carrier," G. Imarisio and A. S.
     Strub, eds., Reidel Publ. Co., Dordrecht - Boston - Lancaster,
     pp.16-23 (1983).

Prof. E. Gileadi
Tel Aviv University, Israel (communicated):

1.   An important factor, at least in the early stages of implementation
     of the hydrogen economy, will be the large quantities of hydrogen
     obtained as a by-product of chlorine production. This amounts to
     close to one million tons of hydrogen produced annually worldwide.
     At present most of this hydrogen is used in the plants as a fuel, for
     lack of proper markets and means of transportation. However, in the
     framework of the hydrogen economy this will no doubt be the cheapest
     source of hydrogen.

     In this context one doubts the need for oxygen depolarized
     cathodes in the chlor-aklali industry. If successful, such designs
     will save about 0.8 V of cell voltage at the cost of producing no
     hydrogen. At such a low voltage, and with no extra investment cost,
     hydrogen production should be viable even for its calorific value, let
     alone its possible uses in internal combustion engines and fuel cells
     or in the production of synthetic fuels.

2.   A possible future use of hydrogen as a cryogenic medium has not been
     mentioned. The best superconducting materials developed so far have a
     transition temperature close to the boiling point of hydrogen. With
     the development of new materials having transition temperatures higher
     than the boiling point of hydrogen, a whole new technology of electro-
     mechanical devices can be developed, which may require significant
     amounts of liquid hydrogen for refrigeration.

SECTION II
ELECTROCHEMISTRY AND THE ENVIRONMENT

ANALYSIS AND REMOVAL OF POLLUTANTS

R. Kalvoda

The J. Heyrovský Institute of Physical Chemistry
and Electrochemistry
Czechoslovak Academy of Sciences, Prague

INTRODUCTION

One of the most pressing tasks for mankind during the scientific and technical revolution, in addition to efforts to preserve peace, is the protection of the environment and creation of conditions for modification of the environment in an ecologically optimal way for the further development of our life on this planet.

From the historical point of view it can be mentioned that the problem of environmental protection is not new: These questions were discussed already by Hippocrates (460–377 B.C.) in his book on Air, Water and Environment.

The constantly increasing pollution of the biosphere is a result of increased energy consumption, rapid industrial development and transportation requirements. In energy production based on thermal power using fossil fuels, pollution arises primarily from the ash and sulphur oxides which is also produced by gaseous and liquid fuels, especially heavy oils. These sulphur oxides are also responsible for the problem with the so-called acid rain and its consequences. In large cities traffic is a major source of pollution. Carbon monoxide is particularly harmful, as are uncombusted polycyclic hydrocarbons and nitrogen oxides. The problem of lead compounds used as fuel additives is also important in this connection. Long-term use of these substances leads to an increase in the amount of lead, particularly in the vicinity of highways, in the air, in agricultural plants, water, etc. Further the chemical, metallurgical and food industries have a bad influence on the biosphere. Agricultural development, especially that involving the use of artificial fertilizers, pesticides, growth factors, etc. produce similar problems. Another important problem is the purity of water, i.e. the situation in water reserves. Heavy metals coming from industrial production are an important problem in ground water bodies exploited for water supply. Also the purity of river and seawater is important for all living organisms.

It is thus imperative that modern societies find means of limiting pollution of the biosphere. A great variety of branches of science participate actively in the solution of environmental problems, including medicine, biology, physics, chemistry and also electrochemistry. The latter is important in several reports. In addition to the analytical aspects, electrochemistry is used in the removal of toxic substances or other products and wastes polluting the environment as a result of industrial activity, from the point of view of agriculture or of communal hygiene. Electrochemistry can also help in decreasing or eliminating the occurrence of some substances, e.g. by wasteless techniques, especially in the chemical industry and related fields. Moreover, electrochemistry can, and most probably will, affect energy production and all kinds of transport, which are among the major polluters of the environment.

This article summarizes the potentialities offered by electrochemistry from an analytical point of view, in monitoring various substances polluting or affecting the environment as well as from the aspect of removal of pollutants of any kind - toxic compounds or other types of wastes influencing the environment and the health of the population. For more details see[1,2,3].

ENVIRONMENTAL ELECTROANALYSIS

Analytical chemistry plays an important role in the protection of the environment. This branch of chemistry finds application in the determination of pollutant concentration - both qualitative and quantitative - in the biosphere, in determining the pollutant pathway from the source to man (or other object of interest) and also in elucidating further transformations into other substances along this pathway, e.g. as a result of the interaction among various pollutants, as a result of metabolism etc. Analytical chemistry is also a means for evaluation of the effectiveness of various processes that prevent the formation of pollutants, or that remove those already formed.

Most applications of environmental analysis involve trace determinations, often at a ppb level or lower. High sensitivity of the methods must be accompanied further by sufficient selectivity, precision and accuracy. Easy sample treatment and rapidity of the analytical procedure are also desirable. Because series analyses are often required, methods that are easy to automate are advantageous. In the selection of the method, the cost of instrumentation that must be available in a great number of laboratories is also important. Measurements must be often carried out in the field and thus large apparatus is excluded even if it fulfills all other criteria. It need not be emphasized that microanalytical instruments should be applicable to a wide range of substances (provided that they are not single-purpose analyzers or monitors) and that it is advantageous if several components can be determined simultaneously. Understanding of this theoretical bases of the particular method is clearly a requirement for their successful use in practice.

All above mentioned criteria are fulfilled by electrochemical methods. Electrochemical methods are attractive for ecological research also because they enable immediate measurement of changes in the component concentrations and because they can be frequently used for continuous monitoring and are suitable for field work, where the systematic error caused by transport and storage of the sample is avoided. It is true that there are a number of much more sensitive physical methods (X-ray electron difraction, neutron activation analysis, mass spectrometry, etc.) which, however, cannot be used in the field and are very expensive.

Practically all commonly employed electroanalytical methods can be used in environmental analysis: the choice of method depends on the character of the compound to be determined and of the matrix in which it occurs, as well as on sensitivity and selectivity requirements. The principal methods are voltammetry and polarography, potentiometry, coulometry and conductometry.

Polarography and Voltammetry. Of the various polarographic methods, differential pulse polarography (DPP) finds the greatest use in environmental analysis because of its sensitivity enabling determination of concentrations down to 10^{-6} or 10^{-7} mol^{-1} (approx 100 - 10 ppb). The DPP can be used to determine inorganic ions and compounds as well as organic compounds with an electroreducible group or compounds which can be oxidized at the electrode. DPP can be used even for a determination of electro-inactive organic compounds which exhibit surface active properties. In this case are followed the s.c. tensammetric peaks formed in the region where the adsorption/desorption process takes place.

Especially for heavy metal ions at concentration below the ppb level anodic stripping voltammetry (ASV) can be used, where the test metal is electrolytically deposited on the electrode and then is stripped off under voltammetric control. In addition to this electrolytical accumulation mode the adsorptive accumulation of surface active compounds at the mercury electrode is becoming important in trace analysis. Adsorptive stripping voltammetry (AdSV) enables not only the tensammetric determination of polarographically nonreducible organic compounds (like detergents, drugs, polychlorinated biphenyls, etc.) but yields significant possibilities in voltammetric analysis of many electroreducible compounds with surface active properties like complexed metal ions (e.g. Ni, Co, U) and organic compounds in the ppb and sub ppb concentration range. The two types of stripping methods in combination with DPP are of the most sensitive analytical methods. Because these stripping methods can be used even with d.c. polarography - which can be implemented by extremely simple instrumentation (consisting of 2-3 operational amplifiers), stripping voltammetry belongs among methods using typically "low cost" instrumentation.

From voltammetry and polarography are derived also many single purpose analyzers mainly for monitoring various gases (O_2, SO_2, H_2S, CO, etc.). Polarographic and voltammetric detectors find important applications in high-performance liquid chromatography (HPLC). This combination of the highly selective separation method with sensitive polarography leads to a considerable broadening of the scope of substances which can be determined by the chromatographic method. It can be expected that this advantageous combination will become one of the most frequently used analytical methods even in environmental chemistry.

Potentiometry. From the point of view of environmental protection the ion selective electrodes (ISE) find the most applications for monitoring the activity of ions in solution. One of the important measurements for the characterization of the environment is the measurement of the pH, which is now mostly carried out using glass electrodes. A glass electrode covered with a semipermeable membrane having an electrolyte film is also the basis of various sensors applicable to analysis of waters and the atmosphere (NH_3, SO_2, NO_x, H_2S, etc.). The development of ISE is one of the greatest successes of analytical chemistry in the sixties and seventies. ISE's for many anions and cations are now common. These electrodes can be used for measurements at concentrations from cca 10^0 or 10^{-2} down to 10^{-6}M. Measurements with ISE's are rapid, simple and easily applicable to continuous measurements. Also sensors based on MOSFET semiconductors (ISFET, CSFET) should be mentioned. Selectivity is attained by choice of a suitable membrane or a layer deposited on the transistor gate. Semiconductor sensors are the basis of monitors of various gases and vapors. Miniaturization of the sensors makes it possible to locate different sensors on a single support in the form of a "mosaic", thus yielding an analyzer for a whole range of compounds: the signals of the individual sensors can be then periodically recorded in a measuring center.

Coulometry. Two methods of coulometry are used: coulometry at controlled potential and coulometric titrations. The main advantage of the coulometric method is the elimination of the necessity of standardization as the Faraday constant is a standard. In analysis of complicated samples encountered in environmental analysis the coulometric titrations are more advantageous where 100% current efficiency can be more readily attained by suitable choice of the reagent-solvent system. Coulometric titrations are suitable for determining the amount of substance in the range 0.01 to 100 mg (and sometimes below 1 µg). Under optimum conditions these titrations can be carried out with a precision and accuracy of 0.01%. Automatic coulometric analyzers for the determination of gaseous pollutants (SO_2, O_3, NO_x, etc.) have proven to be useful in environmental chemistry.

Conductometry. This rather unspecific method is used most frequently in the control of industrial chemical processes; the output controls of particular technological processes are important in the pollution control, mainly of flowing waters and the atmosphere. The very low conductivity of pure water permits this method to be used to track the total content of pollutants, which is frequently sufficient. A typical example of the use of conductometric methods for environmental protection are analyzers of detergents in waste waters, of concentration of synthetic fertilizers in irrigation waters, of the quality of potable waters, etc. In addition to classical conductometric methods, high frequency methods (oscillometry) are also used, in which the electrode system is not in direct contact with the sample.

Application of Electrochemical Methods in Analysis of Pollutants

Chemical analysis of waters. One of the tasks of environmental analysis is to determine the level of various chemicals present in water and decide whether their concentration is within the internationally approved limits.

Oxygen is one of the most important components of the atmosphere with enormous importance in the life cycle. In drinking and natural water evaluation one of the most important questions is the determination of the dissolved oxygen in it. The lack of oxygen stimulates the action of anaerobic biological systems that depreciate potable water. The decrease in concentration of oxygen in natural water indicates the contamination of water with organic of any kind. For measurement of content of dissolved oxygen in water, the most used method is the amperometric one using the Clark type sensor. Many techniques for this measurement were proposed also for field measurements[4]. The oxygen sensors are used also for research of photo-synthetic activity of plants and testing the effect of pollutants on it. The same holds for hydrobotanical and limnological research where the stratification of the oxygen concentration in water reservoirs must be known. On the market many types of instruments for oxygen measurement are available, e.g. Orbisphere Laboratories (Geneva) offer a set of instruments for measurements in the ppm and ppb region, which can measure simultaneously the temperature of water and compensate for the presence of Cl^- ions. The distance between the sensor and measuring device can be as much as 500 m. An advanced device for determining the (BOD) (biochemical oxygen demand) of waste waters is marketed by Leeds and Northrup. The measurement of oxygen concentration with a Clark sensor can serve for detection of toxic substances (waste, chemicals used in industry, agriculture, households, etc.) in water using their inhibition effect on aerobic degradation of organic substances by a culture of indicator organisms[5].

One of the other most important methods in chemical analysis and particularly in environmental measurements is the determination of the pH values of water which is now mostly performed potentiometrically with the glass electrode. The glass electrode covered with a semipermeable membrane and a film of electrolyte is also the basic arrangement for various sensors used in water and gas analysis. The NH_3 and SO_2 sensors working in the ppm range are widely used. The presence of ammonium ions in water signals the pollution with mainly agricultural wastes. For the determination of various anions in water like NO^-_3, F', Cl', etc. ISE's are available, enabling simple measurements in the field. Examples can be mentioned for the characterization of water wells and water sources e.g. for desert districts in Somaliland where complex electroanalytical measurements of pH, dissolved oxygen, Na^+, k^+, Cl^-, Fe^-, Ca^{2+}, NO^-_3 and NH^+_4 ions content were performed[6].

An unspecific but frequently performed measurement is the determination of the redox potential which monitors the content of the concentration of reducing and oxidizing compounds in water. The total amount of impurities in waters can be measured by conductometry, which is of course also unspecific. Typical is the measurement of the content of fertilizers in waters, detergents in waste water and of different salts.

The determination of surface active compounds in water is important. The presence of these compounds signals polluted water. Surfactants, due to their hydrophobic nature and to their lipophilic action are dangerous for river, lake and marine organism because they dissolve in the cellular membrane and may destroy it.

Surface active compounds due to their adsorption on the electrode-solution interface change the properties of the electrode double layer which can be measured and related to the concentration of these surface active compounds in the bulk of the solution examined. There are many electrochemical methods which for such measurements can be used. Many of them are based on polarography or voltammetry. From the historical point of view it may be mentioned that one of the first applications of polarography was just in environmental protection[7]. It was a kind of "polarographic adsorption analysis" based on the phenomenon of suppression of the polarographic maxima due to surface active compounds in water. An automatic device for evaluation of the water purification process by coagulation worked on this principle before World War II at the water tower of Prague. The sensitivity of this method is about 0.01 mg l^{-1} in surface active compounds. Problems in practical application of the polarographic maxima are the lack of specificity and dependence of the maxima height on many parameters like composition of the electrolyte, Hg-electrode properties, etc. The maximum suppression method yields only a measure of the total surfactant activity of the sample.

For the determination of the surfactant activity of seawater various methods based on suppression of the polarographic maximum or on measurement of the polarographic charging current have been developed. The surfactant activity of the seawater sample is evaluated by their comparison with calibration curves obtained with artificial seawater containing the surfactant Triton X-100. Thus surfactant activities equivalent to, from 0.1 mg to 3 g Triton X-100 per liter of seawater could be determined[8]. Also seawater samples at coastal stations after an oil spill were measured[9]. The suppression of the polarographic maximum was used also for the determination of detergents in waters including waste waters, e.g. in fresh laundry effluents, in sewage samples from the sewage collector of a hospital[10], etc.

A frequently used method for determination of surfactants of various types is tensammetry. A large variety of nonionic surfactants was analyzed by means of derivative chronopotentiometry mainly for water purity determination in respect to fisheries[11]. The concentration range for most of the surfactants investigated is within the range of 1-200 mg l^{-1}.

A sensitive method for the estimation of different surface active compounds (tetraalkylammonium salts, dextrans, crude oil components) in water is based on the measurements of the depression of the electro-capillary curve under conditions where the transport of the compound to the dropping Hg-electrode is accelerated by stirring of the solution being examined[12]. The detection limit is in the range of 10-100 µg l^{-1}. More details on "adsorptive polarographic analysis" are given in[13].

An important problem is the pollution of waters with traces of heavy metal ions. This problem is discussed from the analytical point of view in the following paper in this volume.

Organic compounds which are harmful for the environment can be divided into two groups. In the first one are various "useful" chemicals, mainly agrochemicals, necessary for improving food yield – like pesticides – or applied as growth stimulators to animal feeding. They are like good slaves but with their toxic side effects are bad masters. On the other hand there are organic chemicals that due to industrial activity are only discharged into the environment as "wastes".

Among "useful" organic compounds, mostly pesticides, herbicides, insecticides and other agrochemicals are discharged into aquatic systems prevalently in the runoff from agricultural areas. Because of their high toxicity an important task of environmental analytical chemistry is to monitor their concentration in waters and soil leaches but also in plant and animal tissues. The most frequently used analytical method are gas chromatography and spectrophotometry. But, as many of these compounds have in their structure an electroreducible functional group, polarography can be used for their determination. The application of DPP enables determinations in the ppb concentration region and the combination of DPP with adsorptive stripping analysis even in the sub ppb region. The main problem is determination of pesticides is their isolation from the matrix, separation from other compounds, etc. Therefore, frequently electrochemical detection is combined with separation techniques like HPLC. From the enormous quantity of published papers on these topics only some reviews in monographs [14-17] will be selected for further information.

In a discussion of the possibility of determination of organic compounds polluting water mainly due to industrial activity, it can be generally stated that DPP is one of the most sensitive analytical methods. In order to find out whether the respective compound can be determined polarographically, the best way is to consult a comprehensive monograph on organic polarography[18], a bibliography of polarographic papers[19], or tables on electrochemical properties of organic compounds[20].

From the many possibilities of polarographic determination of organic compounds for the sake of brevity the following few substances may be enumerated: acetaldehyde (and aliphatic and aromatic aldehydes,) aceto-phenone, acridine-compounds, aromatic amines, aniline, benzidines, benzal-dehyde, benzene and its homologues (after nitration) chlorphenols, nitro-benzene and nitro compounds in general, phenols (after nitration,) phthalic acid, polychlorinated biphenyls, methacrylates, naphthylamines, etc.

<u>Determination of pollutants in the air.</u> For the determination of SO_2, H_2S, NO, NO_2, CO, O_2, O_3, HCN, HCl, HF in the air many electroanalytical methods were published. Commercially produced instruments based on poten-tiometry, coulometry, voltammetry, conductometry, or using as sensor an electrochemical cell are available on the market. The application range is mostly from 0.1 to 10 ppm are available. Pocket size "personal" dosimeters for O_2, CO, HCN, NO_2, H_2S which signal acoustically if the takehold concen-tration has been reached. It can also be expected that personal monitors

will be available for the public (e.g. for SO_2 to indicate whether a building should be ventilated at a given moment, in relation to the SO_2, concentration in the outside air). More details are in [1,21].

Several types of semiconductor sensors were developed for analysis of toxic gases and vapors. They serve for detection and determination of hydrocarbons, alcohols, ethers, ketones, esters, nitrated compounds, ammonia, carbon monoxide, hydrogen, methane, etc. Their detection limit is often lower than 0.1 ppm. More details are given in [22]. The company Sema Electronics (Irvine, England) has sensors for about 200 different compounds on its production list and is prepared to "tailor" a sensor on special demand.

Polarographic also can be utilized for the determination of many organic compounds present in the air, e.g. in the vicinity of factories, in the industrial atmosphere, etc. In [23] are listed compounds like formaldehyde, acrolein, acetaldehyde, furfurol, hexachlorbutadiene, and nitrocyclohexane. The tested air is passed through a trap containing the supporting electrolyte. Application of electrochemical detection in HPLC to the measurement of toxic substances in air is the subject of a paper[24]. Derivatives of phenols and amines were chosen as examples.

Electroanalytical Chemistry and Environmental Carcinogenity

The percentage of cancer diseases caused by the effect of chemical compounds has been estimated to about 80 (UNESCO, IARC). It is thus important to collect data on the relationship between carcinogenity and physicochemical properties of the most wide spread environmental carcinogens. The population comes into contact with chemical carcinogens in different ways by inhalation of polluted air and/or by smoking cigarettes (polycyclic hydrocarbons, formaldehyde, nitrosamines, etc.) drinking contaminated water (traces of Cd, Pb, Cr, Ni ions, chlorinated aliphatic compounds of the chloroform type, etc.) consuming food containing carcinogenic compounds (nitrosamines, pyrolytic products of proteins, mycotoxines, growth stimulators added to feedstuffs, residues of pesticides, thiourea, etc.) and with application of some remedies (phenacetine, reserpine, etc.).

The aim of analytical chemistry is to signal the presence of chemical carcinogens in the environment and their determination in various matrices. Analytical chemistry has also to check the efficiency of the various decontamination and destruction processes of these hazardous compounds.

Analytical chemistry further studies the reactive mechanism of these compounds with various enzymatic systems or with other compounds, which can be an analogy of reactions occurring in vivo. Also the study of interaction of carcinogens with DNA can be emphasized.

About 70% of all chemical carcinogens can be determined polarographically or voltammetrically, e.g. amines and phenols, such as benzidine, o-dianisidine, naphthylamine, diphenyl, amines, p-chlorophenol, etc. Attention has been paid mainly to various polycyclic substances such as benzpyrene[25] (down to a concentration of 10^{-8} and 1^{-1}).

From the aspect of chemical carcinogenity the polarographic determination of N-nitrosamines become important. These compounds may be present in food, beverages and atmosphere, occasionally they can be formed in the body by nitrosation of amines, especially in the stomach. There were many papers published describing the determination of various nitrosamines (derivatives of proline, pyrolidine, piperidine, etc.) in different matrices[26]. The most important problem in their analysis (as with the determination of other organic compounds), is the isolation of test substances from the matrix, nevertheless isolation procedures are required for all measuring techniques. An advantage of polarography, especially DPP is the possibility of using small samples. For separation purposes often HPLC is suitable followed with electrochemical detection.

Polarography can be supplied also in the analysis of other chemical carcinogens mentioned in the 3rd Annual Report on Carcinogens[27] like benzene, azobenzenes, acridine compounds, formaldehyde, thiourea, polychlorinated biphenyl, phenytoin, reserpine, sacharine, substances used in manufacture of polymers, like acrylonitrile. Polarography has found use in the determination of other carcinogens, such as aflatoxins, trichocene toxins, etc. An important task of analytical chemistry has recently appeared in the determination of growth stimulators used in animal breeding, such as various biofactors and food mixtures and in the determination of residues and metabolites in bowels and muscles of animals bred for meat production. The nitrovin growth stimulant can be determined down to a content of 1 mg/kg in the feed mixtures. Residues of carbadox and its metabolites have been determined in the liver, kidneys and muscle tissues of calves. The determination of traces of estrogenic growth promoting hormone (estriol, estrone, estradiol, diethylstilbestrol, etc.) in meat can be carried out using HPLC with voltammetric detection[17,28].

In addition to purely analytical applications of polarography, the method can also be used for the screening of various compounds in vitro for the mutagenic and carcinogenic effects, as a result of their interaction with nucleic acids. For example, 7-methylguanine was determined in acidic hydrolyzates of nucleic acids after alkylation[29]. Alkylation agents, considered to be mutagenic and carcinogenic, are capable of affecting DNA and RNA through alkylation of the guanine N (7) atom. The conversion of guanine into 7-alkylguanine can thus be used as a measure of the alkylation of the nucleic acid. To determine the degree of methylation of DNA, the ratio of 7-methylguanine to guanine is measured. Thus the method enables direct determination of the concentration ratio of the two substances after acid hydrolysis, without separation.

The DPP method may be used to demonstrate structural changes in natural DNA caused by physical effects (temperature, irradiation) or by the action of chemical reagents. These interactions are manifested by a change in the height of the peak of natural DNA and possibly by the appearance of another peak corresponding to denatured DNA[30].

For testing on mutagenity of chemical compounds an electrochemical modification of the Ames test was developed[31]. In the original Ames test a strain of Salmonella typhimuri is used, which requires histidine in the nutrient medium for its growth. However, mutagenes produce a form that can procreate even without histidine; the cultivation lasts two days and down to 10 µg of a mutagens can be determined in 1 ml of solution. In the electroanalytical method the microorganism is placed in a suspension on a membrane filter that is fixed to an oxygen sensor membrane and its growth may be detected in a buffer solution saturated with atmospheric oxygen. After 10 hours a decrease in the sensor signal due to oxygen consumption by the growing microorganism is detected in the presence of as little as 0.001 µg mutagene in 1 ml solution.

ELECTROCHEMICAL REMOVAL OF ENVIRONMENTAL POLLUTANTS

As mentioned in the introduction, the detection of the pollutant in the environment is only the first necessary step in solving the problem of a cleaner environment. A more important problem is the removal or destruction of the pollutants. (Of course the best way would be the prevention of any contamination of the biosphere by more advanced technologies). Here again electrochemistry plays an important role. One of its advantages is the fact that electrons released from or consumed by the electrode are a clean reagent that does not contribute to a further increase in the amount of chemicals in the environment as is the case in other chemical processes.

Electrochemical methods used in pollutant removal are divided in the following groups:

1) Electrolytical metal removal and recovery from waste waters.
2) Electrolytical decomposition and/or destruction of organic and toxic compounds.
3) Electrolysis, electrodialysis, electrification and similar methods in municipal waste water treatment.

The principal industries and processes responsible for metal ions pollutants are chemical industry, metallurgy, mining operation, and electro-technologies like metal plating and finishing. Perhaps the last mentioned metal finishing industry is in many countries the biggest supplier of pollutants such as cadmium, lead, nickel, hexavalent chromium and cyanides. The removal of these toxic substances in waste waters is thus an important necessity and is often combined with the effort of their recycling for further use, even also for powder metallurgy.

In electrolytical treatment of waste waters and of diluted solutions in particular some difficulties are encountered - mainly in low electrical conductivity of the electrolyzed solution and/or low concentration of the compound which has to be removed. The problem of low conductivity can be solved by addition of indifferent electrolytes, decreasing the distance between electrodes, etc. The low concentration of the depolarizer results in poor current efficiency or a long lasting electrolysis. The use of classical types of electrolyzers with planar electrodes is ineffective. Thus various types of electrochemical reactors with forced flow of depolarizer to the electrode surface, with large surface electrodes minimizing the interelectrode gap, generators with three dimensional electrodes, etc. have been devised.

From the many systems that have found practical use in the industry the following ones can be mentioned. [For more details see 1, 32].

In fluidized bed reactors the solution is forced vertically through the bed consisting of electrode particles formed of metallic or metal coated glass spheres. These systems differ in the way that the electric contact with the bed particles is provided and in the the way that the particles are separated in order to recover the electrolytically deposited metal from them. This principle is used in e.g. the commercially produced reactor by Ako-Zout or the electrolyzer Eco-Cell. A parallel plate cell system is used in the "Swiss roll" arrangement. The cell is similar to spirally wound batteries consisting of large surface electrodes with a separator cloth between them. The solution is pumped through the system along the central axis[33]. There are also reactors using carbon fiber electrodes. One of their important properties is the enormous electrode surface area. The reactor produced by HSA Reactors Ltd[34], mainly for use in metal finishing industry works in a closed loop configuration; the percentage of removal of Cd(II), Cu(II), Ni(II), Zn(II), Pb(II), Cr(II), CN$^-$ ions is about 95 to 99%. Deposited metals are dissolved periodically by an anodic process forming a concentrated metal ion solution. There are also procedures using catalytic active porous electrodes[35-37]. The whole system works without an external current supply and is based on the principle that electrochemical oxidation and reduction processes take place simultaneously on a suitable catalyst. The pollutants are oxidized by an oxidizing agent like oxygen, air, etc. or reduced by a reducing agent like hydrogen, thus converting them into harmless products. The system for catalytic recovery of metals from aqueous solution uses candle-like porous tubes coated with a catalyst on the wall which is in the contact with the solution to be purified. A stream of hydrogen penetrating through the tube is oxidized to protons on the catalytic surface, simultaneously metal ions like $Cu(II)$, $Ag(I)$, $Hg(II)$ are reduced to the respective metal, on $Cr(VI)$ to $Cr(III)$ ions. After metal deposition, oxygen or air instead of hydrogen is bubbled through the tube and the metals are dissolved. On the catalytically active surface CN^- to CNO^-, SO_3^{2-} to SO_4^{2-}, NO_2^- to NO_3^- may be oxidized by air. On a similar principle a closed loop system for etching of printed circuit boards for electronic industry was proposed, where the dissolved copper is recovered. The process can be utilized also for etching of other metallic materials, e.g. for surface treatments.

Electrolytic methods can be used also for removal or destruction of other types of pollutants. Often indirect oxidation processes are used, like in the case of cyanides which can be destroyed by electrogenerated chlorine (or hypochlorite)[38]. Similarly phenols, thiocyanides, sulphides, etc. can be destroyed. Ammonium in waste waters is removed by oxidation to nitrogen[39]. Electrolytic processes are proposed for treatment of radioactive materials and for purification of radioactive waters. Electrolytic methods serve also for removal of gaseous pollutants. In United Technologies system[40] sulphur dioxide is removed from fume gases using suspended carbon slurry as scrubbing agent as well as electrode material at which sulphuric acid is formed. The acid is then concentrated, hydrogen used for other purposes and the slurry regenerated. For removal of SO_2 from gases, also the catalytic method mentioned above working without external power supply was proposed. Sulphur dioxide is oxidized to sulphate ions by atmospheric oxygen on activated carbon particles. A 25 dm^3 volume column which can purify 0.5 m^3 of air per hour (the concentration of 10^3 ppm of SO_2 is reduced to less than 0.1 ppm) was presented at the Achema Exhibition 1979 in Frankfurt/M. For electrochemical removal and concentration of hydrogen sulfide from hot coal gas a membrane cell arrangement was proposed[51]. The device consists of high temperature molten sulfide electrolyte cell ($\sim$ 700° – 1000°C) with porous carbon electrodes. The process-gas is passed through the cathode chamber, where H_2S is removed by reduction; the anion migrates to the anode, where elemental sulphur is formed.

Electrochemical methods play an important role in municipal waste water treatment[38,41,42]. The waste is mixed with seawater (or saline brine solution) and electrolyzed with evolution of chlorine. The magnesium salts present in seawater are exploited for the formation of magnesium hydroxide from the CH^- ions produced at the cathode. This magnesium hydroxide is an important means for the flocculation and sedimentation of suspended particles, adsorption of soluble phosphates and formation of insoluble phosphate compounds. Chlorine reacts with hydroxyl ions to form hypochlorite serving for sterilization purposes. Using a graphite anode and stretched iron cathode cell 60% in BOD, 99% suspended solids and 90% phosphate were removed from the waste water. Several pilot plants based on this principle were developed, e.g. on the island of Guernsey[43], town of Sorrento, etc. For disinfection processes (e.g. in swimming pools, etc.) the commercially available systems (e.g. Cronzio De Nora, Milano) for production and distribution of hypochlorite solution has found widespread use.

The eutrophication river and lake waters mainly due to the high content of phosphates, is an important problem. Using the Electro-N-system the amount of phosphates was reduced from 75 ppm to 0.3 ppm[44].

Some future trends in sewage treatment are mentioned in[45]. In sewage the solid material is about 60% in cellulose which may be converted into CO_2 by electrolysis. At a mercury electrode CO_2 can be reduced to formaldehyde, at a cadmium electrode to methanol. Formaldehyde could be the starting material for protein production.

An important problem is the removal of suspended and colloidal particles of any kind (fat particles, mixtures oilwater, etc.) from waste waters in different industries (food, paper, petrochemistry, etc.). In such cases electrification is used. In fact electrification is a combination flotation, flocculation and electrolysis. Water which has to be purified is pumped through a reactor with a horizontal electrode system, on which hydrogen or oxygen evolution by electrolysis of water takes place. The stream of gas bubbles, transports the foreign particles to the surface of the reactor, from where they are skimmed off. If Al or Fe electrodes are used the respective metal is dissolved with formation as hydroxide which contributes to the coagulation.

Electrophoretic processes can remove waste waters from colloidal particles of loam, asbestos and even bacteria. Another method, electrodialysis can be used not only for desalination of seawater but in general for removal of different salts of a polluted water stream. For more information, see[1, 46-50].

As an exhaustive treatment of the problem mentioned in the title of this paper can not be given, only some of the principal possibilities of electrochemistry to the achievement of a cleaner environment are listed. It can be emphasized that electrochemistry has its place not only in solving the analytical aspects - mainly in monitoring the polluted environment - but also in "repairing" some of the unfavorable consequences of industrial and other activities - that means removal or destruction of pollutants. Perhaps electrochemistry has in proposing and developing new processes, technologies, sources of energy, means of transportation, etc. which would be less harmful for our environment.

REFERENCES

1. J. O.'M. Bockris, ed., "Electrochemistry of Cleaner Environments," Plenum Press, New York (1971).
2. R. A. Bailey, H. M. Clark, J. P. Ferris, S. Krause, and R. L. Strong, "Chemistry of the Environment," Academic Press, New York (1978).
3. R. Kalvoda, ed., "Electroanalytical Methods in Chemical and Environmental Analysis," Plenum Press, London (1985).
4. L. Šerák, Measurement of oxygen in biological systems, in: "Electroanalytical Methods in Chemical and Environmental Analysis," R. Kalvoda, ed., Plenum Press, London (1985).
5. P. Hofman, Electrochemical analyzers of water toxicity, in: "Electroanalytical Methods in Chemical and Environmental Analysis," R. Kalvoda, ed., Plenum Press, London (1985).
6. M. Mascini and A. Liberti, Ion selective electrodes for measurement in fresh waters, Sci.Total Environ., in print.
7. J. Heyrovský and J. Kůta, "Principles of Polarography," Academic Press, London and New York (1965).
8. T. Zvonarić, V. Žutić, and M. Branica, Determination of surfactant activity of sea water samples by polarography, Thalassia Jugoslavica, 9:65 (1973).
9. Z. Kozarac, B. Ćosović, and M. Branica, Estimation of surfactant activity of polluted seawater by Kalousek Commutator Technique, J.Electroanal.Chem., 68:75 (1976).
10. Z. Kozarac, V. Žutić, and B. Ćosović, Direct determination of nonionic

and anionic detergents in effluents, <u>Tenside Detergents</u>, 13:260 (1976).

11. P. Holmquist, On oscillographic polarography of non-ionic tensides, <u>J.Electroanal.Chem.</u>, 39:470 (1972).

12. L. Novotný and I. Smoler, The method of interrupted convection for measurements of adsorption equilibrium, <u>J.Electroanal.Chem.</u>, 146:183 (1983).

13. R. Kalvoda, Polarographic adsorptive analysis and tensammetry of adsorbable molecules, <u>Pure and Applied Chem.</u>, in print.

14. P. Nangniot, "La Polarographie en Agronomie et en Biologie," J. Duculot, Gembloux (1970).

15. J. Volke and M. Slamnik, Polarography and related methods, <u>in</u>: "Pesticide Analysis," G. Dask, ed., Dekker, New York - Basle (1981).

16. W. F. Smyth, "Electroanalysis in Hygiene, Environmental, Clinical and Pharmaceutical Chemistry," Elsevier, Amsterdam (1980).

17. R. Kalvoda, Polarographic and voltammetric methods, <u>in</u>: "Electroanalytical Methods in Chemical and Environmental Analysis," R. Kalvoda, ed., Plenum Press, London (1985).

18. M. Březina and P. Zuman, "Polarography in Medicine, Biochemistry and Pharmacy," Interscience Publ., New York (1958).

19. Anon., "Bibliography of Polarographic Literature 1922-1967," Sargent-Welch Scientific Co., Skokie, USA (1969).

20. L. Meites and P. Zuman, "CRC Handbook Series in Organic Electrochemistry," CRC Press Inc., Boca Raton (1977-1984).

21. J. Tenygl, Electrochemical detectors and monitors, <u>in</u>: "Electroanalytical Methods in Chemical and Environmental Analysis," R. Kalvoda, ed., Plenum Press, London (1985).

22. J. Fexa, Semiconductor sensors, <u>in</u>: "Electroanalytical Methods in Chemical and Environmental Analysis," R. Kalvoda, ed., Plenum Press, London (1985).

23. M. D. Manita, R. M. Salikhzhdanova, and C. F. Javorskaya, "Sovremennyie Metody Opredeleniya Atmosfernych Zagryaznenii Naselenykh Mest," Medicina, Moscow (1980).

24. C. J. Purnell and C. J. Warwick, Application of electrochemical detection in high-performance liquid chromatography to the measurement of toxic substances in air, <u>Anal.Proc.</u>, p.151, April (1981).

25. M. R. Smyth and J. Osteryoung, Electroanalysis of environmental carcinogens, <u>in</u>: "Electroanalysis in Hygiene, Environmental, Clinical and Pharmaceutical Chemistry," W. F. Smith, ed., Elsevier, Amsterdam (1980).

26. R. Samuelson and T. Rydström, Pulse polarographic studies on N-Nitrosamines, <u>in</u>: "Electroanalysis in Hygiene, Environmental, Clinical and Pharmaceutical Chemistry," W. F. Smith, ed., Elsevier, Amsterdam (1980).

27. "Third Annual Report on Carcinogen," U.S. Department of Health and Human Services (1983).

28. W. F. Smyth, L. Gold, D. Dadgar, M. R. Jan, and M. R. Smyth, Polarographic and voltammetric methods of environmental analysis, <u>Inter.Lab.</u>, 13:40, (1983).

29. J. M. Séquaris, P. Valenta, and H. W. Nürnberg, Rapid differential pulse voltammetric determination of 7-methylguanine, <u>J.Electroanal. Chem.</u>, 122:263 (1981).

30. E. Paleček, Polarographic analysis of nucleic acids, <u>in</u>: "Electroanalysis in Hygiene, Environmental, Clinical and Pharmaceutical Chemistry," W. F. Smyth, ed., Elsevier, Amsterdam (1980).

31. I. Karube, T. Matsunaga, T. Nakahara and S. Suzuki, Preliminary screening of mutagens with a microbial sensor, <u>Anal.Chem.</u>, 53:1024 (1981).

32. Final Report, UNESCO, Regional Workshop, Electrochemical Removal of

Environmental Pollutants, Padova (1980), edited by UNESCO, Paris (1981).

33. P. M. Robertson and N. Ibl, Electrochemical removal of environmental pollutants, in: (32).

34. S. Gupta, B. Fleet, and I. F. Kennedy, Electrochemical reactors for environmental pollution control, in: "Electroanalysis in Hygiene, Environmental, Clinical and Pharmaceutical Chemistry," W. F. Smyth, ed., Elsevier, Amsterdam (1980).

35. W. Faul and B. Kastening, Elektrochemische leiterplattenätzung mit geschlossenen elektrolyt kreislauf, Technische Information, No.5, Kernforschungsanlage, Jülich.

36. W. Faul and B. Kastening, Katalytische abwasser- und abgasentgiftung, Technische Information, No.14, Kernforschungsanlage, Jülich.

37. W. Faul and B. Kastening, Katalytische metall- rückgewinnung aus wässrigen lösungen, Technische Information, No.17, Kernforschungsanlage, Jülich.

38. A. T. Kuhn, The electrochemical treatment of aqueous effluent streams, in: "Electrochemistry of Cleaner Environments," J. O. M. Bockris, ed., Plenum Press, New York (1972).

39. L. Marinčić and F. B. Leitz, Electrooxidation of ammonia in waste water, J.Appl.Electrochem., 8:33 (1978).

40. H. Maget, SO_2 abatement, Environ.Sci.Technol., 11:225 (1977).

41. L. Mendia and F. Gigliani, Electrochemical possibilities of treatment of liquid effluents: the case of municipal waste waters, in: (32).

42. L. Mendia and E. Buonincontro, Un particulare trattamento delle acque di rifinto cittadine, Acqua industr., 7:8 (1960).

43. H. W. Marson, "Electrolytic Sewage Treatment: The Modern Process," The Institute of Sewage Purification, Annual Conference, Brighton 21-24 June 1966, Conference Paper No.6.

44. K. H. Hartkorn, Elektro-M verfahren zur phosphat entferung aus wässern, Städtehygiene, 9:21 (1973).

45. J. O.' M. Bockris, The electrochemical future, in: "Electrochemistry of Cleaner Environments," J. O'. M. Bockris, ed., Plenum Press, New York (1972).

46. B. Matov, "Elektroflotacija," Izd.Kartya Moldovenjaske, Kishinev (1971).

47. M. G. Granovskij, I. S. Lavrov, and O. Smirnov, "Elektroobrabotka Židkostěj," Izd.Chimija., Leningrad (1974).

48. H. Binder, Electrochemical, electroflocculation and similar methods in sewage-treatment, in: (32).

49. C. P. C. Poon and T. G. Bruckner, Physicochemical treatment of waste-water/seawater mixture by electrolysis, J.Water Poll.Control.Fed., 47:1 (1975).

50. E. Fischerová, Electrochemical treatment of biological wastes (in Czech.), in: "Elektrochemie a Životní Prostředí," J. Balej, ed., Academia, Praha (1982).

51. S. Lim and J. Winnik, Electrochemical removal and concentration of hydrogen sulfide from coal gas, J.Electrochem.Soc., 131:562 (1983).

DISCUSSION

A. A. Orio

Venice University
Italy

CHAIRMAN'S SUMMARY

C. Buess-Herman (Belgium): You have mentioned tensammetry as a method to determine traces of organic compounds. In my opinion this method is not specific at all. From the quantitative point of view if the adsorption-desorption peak is used to analyze the amount of the material present in solution, the amplitude of this peak will depend on the kinetic of the process, the mass transport, the adsorption, etc. Also the potential at which the peak will occur will change with the concentration in a way which is difficult to establish and this will depend on other organic compounds present in solution. Could you please give us an idea of the concentration range which can be explored by this method and also give some comments on my remarks?

R. Kalvoda: Of course, the tensammetric measurement suffers from a big lack of specificity. It can be used directly to determine concentrations as low as 10^{-5}-10^{-6}M. With some stripping preconcentration, one order of magnitude lower can be reached. This method can perhaps give information on the total concentration of a surface active compound. More difficulties are encountered if other surface active compounds are present. I have studied this problem using some electroreducible organic compounds which are adsorbed on the electrode by adsorptive stripping analysis. I have found that if other surface active compounds, such as albumin or other similar substances, are present at a concentration of about 1 mg/l, there are no remarkable effects on the height of the tensiammetric peak. Some difficulties arise at higher concentrations because of competitive adsorption. Preliminary results indicate that it should be possible to eliminate interfering substances by gel filtration.

E. Gileadi (Tel Aviv University, Israel): The use of ensembles of microelectrodes for trace analysis should be considered. Recent studies in our laboratory (H. Reller, E. Kirowa-Eisner and E. Gileadi, J.Electroanal. Chem., 161:247 (1984)) have lead to the conclusion that such devices can extend the limit of detectibility for orders of magnitude, compared to conventional pulse polarography.

A further advantage of ensembles of microelectrodes is the low solution resistance associated with the small diameter of each electrode in the ensemble. This allows electrolysis in poorly conducting solution such as tap water.

R. Kalvoda: I did not forget to talk about this topic. Professor Wightman will speak about this problem using fiber electrodes of the diameter you mentioned. I think that this is a very promising field and that the microelectrodes are very important also for in vivo measurements. For instance the Bioanalytical Systems Company produces or advertizes the so-called "in vivo polarograph" using this type of electrodes with a semi-derivative technique for evaluating the curve. Working with current less than 1 nA, the sensitivity seems to be very high.

R. M. Dell: The following is not a question, but a piece of information. One type of pollutant that you did not mention specifically is radioactive isotopes. We have been interested in the problem of the removal of radio-active isotopes. The metallic radioactive elements are not different from the other metals that you are trying to remove, but two quite difficult specific cases are Sr-90 and Cs-137 because these elements cannot be reduced or plated out. This takes us to ion exchange techniques. But unfortunately these ions are normally present in strong acid solutions and strong acid ion exchangers are expensive, also requiring lots of strong acid to elute them. We have been working on a technique which we borrowed from early work done in the United States which is electrochemical ion exchange. With this method the ion exchanging resin is built into an electrode and this allows us to increase the pH locally and then an acid solution at pH 4 can be passed and eluted by reversing the potential. This method is proving quite promising even if it is still at research stage. We have published a short article on this topic [in the January issue of ACSON] and I will be pleased to send a reprint to anyone interested.

R. Kalvoda: This is a very interesting remark. I think that other elec-troanalytical methods can be adopted to the same purpose. Due to the shortage of time, I dealt very briefly with the problem of the treatment of radioactive wastes in water.

M. Fleischmann: Just a quick comment on the Gileadi remark on microelec-trodes. We have been looking into the oxidation of Hg(O) using micro-electrodes and we have been able to see $10^{-7} - 10^{-8}$ M solutions without addition of electrolytes by straight oxidation. By making the system sufficiently small you are working at pA levels. If this is combined with AC sweep techniques the detection limits are enormously decreased to very low levels indeed. Of course this must be combined with elution chromat-ography.

R. Murray: One aspect that you did not mention is: who does the analysis? The usefulness of an electrochemical procedure depends on where it is done. If it is done by experts, that is one thing. If it is done by low skill people in a routine laboratory, that is something altogether different. I think that electrochemistry often suffers from the complexity of its application in a routine situation when compared to spectroscopy and chromatography. I wonder if you would comment on where you feel that electrochemistry can be applied using the lowest skill level of the routine workers. I believe these problems are different from those of the overall survey.

R. Kalvoda: I think that an important aspect is also the low cost of equipment that I mentioned before. Of course, many problems can be solved with other types of instruments, but field applications are something different. If we have several instruments placed in many stations, the electroanalytical methods are simple and more convenient. Obviously, the people must learn how to use them since they may be a little bit more complicated than just pressing a button. Automation is also applied to the electrochemical instrumentation; nevertheless, a minimum of expertise is required.

V. S. Bagotsky: In connection with the determination of organic compounds in water there is sometimes the problem of determining the total concentration without separating the single compounds. I wish to provide some information about a very simple device, designed in our Institute, based on the absorption of the organic compounds on a smooth Pt electrode and subsequent oxidation of the absorbed species. The absorption can be determined as decrease in the hydrogen absorption from the voltammetric curve or by oxidation of the compounds. The sensitivity is very high, to the level of mg/l. There is also the possibility to analyze three different groups of organic compounds: those which are easily oxidized, those oxidized at normal levels and those which are oxidized with some difficulty. In connection with the last remark, I would like to stress the fact that this device is very simple and all that you have to do for the analysis is to push a button and the screen shows the total concentration of the organic compound in mg/l.

R. Kalvoda: This is a carbon determination.

R. Bagotsky: Yes, this is a carbon determination with a differentiation on the degree of oxidation.

H. W. Nürnberg: I want to make a comment on the point raised by Dr. Murray. The electroanalytical methods are considered by many people as complex and difficult. It is a fact that this is the rumor. Concerning the topic I am going to speak about soon (determination of heavy metals in water), many people say that atomic absorption is simpler. This is a lie because the opposite is true. In trace analysis there is not a royal method. It depends on the problem to be solved and on the level of expertize needed in a certain laboratory. In most countries government laboratories doing environmental surveys can be generously qualified as mediocre for reliable trace analysis. But the point is that you need expertise for the method that you are going to use whether it is voltammetry or atomic absorption, or chromatography, etc. What you need is general trace analysis expertize, because most errors are made before the analysis, during the sampling. Then you have the paradox that with a great effort and precision in the analytical method you have completely irrelevant numbers. All the data reported in the literature before about 1973 must be regarded as unreliable with a few exceptions such as, for instance those obtained in sea water in the laboratory of Dr. Peterson from Caltech. In some cases the data are wrong by orders of magnitude. This was mainly due to the fact that in most cases there was not a general experience in trace metals analysis and not to the analytical method used.

R. Kalvoda: In the journal American Laboratory, there has been reported a comparison between anodic stripping and atomic absorption analysis which indicates that both methods are good depending on the specific type of analysis to be carried out. Therefore, it is true that there is not a universal method.

A. A. Orio: In connection with the last comment I must say that also our experience indicates that in some areas the electrochemical methods are proved to be better than other analytical techniques and on the basis of the many advantages that they have, it is reasonable to foresee a wide diffusion of these methods. A quite important example is the problem of monitoring pollutants in natural waters. Every industrialized country is facing the need of determining the concentration of harmful and toxic species in rivers, lakes, estuaries and sea to establish the environmental endurance. This is normally done by field sampling and subsequent laboratory analysis. However, as pointed out by Dr. Nürnberg, the natural samples can easily be spoiled during the sampling, the transport to the laboratory or the pretreatments with the result that the

application of the most sophisticated analytical methods is totally meaningless.

These disadvantages are eliminated with the use of field stations.
The scientific knowledge of the basic electrochemical principles is deep
and wide enough for very important and useful applications in this field.
Several polluting species can be analyzed with different electrochemical
techniques which, compared to other analytical methods, do not need high
pressure combustion gases for hot flames or delicate optical systems.

The accuracy, the sensitivity, the precision, the reliability, the low
investment cost, the ease of operation and maintenance by non-highly
specialized personnel are the main characteristic of the electrochemical
methods which make them feasible also for automation. Field stations with
electrochemical sensors to monitor non-specific parameters, such as pH, Eh,
DO, salinity, etc. have been sold by companies for about a decade. In the
last few years more technically advanced field stations have become available for the continuous analysis of specific parameters such as NH_3, Cl^-,
F^-, CN^-, $S^=$, Pb, Cd, Cu, Ni, Zn, Hg, Cr, As, etc.

The use of microprocessors make it possible to carry out the entire
analytical cycle as, for instance, cleaning the cell, addition of a standard solution, recleaning, addition of the water sample to be analyzed,
etc. From several stations the data can be transmitted in real time to a
central unit so that a very large region can be simultaneously checked by
one of the very few specialized persons who can suggest or decide the
immediate action to be carried out to eliminate, or at least to minimize,
extemporary environmental hazards. A very large amount of reliable data
can thus be collected and processed by a computer to determine the space
and time distribution of eutrophic or polluting species present in surface
waters such as rivers, lakes, estuaries, lagoons or coastal waters. This
allows the public authorities to detect the human activities responsible
for polluting the environment and in some cases also to locate the specific
source of pollutants. The economical and practical advantages of this
system compared with the traditional laboratory analysis are clearly
evident.

Although in principle about 30 different chemical species can be
determined by ion selective electrodes (ISE) or applying various electrochemical techniques such as pulse polarography, differential pulse polarography, stripping voltammetry, etc. the analysis of single chemical
species in natural waters is sometimes troublesome due to several analytical difficulties. This is the area where new technical developments are
strongly needed. As examples we can consider the already mentioned case
where the presence of surface active substances limit the sensitivity of
the AC methods or the case of saline waters (estuaries, lagoons and sea)
where the high content of chloride is a strong interference for ion selective electrodes. This makes the direct measurement of very important
species such as nitrate and orthophosphates very difficult. For ammonia
the problem has been brilliantly resolved with the gas sensing probe which
has a detection limit and a working range (from 0.1 to 0.03 ppm) useful for
most natural waters. With a proper chemical treatment of the sample, the
same electrode can be in principle used also for the analysis of the
nitrate, but very few field applications have been so far reported. It is
a pity that similar reliable practical applications have not yet been
reported, although few studies have been carried out to this purpose.

The determination of organic substances is another important field
where new developments are needed. The techniques used for the laboratory
determination of organic species (gas chromatography, liquid-liquid chromatography, mass spectrometry GS-MS or HPLC-MS) can be applied to field

measurements with very great difficulties. The favorable properties of electrochemical methods should be exploited for this purpose.

As clearly indicated by Dr. Kalvoda the application of electrochemical processes for the removal of pollutants is mainly concerned with the many sources of toxic heavy metals such as chemical manufactures, metal plating, surface treatment, mining and metallurgical operations. Electrodeposition, electroreduction to insoluble species and the use of sacrificial anodes are the techniques most widely used in this field. The production of chlorine from saline waters in urban or industrial wastewater treatments must also be quoted as a very interesting and practical example of an electrochemical application to the disinfection of different types of water.

As a short closing remark I think that also from this very interesting lecture provided by Dr. Kalvoda and from the very stimulating discussion which followed, the importance and the usefulness of electrochemistry in solving a large number of environmental problems is clearly evident. Further improvements are needed and can certainly be expected in years to come, and I hope to be able to meet all of you again in future meetings, dealing with new and even more interesting applications of electrochemical processes and techniques to solve the many environmental problems.

APPLICATIONS AND POTENTIALITIES OF VOLTAMMETRY

IN ENVIRONMENTAL CHEMISTRY OF ECOTOXIC METALS

H. W. Nürnberg

Institute of Applied Physical Chemistry
Nuclear Research Center (KFA), Julich
West Germany

INTRODUCTION

A large and still increasing amount of potentially and actually hazardous chemicals is emitted from a variety of anthropogenic sources into the environment. Among these environmental chemicals a number of heavy metals and some metalloids have gained particular significance with respect to their ecotoxicity[1,2]. Some metals, e.g. Cd, Pb, Hg and As (III), have a significant toxicity per se and constitute thus always an ecotoxic risk, if they are abundant in an ecosystem above their natural base levels. Other metals, e.g. Cu, Zn, Ni, Co, Se have for plants, organisms and man indispensable essential functions below threshold levels, which depend on the metal and organism, but will also exert toxic effects above the respective threshold levels[1,3,6,].

A common specific ecochemical feature of metals is, that they undergo not biodegradation, as do ultimately most hazardous organic environmental chemicals. Instead metals participate in biogeochemical cycles with different residence times in various environmental compartments. The global biogeochemical cycle has a distinct tendency to transfer metals from the land to the sea (see Figure 1) where their residence times are usually rather long. The two main pathways of heavy metals in this global biogeochemical cycle go through the atmospheres and with the rivers and continental run off water into the sea. A substantial part of the metals emitted from anthropogenic and natural sources on land into the atmosphere is after atmospheric transport over certain distances already again deposited into terrestrial ecosystems.

Man and animals usually take up hazardous metals predominantly with food[6]. The particularly toxic metals mentioned above are insidious poisons. Although a substantial portion ingested with food is excreted again, a certain amount remains in the body, where the toxic metals accumulate in certain organs building up there a growing stock and therefore exerting chronic toxic effects growing progressively with time.

In this context it is emphasized that environmental research and protection (besides provision of sufficient energy and food, improvement of the health situation by conquering major diseases, and overcoming of problems connected with shortages in a number of raw materials) is one of the five major global problems to be solved and thus one of the present

121

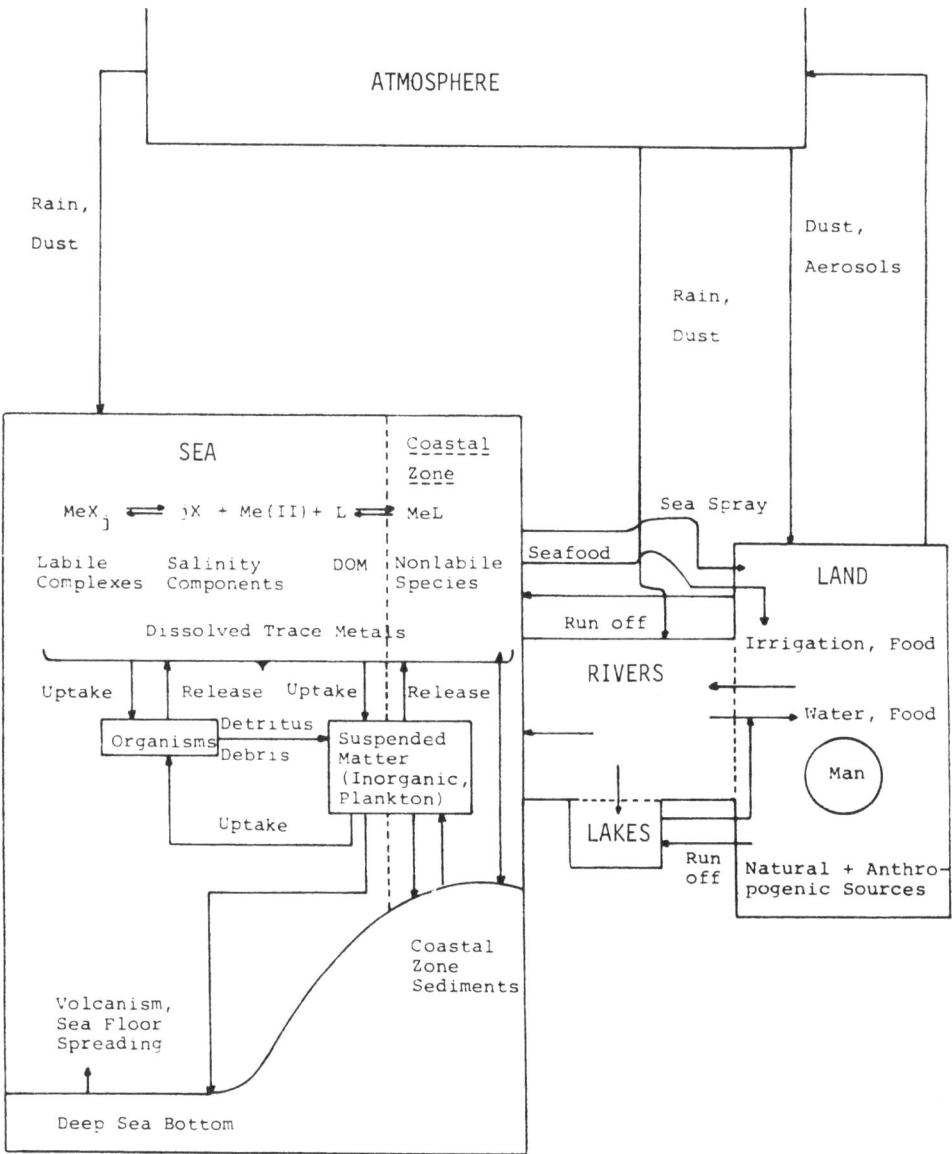

Fig. 1. Global biogeochemical cycle of metals. For the sea, given
details on the interactions and concentration regulation, refer
respectively also to rivers and lakes.

major challenges to natural science and technology. As for the other four
major problem areas mentioned, electrochemistry has the capacity for major
contributions for many tasks in environmental research and protection.

It is obvious, that the environmental burden caused by hazardous
metals, particularly with respect to the per se toxic metals, requires
surveillance of the emissions and immissions for the introduction of
appropriate environmental protection regulations as well as intensive
ecochemical and ecotoxicological research to deepen and expand the knowledge on the fate, behavior and effects of ecotoxic metals in the various
environmental compartments (atmosphere, hydrosphere, terrestrial eco-

systems), in food and in man and animals. For these tasks in monitoring
and research, which constitute in terms of chemical methodology demanding
topics in trace metal chemistry and trace metal analysis, suitable analytical methods are of key significance. Several errors in the data to be
measured would cause severe ecological, economical, social and scientific
penalties.

GENERAL ASPECTS OF THE APPLICATION OF VOLTAMMETRY IN THE ECOCHEMISTRY
OF METALS

Within the last decade it has been established, that voltammetry is
one of the few instrumental trace methods which fulfill in a comprehensive
manner the requirements in environmental trace chemistry of the ecotoxic
metals mentioned afore. Therefore, suitable voltammetric methods, in the
first place differential pulse voltammetry, have become one of the preferred approaches in environmental trace analysis and ecochemical research of
ecotoxic metals and metalloids[7-10].

Voltammetry combines extraordinary determination sensitivity with
inherently high accuracy, i.e. small tendency for systematic errors, and
good precision. These important properties are connected with the fact,
that this electroanalytical approach is based on the Faraday law according
to which 1 mole of a substance undergoing an electrode process is equivalent to the transfer of the enormous electrical charge of n x 96500 C
through the interface electrode/solution. The quantity n is the number of
electrons transferred in the elementary step per ion or molecule and for
trace metals corresponds frequently to 2.

Further important advantageous features of voltammetry are the following: Voltammetry enables the simultaneous determination of groups of
heavy metals e.g. Cu, Pb, Cd, Zn, Se(Iv) or Ni, Co or Cu, Hg (Figure 2).
It has therefore been termed an oligomethod. Consequently it has also a
reasonably high determination rate. If more than one metal has to be
determined, as is usually the case in environmental samples, voltammetry
equals or even overcomes in analysis rate (with respect to this aspect in

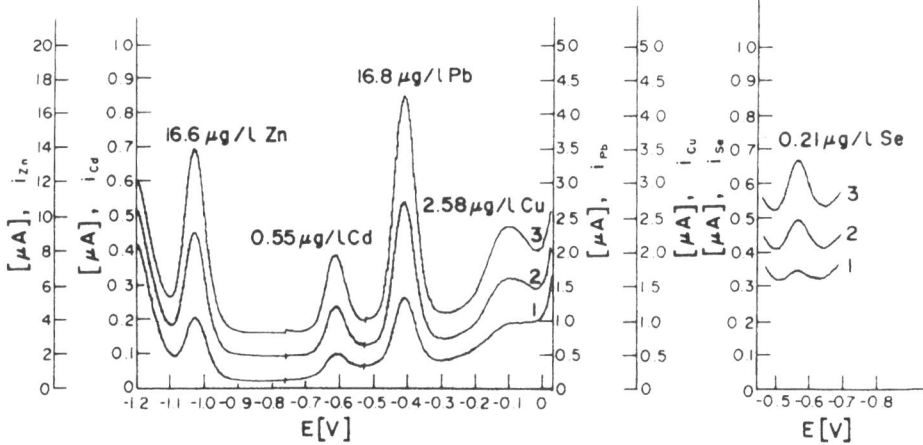

Fig. 2. Simultaneous determination of Cu, Pb, Cd, Zn by DPASV at HMDE,
subsequently of Se(IV) by DPCSV, in rain water acidified to pH 2;
preconcentration times: 3 min at -1.2 V for DPASV and 5 min for
DPCSV. 1 original analyte, 2, 3 after first and second standard
addition. Total analysis time with 2 standard additions 30-40
min.

the past particularly overvalued) atomic adsorption spectrometry AAS, which is a single element technique, being moreover in the electrothermal mode (ETAAS) inherently inferior to voltammetry with respect to determination sensitivity and accuracy[9,10]. The present introduction of automation into voltammetry will further enhance convenience of application and in routine analysis the determination rate[11]. This is particularly the case with automatic devices which include all sample pretreatment steps in the automation also. Even then voltammetry retains two further important advantages. The instrumentation remains very compact and can be therefore easily used in field studies on ships or in mobile terrestrial laboratories. An important practical aspect for the user is, that reliable instrumentation produced by reputed companies is available. As Table 1 shows, the costs to be invested for instrumentation are significantly lower than for all other alternatives in trace metal analysis. This makes voltammetry very attractive for the numerous institutions and laboratories charged with the environmental monitoring of the levels of ecotoxic metals in natural waters, atmospheric precipitates, waste water, biological materials, food and body fluids.

Finally it has to be explained, that voltammetry is essentially a substance specific and not just an element specific method like the other nonelectrochemical alternatives listed in Table 1. This opens to voltammetry the important field of speciation investigations on heavy metal complex species in all types of natural waters.

As a rule the ecotoxic metals exist at trace or even ultra-trace levels in the environment. This leads to analysis with concentrations of those metals usually below 1000 µg/l or 10^{-5} M. Consequently the application of conventional polarography is impossible and advanced modes have to

Table 1. Cost Levels of Instrumental Determination Methods for Trace Metal Analysis

Method	Costs 10^3 US $	Remarks
Voltammetry (manual operation)	8-10	Oligo-method suitable for routine
Voltammetric automates	20	Oligo-method, suitable for routine
Graphite tube AAS	25-60	single-element method; costs according to automation level; including hydride system; suitable for routine
Zeeman-AAS	40-60	suitable for routine, direct measurements in many solid sample types without pre-treatment
Atomic emission spectroscopy with ICP excitation	45-200	not sufficiently sensitive for ultra trace range, but multi-element method; suitable for routine
Neutron activation analysis	60-100	multi-element method; additional costs for neutrons and radiochemical laboratory; no routine method

be applied, such as the differential pulse mode, one of the most important and versatile achievements for electrochemical trace analysis from the pioneering work of Barker[12], which is now incorporated in every voltammetric device as the most significant function for analytical practice. Recording of the response from the differential pulse modes applied in polarography and voltammetry is performed according to the principle introduced by Parry and Osteryoung[13].

In the differential pulse mode the test electrode is polarized essentially by a dc-voltage ramp (10 mV/s) on which at adjustable clock times (0.2 to 0.5 s) rectangular voltage pulses (height 25 - 50 mV; duration 20 - 60 ms) are superimposed[8,14,15]. In this manner the charging current of the double layer capacity, limiting the utility of voltammetry, can be efficiently separated from the faradaic current component, which is due to the electrode process of the substance being analyzed and thus proportional to its concentration in the analyte. The resulting substantial improvement in signal-to-noise ratio has opened to voltammetry the area of trace and in conjunction with electrochemical preconcentration of ultra trace analysis down to the range of less than 10^{-3} µg/l or 10^{-11} M. These are by the way practical detection limits caused by the low blank levels which remain even with appropriate precautions, whereas with the electrochemical preconcentration mentioned below, voltammetry could determine in principle even substantially lower concentrations, if the blanks could be further reduced. In this context it has, however, to be emphasized that metal concentrations in the analysis typically below 100 µg/l or 10^{-6} M cannot be determined reliably by direct differential pulse polarography (DPP). Preconcentration is required. The basic advantage of voltammetry is, that this preconcentration can be performed electrochemically and therefore without any additional contamination risk and corresponding loss of accuracy. For this electrochemical preconcentration, which has to be achieved at the interface of the test electrode, two approaches can be distinguished.

One is the stripping method used already for many years in voltammetry (Figure 3). It is however, restricted to those metals which form stable amalgams if, as is often the case with respect to many other advantages, a mercury test electrode is used. Among the ecotoxic metals considered a

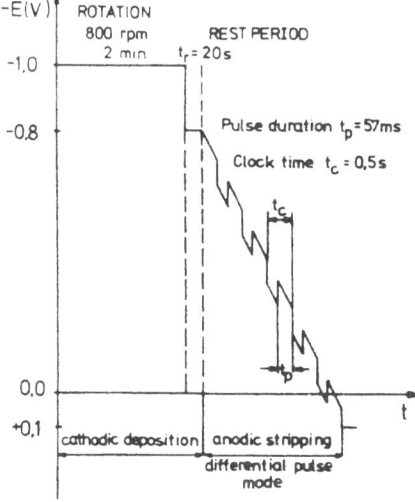

Fig. 3. Principle of differential pulse anodic stripping voltammetry (DPASV).

mercury test electrode can be used for Cu, Pb, Cd, Zn, Se(IV)[8,16] while Hg[13,18] or As(III)[19] require an activated gold electrode. Often the trace metal is accumulated at the test electrode by cathodic deposition and then determined by anodic reoxidation in the differential pulse mode. This termed differential pulse anodic stripping voltammetry (DPASV). For certain cases, e.g. Se(IV), differential cathodic stripping voltammetry (DPCSV) is applied[16] (see Figure 2).

Other metals, e.g. Ni and Co and now also Cr[20], but also those accessible to stripping voltammetry, can be very sensitively determined by the recently introduced adsorption differential pulse voltammetry (ADPV)[21,22]. To the analyte a suitable organic chelating ligand is added, e.g. dimethylglyoxime (DMG) for Ni and Co. The electrochemical preconcentration is then achieved by setting a suitable adsorption potential at the test electrode. The trace metal chelates adsorb and are thus accumulated at the interface. Subsequently the potential is scanned in the differential pulse mode into the potential range where the electrode process of reduction for the trace metals can occur. This electrochemical preconcentration approach is also not affected by any additional contamination risk and has still a wide potential application area to be explored for further metals. Recently also a very convenient ADPV-procedure with an appropriate ligand for Zn, which can also be determined by DPASV, has been reported[13].

It is the combination of the electrochemical preconcentration versions outlined with the differential pulse mode which has opened the lower trace and the ultra trace concentration range to the voltammetric determination and investigation of ecotoxic metals and metalloids.

The determination limits hitherto attainable in natural waters are listed in Table 2. These low concentration levels could also be reached in aqueous analytes resulting from digested biological material and food, but in practice these matrices always contain higher metal levels than natural waters in the dissolved state and consequently also yield higher levels in the analyte than the limits given in Table 2. It has further to be explained that the ultimate given practical determination limits given are only attainable by laboratories with special expertise in voltammetry and generally in ultra-trace chemistry using clean rooms and all precautions to minimize contamination[14]. Nevertheless, the values are practical determination limits and not yet the ultimate methodological limits inherent to voltammetry itself. Thus, the determination limits can be lowered further by particular efforts, if the necessity exists, e.g. for base line studies on heavy metal levels in arctic[25] or antarctic snow and ice.

APPLICATIONS

The applications of voltammetry in environmental chemistry of ecotoxic metals have become extensive and manifold and comprehend all types of ecosystems and environmental materials including man and his food. On the basis of reliable accurate analytical data, this electrochemical approach has contributed significantly to the expansion and deepening of the knowledge on the abundance, levels, behavior, transfer, pathways and fate of ecotoxic metals in the atmosphere, hydrosphere and terrestrial ecosystems and their ingestion by man and animals. The great potentialities of voltammetry in these areas of environmental research and monitoring will be demonstrated by a number of examples taken mainly from the extensive environmental research program of the author's institute which is one of the laboratories that has used voltammetry successfully for these tasks for more than a decade.

Table 2. Ultimate Practical Determination Limits (in µg/l; ppb) of Differential Pulse Voltammetric Methods for a Precision of 20% RSD (in brackets ≤10% RSD)

Method	Cu	Pb	Cd	Zn	Ni	Co	Hg	As(III)	Se(IV)
DPSV/HMDE	0.05 (0.10)	0.02 (0.30)	0.02 (0.10)	0.02 (0.50)	–	–	–	–	0.1 (1.00)
DPSV/MFE	0.007 (0.05)	0.001 (0.0015)	0.003 (0.0015)	n.d.	–	–	–	–	–
DPSV/Au	0.02 (0.10)	n.d.	n.d.	–	–	–	0.04 (0.20)	0.1 (2.0)	–
ADPV/HMDE	–	–	–	n.d	0.001 (0.02)	0.001 (0.02)	–	–	–

The area where voltammetry offers most particular potentialities and is frequently the most superior method of determination and investigation is the chemistry of ecotoxic metals in all types of natural waters[15]. This also includes the atmospheric precipitates, ratio and snow[16,26].

The scheme of the complete analytical procedure for natural waters, adapted to the utilization of voltammetry in the determination stage, is shown in Figure 4. High performance determination methods such as voltammetry in the differential pulse mode, can only display their full potentialities, if they are combined with adequate sampling and sample pretreatment procedures, adapted to minimize or exclude the contamination risks lurking everywhere, particularly in the ultra trace ranges encountered with levels usually between 0,5 and 10^{-3} µg/l in the oceans[27,28] or in snow and ice from polar regions[25].

Sampling is often the most critical stage. Systematic errors introduced here can make the whole analysis useless and may create the paradoxical situation, that with effort and diligence completely irrelevant, because inaccurate, data will be determined very precisely. Thus, sampling for trace metal chemistry of natural waters requires specially designed techniques and sampling devices. The technical requirements are much more stringent for deep water. Suitable and reliable techniques have been developed meanwhile for surface waters[14,29] and deep waters[30,31].

A comprehensive account of all stages of the complete analytical procedure for natural water samples has been recently given[15]. With respect to sample pretreatment here, only the following will be mentioned. For certain water types only a restricted pretreatment indicated by the dotted lines in Figure 4 is necessary and recommendable. Thus, in many regions of the oceans the content of suspended matter is so low that filtration can be omitted. In many oceanic regions also the biological productivity is so small that the level of dissolved organic matter (DOM) remains negligible and the UV-irradiation step can be omitted if proved to be unnecessary[28]. However, usually in coastal waters, estuaries, rivers and lakes the existing DOM-level requires the UV-treatment, because in this manner all DOM components are decomposed by photolysis without additional contamination risk. They would otherwise bind a part of the dissolved heavy metals in strong complexes rendering them not readily accessible to voltammetric determination. pH-Adjustment of the analyte to optimal values for voltammetric determination is always necessary, if the ultimate determination limits given in Table 2 are to be attained. For speciation investigations a special pretreatment scheme has to be applied (see Figure 8).

Mandatory general trace analytical requirements are the scrupulous cleaning of all labware, sampling flasks and filters according to established procedures, the handling of the samples on clean benches with filtered laminar air flow, for work at the utmost ultra trace level (10^{-3} to 10^{-4} µg/l) even in clean rooms, and in general a trace chemical working philosophy of the staff taking greater precautions against contamination interferences, the lower are the trace levels of the ecotoxic metals to be determined[24].

The essentials of the voltammetric determination procedures are indicated in Figure 4.

Cu, Pb and Cd are usually determined simultaneously by DPASV using as test electrode the mercury film electrode (MFE) on a glassy carbon substance[32]. Zn can also be determined, yet at the MFE not at pH 2 but

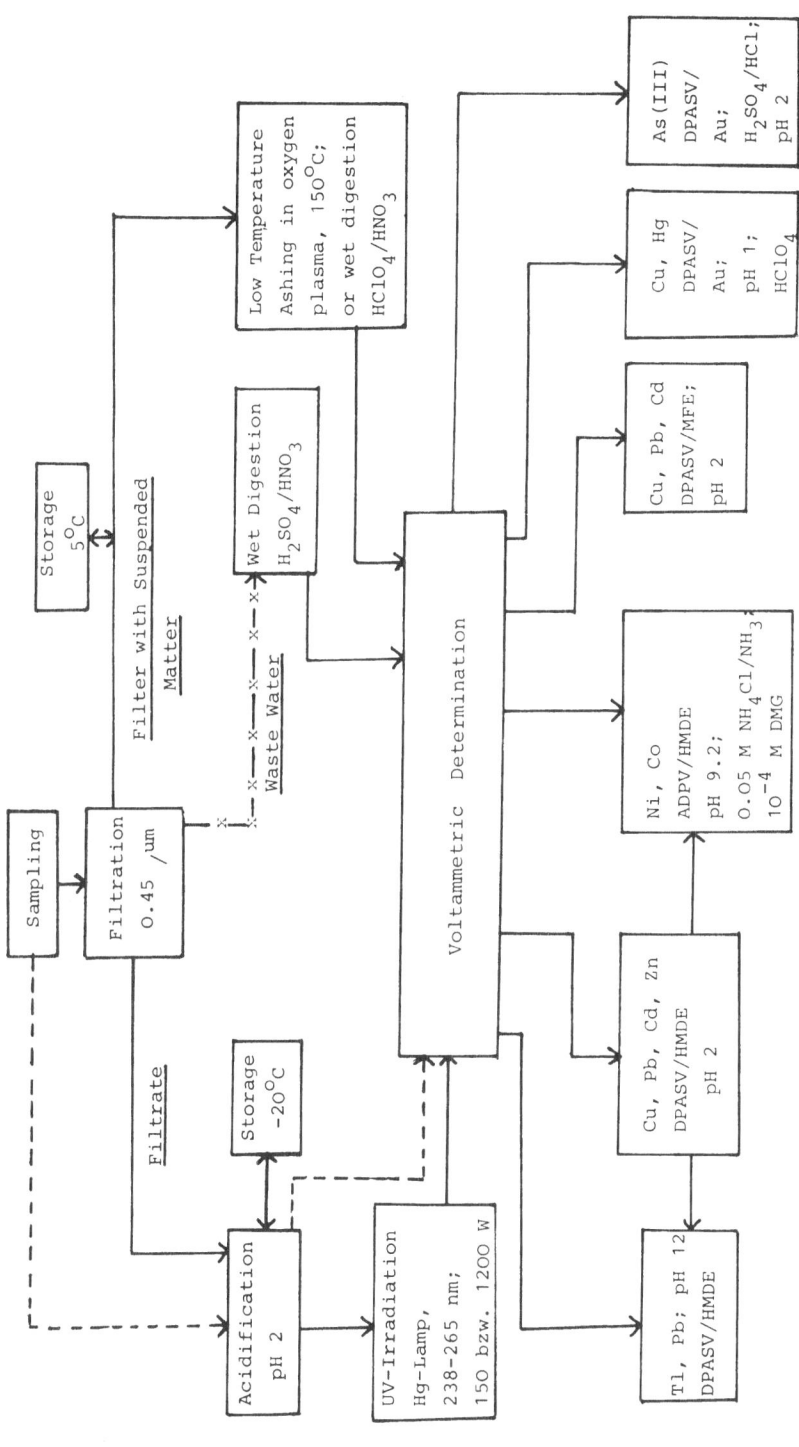

Fig. 4. Flow chart of complete analytical procedure with voltammetric determination for various types of water.

separately in an acetate buffer at pH 4,5, while for trace metal levels above typically 0.1 μg/l simultaneous determination of Zn with Cu, Pb and Cd at the HMDE is even possible at pH 2[16].

Ni and Co are determined simultaneously after adjustment of the analyte or in a separately prepared aliquot of the sample by ADPV at the HMDE[21,22]. A highly sensitive ADPV-determination for Cr has just been developed[20].

Cu and Hg are determined simultaneously very conveniently by DPASV at a specially activated gold electrode[17]. For very small Hg-concentrations, typically below 3×10^{-5} μg/l, as they occur frequently in seawater, DPASV in the subtractive mode at a twin gold disc electrode has to be used[18].

Also, the relatively most toxic As(III) can be specifically determined by DPASV at an activated common gold disc electrode[19].

Oceans and Coastal Waters

A challenging task in chemical oceanography of trace metals is the elucidation of their distribution in the oceans and of their vertical concentration profiles with depth in correlation to the oceanographic parameters (currents; mixing; stratification, age and transfer of water masses; upwelling and advection of water, etc.) aeolian input and biological productivity (predominantly phytoplankton). Up to now the distribution contours for Cu, Pb and Cd and partially also for Ni and Co in the surface waters of the North Atlantic[27,28], the eastern Arctic Ocean[30,33], the western and central Pacific[27,28], including the Tasman Sea[27] and the Peru Basin[27] and in the Scotia and northern Weddell Sea[27] have been established. Examples of the ranges in surface waters are given in Table 3. Although the heavy metal levels are all in the ultra trace level, significant differences exist for different geographic regions as function of the before mentioned parameters. Well defined correlations have been found for the depth profiles of nutrient-like metals, as Cd and Cu, and the depth profiles of the nutrients phosphate, nitrate and silicate[27,28,30,34]. For silicate a new determination alternative by ADPV has been developed[35]. The depth profiles of nutrient-like heavy metals always remain of the same general type but show pronounced diversity in their quantitative pattern (see Figure 5) for different geographical positions according to the respective variations in the above mentioned parameters[34]. While the depth profiles of nutrient-like heavy metals reveal, that their concentration regulation in surface waters is, depending on the oceanographic parameters, more or less strongly affected by cycling in the water column a completely different behavior is observed everywhere for Pb, which has no correlation to the depth profiles of the nutrients[36,37]. Recently the particular effects of bisulphide and polysulphides in strongly anoxic deep waters could be established from depth profiles for Cd, Cu and Pb in the Black Sea[38].

A recent study in surface waters has revealed the distribution and transport with the currents of the Cd-pollution (see Figure 6) from the major fluvial input areas at the North Sea Coast (estuaries of Scheldt, Rhine, Weser and Elbe) over the North Sea and Norwegian Sea up to the Barents Sea and the East Greenland Strait[34]. Detailed studies have contributed significantly to establish reliable data for the levels of Cd, Pb and Cu in the dissolved and suspended matter phase of coastal waters along the Belgium, Dutch and German North Sea Coast including the particular situation in the Wadden Sea and the distribution in the German Bight[39,40]. In the latter area also the Hg-levels have been determined[18].

130

Table 3. Ranges in μg/l of Heavy Metal Concentrations in Surface Water of the Oceans

Region	Cu	Pb	Cd	Ni	
Central North Atlantic 20°-45°N; 20°-60°W	70-150	40-60	3-7	–	
North Sea, Norwegian Sea, Island Sea, Barents Sea	–	22-50	8-30	–	
Eastern Arctic Ocean 79°-82°N; 10°W-40°E	50-200	4-30	4-9	50-200	
Western Mediterranean	140	62	19	–	
Eastern Pacific 15°N-15°S; 90°-130°W	40-80	7-18	2-5	–	
Aitutaki Passage Cook Islands 20°S; 158°W	40-60	4-7	0.1-1.7	–	
Peru Basin 5°-10°S; 90°W	94	10	7	230	
Tasman Sea 28°-33°S; 152°-154°W	60-100	8-24	3-6	–	
Scotia and Northern Weddell Sea 67°-62°S; 44°-53°W	100-200	3-13	17-54	–	(upwelling effects)

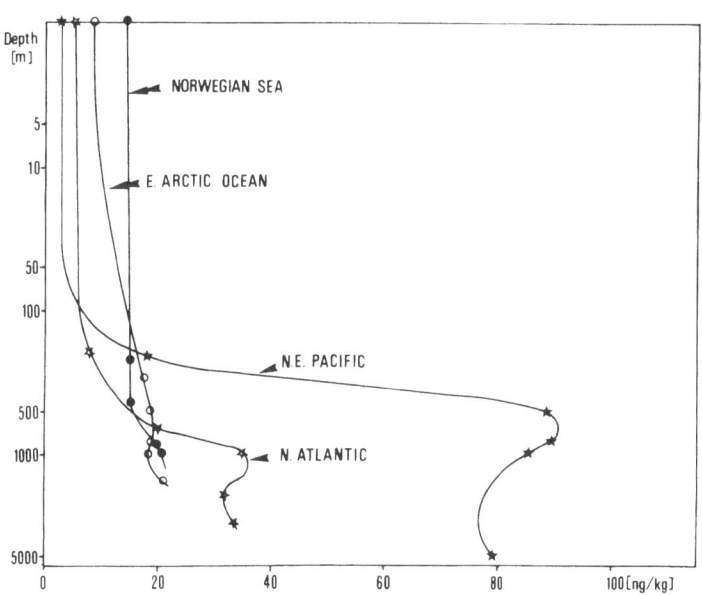

Fig. 5. Diversity of depth profiles for cadmium in various parts of oceans. The enhancements at depths around 1000 m are caused by the release of Cd from decomposed phytoplankton detritus and decomposed fecal pellets of zooplankton.

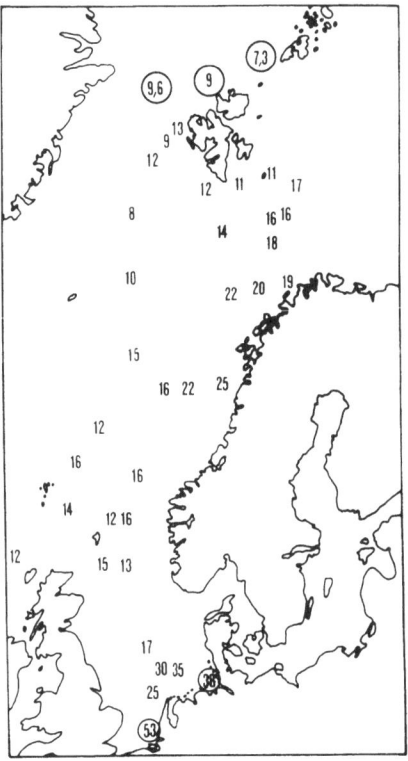

Fig. 6. Distribution of Cadmium in surface waters in North Sea, Norwegian Sea, Island Sea and Barents Sea and for comparison in eastern Arctic Ocean. Levels in 10^{-3} µg/l. Encircled values indicate average levels in corresponding area.

A similar extended study has established the heavy metal levels in Ligurian and Tyrrhenian coastal waters[27,39,41].

Estuaries

In estuaries trace metal chemistry intensive studies have been devoted to the distribution of heavy metals between dissolved and suspended matter phase, the special influence of phytoplankton population and the complicated heavy metal concentration regulation in the estuarine water column affected by the salinity gradient and resuspension and solubilization phenomena. Detailed long-term studies are investigating the situation and its seasonal dependence in various parts of the Oosterscheldt in connection with effects exerted by the completion of the Dutch Delta Plan[42,43,44]. The influence of the dry and wet season have been studied in an African tropical estuary[44].

For the Tyrrhenian estuaries, which belong the the practically non-tidal Mediterranean Sea, it has been shown, that the overwhelming part of the fluvial heavy metal freight is sedimentated in the estuarine zone and consequently the fluvial heavy metal input to the Mediterranean Sea by Arno, Tiber and the smaller rivers remains insignificant[45]. An example from this study is shown in Figure 7. When present studies in progress for the Rhone and Ehro[46] estuaries have been completed, a relevant basis will exist to assess the relative significance of fluvial and aeolian input of ecotoxic metals into the western Mediterranean Sea.

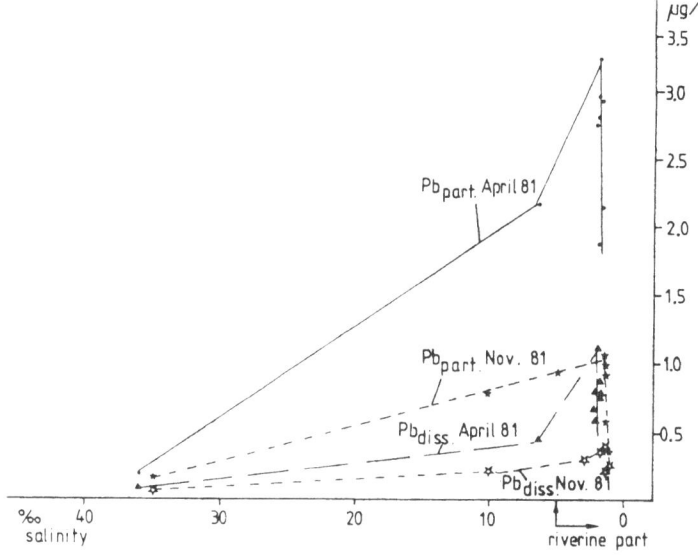

Fig. 7. Concentration regulation of lead in the dissolved and particulate
phase in the Arno Estuary as function of the salinity and season.
Peak in riverine part indicates resuspension and redissolution
effects. Common Pb-levels at begin of sea water region (38°/oo
salinity), 500 m seawards from river mouth correspond to coastal
water Pb-levels[27] and indicate almost complete sedimentation of
fluvial Pb-freight in estuarine zone.

Rivers and Lakes

Among the fresh water systems an extended study for 1979 has revealed
the pattern of the heavy metal burden (Cd, Pb, Cu) in the River Rhine and
the contributions from its tributaries[47]. The suitability of the voltammetric approach for river studies are emphasized by a recent methodological inter-comparison[48].

As the first part of a program on large European lakes the main
features of the concentration regulation for Cd, Pb and Cu in the Lake of
Constance have been clarified[50,59]. At present the somewhat different
situation in Lake Zurich is under investigation[51].

Speciation

Ecotoxic metals exist in the dissolved state in seawater and natural
fresh waters to a significant or even overwhelming extent not as free
hydrated ions but as complexes. Usually the complexes with inorganic
ligands (in seawater Cl^-, OH^-, CO_3^{2-}) are rather labile while certain
organic monomeric and polymeric ligands originating from DOM can form
rather stable complexes. This speciation in the dissolved state is of
great significance in aquatic ecosystems, as it determines the interactions
of the dissolved ecotoxic metals with the interfaces of suspended particles
and bottom sediments and also with biological interfaces in the uptake by
the phytoplankton. This lowest trophic level is very significant for the
entrance of the metals into the aquatic food webs, as well as with respect
to nutrient-like metals for the recycling between deep and surface waters
along the water column in the ocean mentioned above.

Voltammetry provides a versatile and convenient methodological tool both for diagnostic exploratory investigations as well as for detailed studies on the particular speciation of a given metal with a given inorganic or organic ligand in a certain water type. For a more detailed treatment of the applications and potentialities of voltammetry in speciation studies References 52-58 should be consulted.

Diagnosis of Species Categories

The analytical scheme in Figure 8 provides first diagnostic information on the distribution of dissolved heavy metals between various groups of complexes with increasing stability[53,57]. At natural pH the very labile complexes MeX_j formed usually with inorganic ligands ($X = Cl^-$, OH^-, CO_3^{2-}, HCD_3^-, SO_4^{2-} and the free hydrated metal ions produce voltammetric signals. After acidification to pH 2 also the metal amount bound in organic species of medium stability is exchanged with H^+-ions and becomes determinable while the decomposition of very stable inert organic complexes requires UV-irradiation.

Complexation Capacity

A characteristic diagnostic speciation parameter of a natural water sample from a certain aquatic ecosystem or a geographical region in the sea is the complexation capacity $\Sigma[L]$ for a certain heavy metal, e.g. Cu(II) or

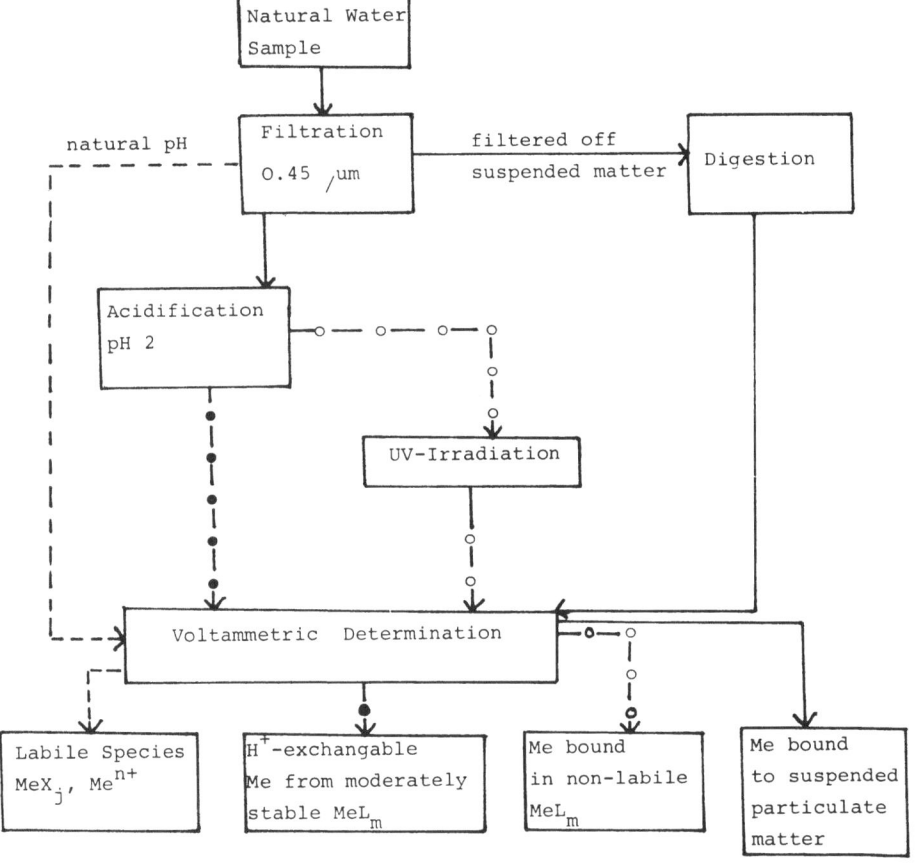

Fig. 8. Scheme for diagnostic studies on categories of heavy metal species types in natural waters.

Pb(II)[52,53,55,56,59,60]. The complexation capacity corresponds to the sum of the concentrations of all ligands $\Sigma[L]$, which form with the titrant heavy metal nonliable complexes MeL_m. These complexation capacities can be conveniently determined by a voltammetric titration of the respective water sample with the chosen titrant metal. In this manner relative scales for the contents of nonlabile heavy metal complexes forming ligands L in a natural water or a region of the sea can be determined and the resulting complexation capacities may be correlated to the phytoplankton density or DOM-level[52].

Studies on Defined Species and General Speciation Parameters

The basic property of substance specificity provides for voltammetry a particular potential for very detailed studies on individual heavy metal complexes and on general parameters governing metal speciation in natural waters.

Thus, the stoichiometric stability constants and the ligand numbers j of labile complexes MeX_j can be determined in the given water type as function of the adjusted concentrations for the ligand studied [X] at realistic levels of the heavy metals studied by the pseudo-polarogram approach[61] using anodic stripping voltammetry. From the resulting coordination chemistry data determined for all those ligands X predominantly present in the water type studied, the distribution of the heavy metal studied over the various labile complex species MeX_j can be evaluated. In this manner the species distribution of dissolved Cd and Pb valid for large parts of the oceans, where the DOM-level is small, has been evaluated[57,62]. In many coastal waters, in estuaries, rivers, lakes and ponds, however, the DOM-level will be sufficient to alter the speciation pattern of both metals significantly by consuming a certain amount of the overall dissolved concentration of heavy metals in the formation of strong nonlabile complexes MeL_m with organic ligands.

Very illuminating model studies on the general parameters governing the formation of nonlabile strong organic complexes and chelates with natural ligands provided by DOM in seawater and hard fresh water have been performed with NTA[63-66]. The measurements are based on the voltammetric determination of that amount of the heavy metal studied which remains uncomplexed for a given adjusted NTA-Concentration. In this manner the investigations could be performed at realistic overall concentration of the heavy metal studied, due to the high sensitivity of voltammetry. Investigations on the NTA-concentrations required to achieve a certain complexation degree with Cd, Pb and Zn have shown that the complexation by strong organic ligands is strongly influenced by specific effects exerted by certain salinity constituents present in substantial success compared with the trace level of the studied dissolved heavy metal. Of prime importance in this context the competition of Ca and Mg ions for the ligand. This is a common feature in seawater and hard fresh waters for all complexing DOM-components. Therefore, substantially higher ligand concentrations are required, as would be expected, then if the heavy metal complexation were affected only by the general salt effects corresponding to the ionic strength adjusted by the salinity of the respective water type.

Furthermore, important prognostic conclusions emerged from the investigations of the complexation of Cd, Pb and Zn with NTA. It could be predicted[64-66], that the levels of dissolved humics are too low in the open sea to affect the speciation pattern of the three heavy metals studied. Recent studies (see Figure 9) with marine humic materials in seawater have confirmed those prognostic conclusions directly[67]. Analogous investigations with amino acids have also confirmed the expectation, that the amino acid levels existing in the sea are too low with respect to their

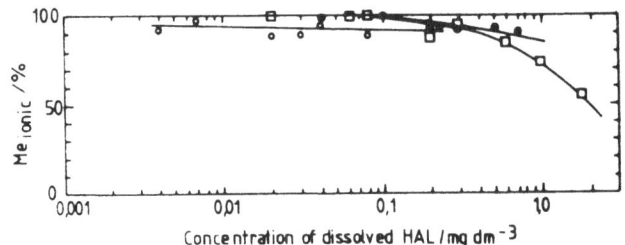

Fig. 9. Decrease of uncomplexed Cd(o), Pb(o) and Zn (□) measured by DPASV
as function of increasing concentration of dissolved humic acid
(HAL) in sea water, pH 8. Total concentrations of trace metals:
3.3×10^{-8} M Zn(II); 3×10^{-9}M Cd(II); 1.9×10^{-8} M Pb(II).
Below 0.3 mg/l humic acid trace metal complexation remains
insignificant. In the oceans humic acid concentrations hardly
exceed 0.1 mg/l.

moderate stoichiometric stability constants to exert any significant
contribution on the speciation pattern of Cd, Pb and Zn in seawater[68-70].

Of technological and environmental relevance are studies on the
speciation of Cu(II) and Cu(I) by organic anticorrosion additives in the
coolant water of power stations[71,72].

Kinetic studies with EDTA, as a well-defined model ligand for strong
complexes forming natural ligands, on the chelation of Cd, Pb and Zn have
revealed the typical formation rate constants k_f and the mechanism oper-
ative in the formation of stable complexes of heavy metals with DOM-
components in seawater and hard fresh waters[73,74]. It consists in ligand
exchange according to:

$$CaL + MeX_j \xrightarrow{k_f} MeL + Ca^{2+} + jX^- \ .$$

Due to the Ca(II) excess, Ca(II)-complexes are first formed with the strong
organic ligand L, (here EDTA). The analogous heavy metal complexes MeL are
then formed if a free heavy metal ion or a labile inorganic heavy metal
complex MeN_j encounters the Ca(II)-complex CaL. The reason for this ligand
exchange is, that the heavy metals form much more stable complexes with the
respective natural organic ligands L than Ca(II).

Another important practical result from these speciation studies with
NTA and EDTA is a new very sensitive voltammetric procedure for the simultaneous determination of both ligands, down to trace levels in natural
water[75]. The method is based on a voltammetric titration with Bi(III).
As NTA is used in laundry detergents as a substitute for phosphate to avoid
eutrophication in rivers the control of NTA has gained great significance,
because intolerable levels could cause substantial solubilization of heavy
metals from the significant deposits of ecotoxic heavy metals in the bottom
sediments of the rivers.

Wet Deposition of Ecotoxic Metals from the Atmosphere

Substantial emissions of ecotoxic metals into the atmosphere occur in
the heavily industrialized and highly populated regions of the world. To
a large extent the ecotoxic metals are bound to fine aerosol particles and
therefore, undergo mesoscalic transport and distribution with the wind or,
after advection to the upper troposphere, even hemispherical transfer.
For Cd, Pb, Cu, Zn wet deposition with rain and snow are the prevailing

transfer mode from the atmosphere into terrestrial and marine ecosystems.
Therefore, reliable analysis of ecotoxic metals in rain and snow has become
a very important application of voltammetry. Again reliable sampling has
been one of the most crucial problems. It was solved by the construction
of an automated sampler, which opens within 2,5 s at the beginning of
rain- or snowfalls and remains closed during the dry periods[16,26,76].
This sampler is now becoming part of a German regulation. A network of
20 stations distributed over the Federal Republic from the North Sea coast
to the Alps has operated since 1980, covering typical rural regions without
strong particular emission source, urban agglomerations, heavy industry
zones and particular problematic locations where lead smelters and accumulation of metallurgical industry (Stolberg and Goslar) are situated.
Usually weekly samples are collected and after filtrations, acidification
and UV-irradiation of the rain or molten snow, Cu, Pb, Cd, Zn are determined simultaneously by DPASV at the HMDE and subsequently Se(IV) by DPCSV.
This German Wet Deposition Program[26,76,77] carried out since 1980 by the
institute, have revealed the main contours of the deposition burden with
ecotoxic metals in the various regions of the country (Table 4). This also
refers to the deposition of acid rain which is monitored via electrometric
pH-measurements as well. From the wet deposition data an approximate
assessment of the annual burden of the FRG by deposition of ecotoxic metals
from the atmosphere has been possible (Table 5)[26].

Detailed regional research campaigns have further clarified the situation in problem areas with respect to the burden from wet deposition of
ecotoxic metals[78]. At present the investigations are extended to the
analysis of fog and clouds from which forest trees intercept pollutants
very efficiently, particularly at the crest of hill ridges. Toxic heavy
metals are suspected of playing a significant role in the death of the
forests, besides the obvious damaging effects of SO_2 and NO_x. In this
context a very reliable low cost voltammetric SO_2-monitor to follow the
SO_2-content in stack gases, has continuously been developed and commercialized[79].

Table 4. Average Daily Wet Deposition (in $\mu g/m^2/d$) of Ecotoxic Metals with
Rain and Snow in the Federal Republic of Germany

Region	Cd 1980	Cd 1981	Cd 1982	Pb 1980	Pb 1981	Pb 1982
Rural regions (90% of territory)	0.4–1.0	0.5–0.9	0.3–0.7	20–40	20–40	9–38
Hamburg	1.0	0.7	n.d.	48	38	n.d.
Frankfurt	1.2	1.4	n.d.	50–60	50–80	n.d.
Essen	2.35	2.3	1.9	116	158	85
Dortmund	2.2	1.8	1.1	78	96	56
Stolberg 4 km from lead smelter	1.7	4.0	2.4	91	196	130
Stolberg lead smelter location	n.d.	23.3	9.7	n.d.	645	447
Goslar	6.5	9.0	5.3	88	152	92
Tolerances of German Regulation (1982) for Sum of Wet and Dry Deposition		5.0			250	

Table 5. Approximate Annual Wet Deposition Burden (in t/y) by
Ecotoxic Metals of Federal Republic of Germany

Metal/Year	1980	1981	1982
Cd	82	79	68
	(61)	(61)	(54)
Pb	3350	3630	2450
	(2450)	(2450)	(1810)
Cu	842	800	590
	(570)	(570)	(453)
Zn	5600	5900	2900
	(3620)	(3620)	(1810)

The values in brackets refer to the annual burden of 90% of the
territory without particular emission sources and zones and are
mainly caused by mesoscalic transfer from the main emission areas.

The research on wet deposition of ecotoxic metals has also established
several findings of general significance. Over 90% of the wet deposited
metals are dissolved in the rain water. Consequently, they reach the
vegetation in a form most favorable for uptake. In this context, it has to
be realized further that wet deposition only constitutes a lower limit for
metal uptake by the vegetation, because the rain leaches and dissolves
further metal amounts from the dust particles deposited in dry periods onto
the leaves[16,26].

Time profiles of the metal concentrations in rain over precipitation
periods (see Figure 10) have established, that in the initial phases of
rainfall the metal concentrations are always substantially higher (typically up to 10- or 15-fold) than the rather stationary values attained after
1-2 h. This indicates efficient rain-out and wash-out in the initial phase
of precipitations[16,26].

An important finding on hemispherical pollution transport to the
remote Arctic has been established during a participation in the great
Swedish polar expedition YMER-80 to the region north of Greenland,
Spitsbergen and Franz-Josef-Land. This region is absolutely free of any
sources of emission. Extreme ultra trace voltammetry analysis, for Cd
down to 10^{-4} µg/kg, of snow samples has revealed for the first time the
seasonal variations in the hemispherical atmospheric transport of ecotoxic
metals[25]. In good correlation with the meteorological situation it has
been established that the metal transfer from the European emission zones
to the Arctic occurs predominantly in fall and winter. Of general significance is in this context, that the metals can be regarded as pilot substances for the hemispherical transport also of the other pollutants.

Waste Water

Municipal and industrial waste waters, discharged treated or even
still untreated, into rivers and coastal waters are the further important
source of ecotoxic metals particularly with respect to the burden of
aquatic ecosystems. Voltammetry offers an efficient and convenient way for
control of the levels of ecotoxic metals[80,81]. The necessary pretreatment is simple and consists in filtration (0,45 µm) and subsequent treatment of the filtrate either by rapid UV-irradiation in a photodigestion
device[82] equipped with high intensity mercury lamps (1200 W) with addition of some H_2O_2 or by a fast wet digestion procedure (see Figure 4).
Then follows the simultaneous determination of Cu, Pb, Cd, Zn by DPASV at

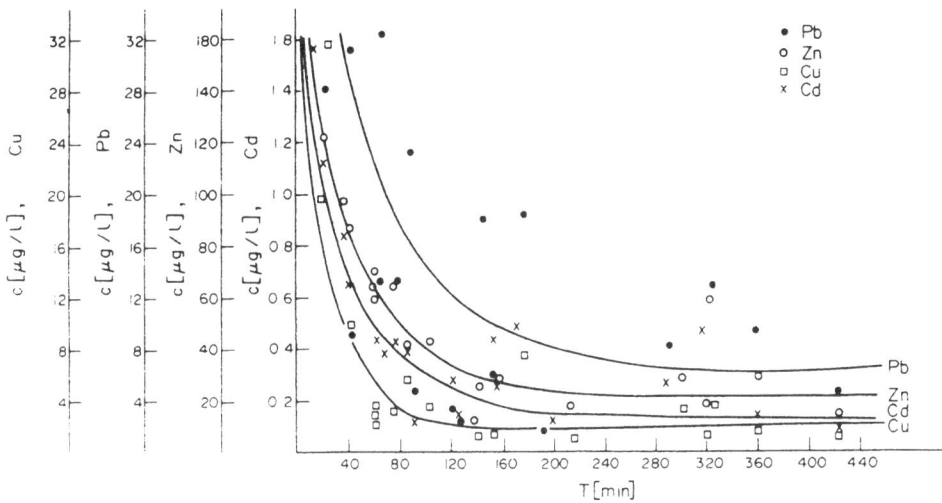

Fig. 10. Typical time pattern of ecotoxic metal concentrations in rain
during precipitation period of about 8 h.

the HMDE[80,81] and of Ni and Co by ADPV[21,22]. Automation of the whole
procedure is possible by combining the new photodigestion device[82], which
includes as an initial step the filtration, with a new fully automated
voltammetric analyzer, the "Voltammat", developed by us[11]. The metal
content in the material filtered off can also be subjected to voltammetric
determination after rapid wet digestion.

Studies on the efficiency of ecotoxic heavy metal elimination by the
usually applied biological waste water treatment plants[80] have revealed,
that the elimination of those metals is rather limited. Moreover, although
the total metal content is decreased compared with the untreated waste
water, the apparently purified water leaving the treatment plant has a
relatively higher or unaltered content of ecotoxic metals in the dissolved
phase as consequence of solubilization from suspended material by organic
complexators in the biological treatment plant.

Drinking Water, Food and Biological Environmental Materials

Drinking water is potentially the most efficient food component for
the ingestion of toxic metals by man. Therefore, the permissible content
of toxic heavy metals (Cd, Pb, Hg, As, Zn and others) is strictly regulated
in many countries. Voltammetry provides a most reliable and convenient way
for toxic metal control of drinking water[9,81,83]. Pretreatment requirements are usually minimum and consist in the adjustment of the appropriate
pH by HCl or a buffer to obtain a suitable analyte. The whole control in
the water works can be automated for continuous control at adjustable
intervals by connecting an automated voltammetry analyzer to a by-pass of
the main water tube[11,84].

Also in all types of solid and liquid food assessment of the levels
of toxic heavy metals has become a significant task in food control for
securing public health, as for the overwhelming part of the population,
which is not specifically exposed by occupation or other reasons such as
metal implants, food is the most important potential and actual ingestion
pathway of toxic heavy metals.

Voltammetry offers here an efficient and convenient way of analysis
particularly since for routine the necessary pretreatment has been harmon-

139

ized by a universally applicable common wet digestion procedure[85,86], slightly modified for meat, fish and fat[87], which provides the necessary complete mineralization of the food samples (Figure 11). In the resulting analyte, voltammetry can then display its advantages of simultaneous determination of the various groups of toxic metals as already described in the determination stage of the procedure for natural waters (see Figure 4). It is obvious, that due precautions against contamination have to be taken also in the sample pretreatment steps[88-90] of dissection, portioning and homogenization before the digestion.

The universal applicability of the analytical procedure developed and adapted to the utilization of voltammetric determination methods (Figure 11) has been featured by extensive investigations on all components of the human food basket[85,86]. Table 6 gives an impression on the magnitude of toxic metal levels to be expected in the various food types.

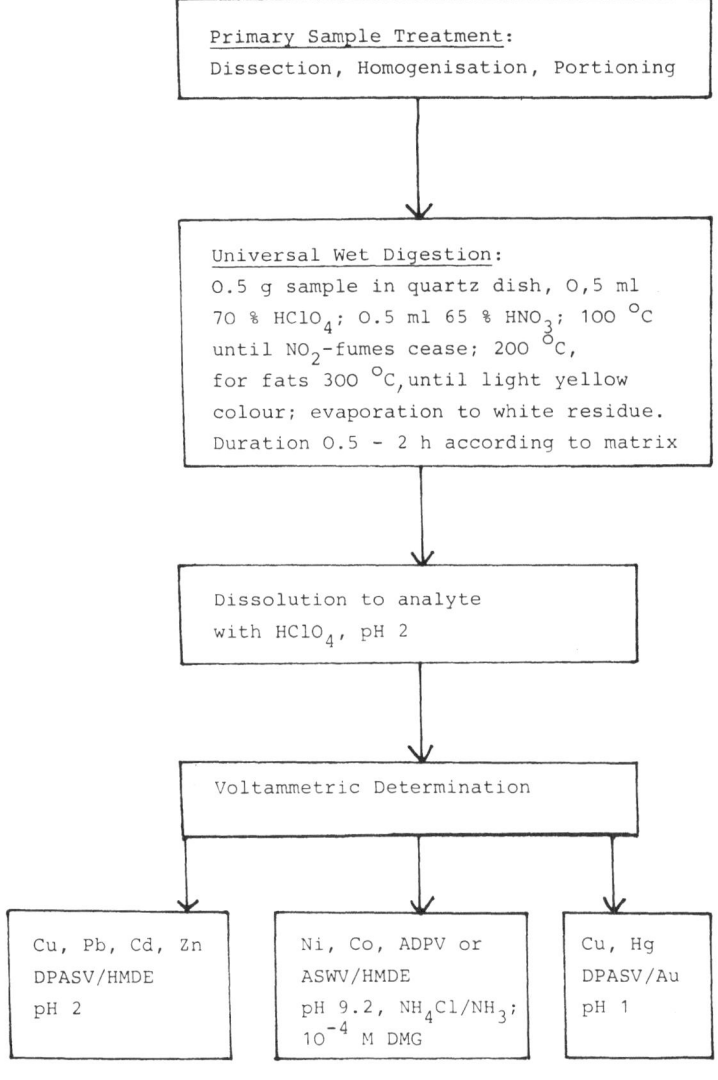

Fig. 11. Flow chart of complete analytical procedure with voltammetric determination for food, biological materials, body fluids and tissues.

Table 6. Typical Heavy Metal Levels (in µg/kg) in Various Types of Food

Food	Cd	Pb	Cu	Zn	Ni
Wheat flour	39	22	822	3860	44
Bread	30	42	2070	18630	88
Rice	5	17.4	1800	11300	160
Apples	2.4	3.3	170	190	33
Beef(FW)	0.4-1	3.6-8.5	1.1-1.8	43-55	1.5-13
Veal(FW)	0.5-1	1.5-1.6	1.2-1.4	35-53	7.6- 7.8
Milk	0.3	1.0	40	3730	4.0
Butter	0.5	7.4	97	1540	25
Margarine	42	76	58	1270	313
Fish(FW)	≤ 1	≤ 10	100-500	n.d.	n.d.
Krill meat	48	50	380	n.d.	n.d.
Krill meal	610	1200	32000	n.d.	n.d.
Beer	0.7	0.9	18	52	7.6
Orange juice	0.85	0.5	183	198	11
Drinking water	0.08-0.2	0.4-0.8	2-4	6.2	n.d.

For marine food it has been established, that significant levels of toxic metals occur in fish only in the storage and detoxification organs, liver and kidneys[88], whereas in the normally consumed muscle meat of fish, only mercury levels[88,91,92] can constitute a problem, particularly in large predator fish, as tuna[92], which are at the top of the marine food chain. Investigations of antarctic krill[93] have shown, that tolerable Cd-levels are obtained only if the krill is processed and the meat is separated for use as food. Similar observations on negligible toxic metal levels as for fish meat has revealed a detailed investigation of slaughter cattle[94]. Also here the metals are accumulated in liver and kidneys. Interesting correlations between Cd and Pb have been found for both storage organs.

Very low toxic metal levels seem to be also common in fruits and with the exception of Zn also in milk and milk products. Higher, to possibly dangerous levels are to be expected in vegetables, grains 'and grain products, wild mushrooms, and certainly in tobacco, particularly in cigarettes, as here the toxic metal content in the paper substantially enhances the levels[86].

For the determination of the rather low levels of certain heavy metals (Cd, Pb, Ni) in certain food types voltammetry constitutes definitely the most superior method. In other cases it will provide just an independent alternative for the necessary analytical quality control. This is mainly the case, if voltammetry can determine just a simple metal and thus cannot display its advantage as an oligo-method for the simultaneous determination of several metals. An example is the Hg-determination in fish and other marine organisms[92], where cold vapor AAS remains the routine determination method[91]. Nevertheless, it is important that voltammetry provides an independent alternative, e.g. for purposes of analytical quality control. Certain liquid food, as most wines and certain fruit juices with low sugar content, require only a much simpler pretreatment[95]. It consists just in UV-irradiation. Beer, however, requires prior wet digestion[86]. Extended studies in German and European wines[96,97] have established the particular suitability of voltammetry as a routine procedure in the toxic metal control of wines and have given an impression of the levels usually to be expected on average (Cd 0,1 - 1; Pb 120; Ni 60 µg/l and Cu covering a wide range between 20 and 2000 µg/l according to the time and frequency of treatment in the vineyards before harvesting).

The procedure presented with the universal wet digestion can also be applied to all types of biological materials from the environment and extensive use of this feature is made in the analysis of samples to be stored in the German Environmental Specimen Bank[98-100], a new tool in environmental assessment and protection operated by the institute for the government. Other important applications to environmental biological materials deal with studies on the uptake of ecotoxic metals deposited with rain by certain types of common plants and by the needles and leaves of forest trees.

Body Fluids and Tissues

Easily and readily available sample types for large scale ecotoxicological studies on the toxic metal burden of population groups or for exposure control in occupational medicine are blood, urine, hair and nails from fingers and toes, while organ samples are usually restricted to postmortem autopsy material. Blood is the transfer medium of toxic metals in the body and reflects the uptake by resorption and eventually also by respiration. Urine is important for measurements on the toxic metal excretion. Hair is in principle a good exposure indicator, provided a sound pretreatment of washing off extraneous adhering metal containing dust particles has been carried out.

The wet digestion procedure with $HNO_3/HClO_4$, as for food and biological environmental materials, can be well applied to human samples[22]. Alternatives for blood are low temperature ashing (LTA) in a microwave induced oxygen plasma or pressurized wet digestion with HNO_3 followed by a mandatory after-treatment of UV-irradiation of the resulting digestion solution[103,104]. The LTA approach[103] provides a particularly high precision. The pressurized wet digestion with UV-aftertreatment is used if for quality control of AAS-determination similar digestion conditions of the blood sample shall be established.

A high performance method has also been introduced for heavy metals in urine[105]. Although direct measurements in urine are possible, prior digestion is to be preferred for unknown samples, where it cannot be excluded that the urine contains metal trapping proteins.

In general, it can be stated, that also in this area of general human and particularly in occupational metal toxicology voltammetry is gaining (due to its high reliability and extraordinary determination sensitivity) more and more important are a trace analytical approach for routine applications as well as for the (particularly in this area) very important analytical quality control to prevent wrong diagnostic medical conclusions caused by irrelevant analytical data[10,104,106,107]. For low levels of Cd and Pb[103] and particularly of Ni and Co by ADPV[22] and by adsorption square wave voltammetry (ASWV)[108] in body fluids and human tissue analytical procedure with voltammetric determination are now becoming the method of choice. This voltammetric approach has become of special importance in investigations of the Ni-burden caused in patients by corresponding Ni-containing orthopedic implants[109].

CONCLUSION

The application areas discussed here, demonstrates that in voltammetry, electrochemistry provides major potentialities and contributes to deepen the knowledge and understanding in ecochemistry and ecotoxicology of metals as well as in the related fields of food investigations and human and occupational metal toxicology. Commonly for natural waters and atmospheric precipitates, but also for a number of cases in food control, biological and human material voltammetry is the method of first choice for routine analysis and monitoring of pollution levels. Doubtless voltammetry has become a mandatory component in the methodological equipment of all laboratories charged with research and routine tasks in trace chemistry and trace analysis of ecotoxic metals.

REFERENCES

1. E. Merian, M. Geldmacher-v. Mallinckrodt, G. Machata, H. W. Nürnberg, H. Schlipköter and W. Stumm, eds., "Metalle in der Umwelt," Verlag Chemie, Weinheim (1984).
2. U. Förstner and G. T. W. Wittmann, "Metal Pollution in the Aquatic Environment," Springer, Berlin - Heidelberg - New York (1981).
3. B. Venugopal and T. D. Luckey, "Metal Toxicity in Mammals," Vol.1, 2, Plenum Press, London - New York (1978).
4. L. Friberg, G. F. Nordberg and B. Vouk, eds., "Handbook on the Toxicology of Metals," Elsevier/North Holland Biomedical Press, Amsterdam - New York - Oxford (1979).
5. A. Vercruysse, ed., "Hazardous Metals in Human Toxicology," Elsevier, Amsterdam - Oxford - New York - Tokyo (1984).
6. O. Hutzinger, ed., "The Handbook of Environmental Chemistry," Vol.1A, Springer, Berlin - Heidelberg - New York (1980).
7. H. W. Nürnberg, Polarography and voltammetry in studies of toxic metals in man and his environment, Sci.Tot.Environm., 12:151 (1979).
8. H. W. Nürnberg, Potentialities and applications of voltammetry in ecological chemistry and pollution control of toxic metals, in: "Futuristic Aspects of Electrochemical Science and Technology," p.89, SAEST, Karaikudi (1981).
9. H. W. Nürnberg, Voltammetric trace analysis in ecological chemistry of toxic metals, Pure Appl.Chem., 54:853 (1982).
10. H. W. Nürnberg, A critical assessment of the voltammetric approach for the study of toxic metals in biological specimens and their ecosystems, in: "Electroanalysis in Hygiene, Environmental, Clinical and Pharmaceutical Chemistry," W. F. Smyth, ed., p.351, Elsevier, Amsterdam (1980).
11. P. Valenta, L. Sipos, I. Kramer, P. Krumpen and H. Rützel, An automatic voltammetric analyzer for the simultaneous determination of toxic trace metals in water, Z.Anal.Chem., 312:101 (1982).
12. G. C. Barker and A. W. Gardner, Pulse polarography, Z.Anal.Chem., 173:79 (1960).
13. E. Parry and R. A. Osteryoung, Evaluation of analytical pulse polarography, Anal.Chem., 37:1634 (1964).
14. H. W. Nürnberg, Differentielle pulspolarographie, pulsvoltammetrie und pulsinversvoltammetrie, in: "Analytiker-Taschenbuch," R. Bock, W. Fresenius, H. Günzler, W. Huber and G. Tölg, eds., Bd.2, S.211, Springer, Berlin - Heidelberg - New York (1981).
15. H. W. Nürnberg, Trace analytical procedures with modern voltammetric determination methods for the investigation and monitoring of ecotoxic heavy metals in natural waters and atmospheric precipitates, Sci.Tot.Environ., 37:9 (1984).
16. V. D. Nguyen, P. Valenta and H. W. Nürnberg, Voltammetry in the analysis of atmospheric pollutants. The determination of toxic trace metals in rain water and snow by differential pulse stripping voltammetry, Sci.Tot.Environm., 12:35 (1979).
17. L. Sipos, J. Golimowski, P. Valenta and H. W. Nürnberg, New voltammetric procedure for the simultaneous determination of copper and mercury in environmental samples, Z.Anal.Chem., 298:1 (1979).
18. L. Sipos, H. W. Nürnberg, P. Valenta and M. Branica, The reliable determination of mercury traces in sea water by subtractive

differential pulse voltammetry at the twin gold electrode, Anal.Chim.Acta, 115:25 (1980).

19. F. G. Bodewig, P. Valenta and H. W. Nürnberg, Trace determination of As(III) and As(V) in natural waters by differential pulse anodic stripping voltammetry, Z.Anal.Chem., 311:187 (1982).

20. J. Golimowski, P. Valenta and H. W. Nürnberg, Voltammetric ultra trace determination of chromium in natural waters, rain and drinking water, Z.Anal.Chem., in preparation.

21. B. Pihlar, P. Valenta and H. W. Nürnberg, New high-performance analytical procedure for the voltammetric determination of nickel in routine analysis of waters, biological materials and food, Z.Anal.Chem., 307:337 (1981).

22. P. Ostapczuk, P. Valenta, M. Stoeppler and H. W. Nürnberg, Voltammetric determination of nickel and cobalt in body fluids and other biological materials, in: "Chemical Toxicology and Clinical Chemistry of Metals," S. S. Brown and J. Savory, eds., p.61, Academic Press, London - New York (1983).

23. C. van den Berg, Direct Determination of Sub-nanomolar Levels of Zinc in Sea Water by Cathodic Stripping Voltammetry, Talanta, in press.

24. L. Mart, Minimization of accuracy risks in voltammetric ultratrace determination of heavy metals in natural waters, Talanta, 29:1035 (1982).

25. L. Mart, Seasonal variations of Cd, Pb, Cu and Ni levels in snow from the eastern arctic ocean, Tellus, 35B:131 (1983).

26. H. W. Nürnberg, P. Valenta, V. D. Nguyen, M. Gödde and E. Urano de Carvalho, Studies on the deposition of acid and of ecotoxic heavy metals with precipitates from the atmosphere, Z.Anal.Chem., 317:314 (1984).

27. L. Mart, H. Rützel, P. Klahre, L. Sipos, U. Platzek, P. Valenta and H. W. Nürnberg, Comparative studies on the distribution of heavy metals in the oceans and coastal waters, Sci.Tot.Environm., 26:1 (1982).

28. H. W. Nürnberg, L. Mart, H. Rützel and L. Sipos, Investigations on the distribution of heavy metals in the Atlantic and Pacific Oceans, Chem.Geology, 40:97 (1983).

29. L. Mart, Prevention of contamination and other accuracy risks in voltammetric trace metal analysis of natural waters. II. Collection of surface water samples, Z.Anal.Chem., 299:97 (1979).

30. L. Mart, H. W. Nürnberg and D. Dyrssen, Low level determination of trace metals in arctic sea water and snow by differential pulse voltammetry, in: "Trace Metals in Sea Water," C. S. Wong, E. Boyle, K. W. Bruland, D. Burton and E. D. Goldberg, eds., p.113, Plenum Press, New York - London (1983).

31. L. Sipos, H. Rützel and T. H. P. Thijssen, Performance of a new device for sampling sea water from the sea bottom, Thalassia Jugosl., 16:89 (1980).

32. L. Mart, H. W. Nürnberg and P. Valenta, Prevention of contamination and other accuracy risks in voltammetric trace metal analysis of natural waters. III. Voltammetric ultratrace analysis with a multicell-system designed for clean bench working, Z.Anal.Chem., 300:350 (1980).

33. L. Mart, H. W. Nürnberg and D. Dyrssen, Trace metal levels in the eastern Arctic Ocean, Sci.Tot.Environm., 39 (1984), in press.

34. L. Mart, H. W. Nürnberg and H. Rützel, Comparative studies on cadmium levels in the North Sea, Norwegian Sea, Barents Sea and the Eastern Arctic Ocean, Z.Anal.Chem., 317:201 (1984).

35. C. S. P. Iyer, P. Valenta and H. W. Nürnberg, A new voltammetric method for the determination of silica traces in water, Anal.Letters, Ser. A, 14:921 (1981).

36. K. W. Bruland, Trace elements in sea water, in: "Chemical Oceanography," Vol.8, Chap.45, Academic Press, London (1983).

37. K. W. Bruland, Oceanic distributions of Cd, Zn, Ni and Cu in the North Pacific, Earth Planet, Sci.Letters, 47:176 (1980).
38. W. Dorten, H. W. Nürnberg, L. Mart and P. Valenta, Depth profiles of Cd, Pb and Cu in the Black Sea, Sci.Tot.Environm., in press.
39. L. Mart, H. W. Nürnberg and P. Valenta, Comparative base line studies on Pb-levels in European coastal waters, in: "Lead in the Marine Environment," M. Branica and Z. Konrad, eds., p.155, Pergamon Press, Oxford (1980).
40. L. Mart, H. W. Nürnberg and H. Rützel, Trace metal distribution in the German Bight and North Sea coastal waters, Mar.Chem., in press.
41. L. Mart, H. W. Nürnberg, P. Valenta and M. Stoeppler, Determination of levels of toxic trace metals dissolved in sea water and inland water by differential pulse anodic stripping voltammetry, Thalassia Jugosl., 14:171 (1978).
42. P. Valenta, H. W. Nürnberg, H. Rützel and A. G. A. Merks, Die belastung der ästuare der wester- und oosterschelde mit ökotoxischen metallen, Jahrbuch "Vom Wasser", 62:235 (1984).
43. P. Valenta, E. K. Duursma, A. G. A. Merks, H. Rützel and H. W. Nürnberg, Dissolved-particulate behaviour of Cd, Pb and Cu in Netherlands Delta estuaries, Sci.Tot.Environm., in press.
44. P. Valenta, H. W. Nürnberg, P. Klahre, H. Rützel, A. G. A. Merks and S. J. Reddy, A comparative study of toxic trace metals in the estuaries of the Ooster and Wester Scheldt and of the Sierra Leone river, Mahasagar, 16:109 (1983).
45. R. Breder, R. Flucht and H. W. Nürnberg, A comparative study on the toxic trace metal situation in the tyrrhenian estuaries, Thalassia Jugosl., 18:135 (1982).
46. W. Dorten, H. W. Nürnberg, X. Modamio and A. Ballester, Distribution of trace metals and nutrients in the Ebro estuary, Rapp.Comm.Int. Mer.Medit., in press.
47. R. Breder, H. W. Nürnberg, J. Golimowski and M. Stoeppler, Toxic metal levels in the River Rhine, in: "Pollutants and their Ecotoxic Significance," H. W. Nürnberg and S. Vigneron, eds., J. Wiley, New York (1984).
48. M. Ihnat, A. D. Gordon, J. D. Gaynor, S. S. Berman, A. Desaulniers, M. Stoeppler and P. Valenta, Interlaboratory analysis of natural fresh waters for copper, zinc, cadmium and lead, Intern.J.Environ.Anal. Chem., 8:259 (1980).
49. L. Sigg, M. Sturm, W. Stumm, L. Mart and H. W. Nürnberg, Schwermetalle im bodensee − mechanismen der konzentrations-regelung, Naturwissenschaften, 69:546 (1982).
50. W. Stumm, L. Sigg, H. Sturm and J. Davis, Metal transfer mechanisms in lakes, Thalassia Jugosl., 18:193 (1982).
51. R. Breder and P. Klahre, Distribution of heavy metals in Lake Zurich, "Proc. 3rd Int. Symp. Interactions Sediments-Water," Geneva, August 1984, CEP Consultants, Edinburgh (1984).
52. H. W. Nürnberg, Features of voltammetric investigations on trace metal speciation in seawater and inland waters, Thalassia Jugosl., 16:95 (1980).
53. T. M. Florence and G. E. Batley, Chemical speciation in natural waters, CRC Crit.Rev.Anal.Chem., 9:219 (1980).
54. H. W. Nürnberg and P. Valenta, Potentialities and applications of voltammetry in chemical speciations of trace metals in the sea, in: "Trace Metals in Sea Water," C. S. Wong, E. Boyle, K. W. Bruland, D. Burton and E. D. Goldberg, eds., p.671, Plenum Press, New York − London (1983).
55. T. M. Florence, The speciation of trace elements in waters, Talanta, 29:345 (1982).
56. P. Valenta, Voltammetric studies on trace metal speciation in natural waters. Part I. Methods, p.46; H. W. Nürnberg, Applications and conclusions for chemical oceanography and chemical limnology. Part

II, p.211, in: "Trace Elements Speciation in Surface Waters and its Ecological Implications," G. G. Leppard, ed., Plenum Press, New York - London (1983).

57. H. W. Nürnberg, Investigations on heavy metal speciation in natural waters by voltammetric procedures, Z.Anal.Chem., 316:557 (1983).

58. H. W. Nürnberg, Potentialities of voltammetry for the study of physicochemical aspects of heavy metal complexation in natural waters, in: "Complexation of Trace Metals in Natural Waters," C. J. M. Kramer and J. C. Duinker, eds., M. Nijhoff/W. Junk Publ., The Hague (1984).

59. J. C. Duinker and C. J. M. Kramer, An experimental study on the speciation of dissolved zinc, cadmium, lead and copper in the River Rhine and North Sea water by differential pulse anodic stripping voltammetry, Mar.Chem., 5:207 (1977).

60. A. Neubecker and H. E. Allen, The measurement of complexation capacity and conditional stability constants for ligands in natural waters, Water Res., 17:1 (1983).

61. M. Branica, D. H. Novak and S. Bubic, Applications of anodic stripping voltammetry to the determination of the state of complexation of traces of metal ions at low concentration levels, Croat.Chem.Acta, 49:231 (1977).

62. L. Sipos, P. Valenta, H. W. Nürnberg and M. Branica, Voltammetric determination of the stability constants of the predominant labile lead complexes in sea water, in: "Lead in the Marine Environment," M. Branica and Z. Konrad, eds., p.61, Pergamon Press, Oxford (1980).

63. B. Raspor, P. Valenta, H. W. Nürnberg and M. Branica, The chelation of cadmium with NTA in sea water as a model for the typical behaviour of trace heavy metal chelates in natural waters, Sci.Tot.Environm., 9:87 (1978).

64. B. Raspor, H. W. Nürnberg, P. Valenta and M. Branica, The chelation of Pb by organic ligands in sea water, in: "Lead in the Marine Environment," M. Branica and Z. Konrad, eds., p.181, Pergamon Press, Oxford (1980).

65. H. W. Nürnberg and B. Raspor, Applications of voltammetry in studies of the speciation of heavy metals by organic chelators in sea water, Environm.Tech.Letters, 2:457 (1981).

66. B. Raspor, H. W. Nürnberg, P. Valenta and M. Branica, Voltammetric studies on the stability of the Zn (II) chelates with NTA and EDTA and the kinetics of their formation in Lake Ontario water, Limnol. Oceanogr., 26:54 (1981).

67. B. Raspor, H. W. Nürnberg, P. Valenta and M. Branica, Significance of dissolved humic substances for heavy metal speciation in natural waters, in: "Complexation of Trace Metals in Natural Waters," C. J. M. Kramer and J. C. Duinker, eds., M. Nijhoff/W. Junk Publ., The Hague (1984).

68. M. L. S. Simoes-Goncalves and P. Valenta, Voltammetric and potentiometric investigations on the complexation of Zn (II) by glycine in sea water, J.Electroanal.Chem., 132:357 (1982).

69. M. L. S. Simoes-Goncalves, P. Valenta and H. W. Nürnberg, Voltammetric and potentiometric investigations on the complexation of Cd (II) by glycine in seawater, J.Electroanal.Chem., 149:249 (1983).

70. M. Sugawara, P. Valenta, H. W. Nürnberg and T. Kambara, Voltammetric study on the speciation of Cd (II) with L-aspartic acid in sea water, J.Electroanal.Chem., 180:343 (1984).

71. S. V. Narasimhan and P. Valenta, Determination of the speciation of copper in the presence of morpholine in alkaline solutions by DC and differential pulse polarography, Microchim.Acta, 297 (1983).

72. S. V. Narasimhan, P. Valenta and H. W. Nürnberg, Study on the speciation of copper (I) in the presence of cyclohexylamine in alkaline solution by differential pulse polarography,

Microchim.Acta, in press.

73. B. Raspor, P. Valenta, H. W. Nürnberg and M. Branica, Polarographic studies on the kinetics and mechanism of Cd(II)-chelate formation with EDTA in sea water, Thalassia Jugosl., 13:79 (1977).

74. B. Raspor, H. W. Nürnberg, P. Valenta and M. Branica, Kinetics and mechanism of trace metal chelation in sea water, J.Electroanal. Chem., 115:293 (1980).

75. A. Voulgaropoulos, P. Valenta and H. W. Nürnberg, Indirect trace determination of NTA in natural waters by differential pulse anodic stripping voltammetry, Z.Anal.Chem., 317:246 (1984).

76. H. W. Nürnberg, P. Valenta and V. D. Nguyen, Wet deposition of toxic metals from the atmosphere in the Federal Republic of Germany, in: "Deposition of Atmospheric Pollutants," H. W. Georgii and J. Pankrath, eds., p.143, D. Reidel, Dordrecht-Bosten (1982).

77. H. W. Nürnberg, P. Valenta and V. D. Nguyen, The wet deposition of heavy metals from the atmosphere in the Federal Republic of Germany, "Proc. Int. Conf. Heavy Metals in the Environment," Heidelberg, Sept. 1983, 1:70, CEP Consultants, Edinburgh (1983).

78. H. W. Nürnberg, P. Valenta and V. D. Nguyen, Schwermetallbelastung der Umwelt im Raum Stolberg durch die Feuchtdeposition, in: "Umweltprobleme durch Schwermetalle im Raum Stolberg 1983," Ministerium für Arbeit, Gesundheit und Soziales NRW Hrsg., Anh. III, p.1, Diederichs, Düsseldorf (1983).

79. J. Divisek and L. Fürst, Elektrochemischer Gasanalysator für SO_2 in Abgasen, Z.Anal.Chem., 317:317 (1984).

80. B. Pihlar, P. Valenta, J. Golimowski and H. W. Nürnberg, Die voltammetrische Bestimmung toxischer Spurenmetalle in kommunalen Abwässern und im Ablauf biologischer Kläranlagen, Z.Wasser Abwasser Forschung, 13:130 (1980).

81. P. Valenta and H. W. Nürnberg, Moderne voltammetrische verfahren zur analyse und überwachung toxischer metalle und metalloide in wasser und abwasser, Gewässerschutz-Wasser-Abwasser, 44:105 (1980).

82. W. Dorten, P. Valenta and H. W. Nürnberg, A new photodigestion device to decompose organic matter in water, Z.Anal.Chem., 317:264 (1984).

83. P. Klahre, P. Valenta and H. W. Nürnberg, Ein normiertes pulsepolarographisches verfahren zur Prüfung des Trinkwassers auf toxische metalle, Jahrbuch "Vom Wasser", 51:199 (1978).

84. P. Valenta, H. Rützel, P. Krumpen, H. W. Salgert and P. Klahre, Device for the automated simultaneous voltammetric on-line determination of toxic trace metals in drinking water, Z.Anal.Chem., 292:120 (1978).

85. P. Valenta, P. H. Ostapczuk, B. Pihlar and H. W. Nürnberg, New applications of voltammetry in the determination of toxic trace metals in food, "Proc. Int. Conf. Heavy Metals in the Environment," Amsterdam, Sept. 1981, p.619, CEP Consult., Edinburgh (1981).

86. P. Ostapczuk, P. Valenta and H. W. Nürnberg, The voltammetric approach for the assessment of heavy metal traces in food, Z.Lebensm.Unters. Forsch., in press.

87. H. D. Narres, P. Valenta and H. W. Nürnberg, Insversvoltammetrische bestimmung von schwermetallen in fleisch, leber und nieren von schlachtrindern, Z.Anal.Chem., 317:484 (1984).

88. M. Stoeppler and H. W. Nürnberg, Typical levels and accumulation of toxic trace metals in muscle tissue and organs of marine organisms from different European Seas, Ecotoxicol.Environ.Safety, 3:335 (1979).

89. M. Stoeppler, Analytical aspects of sample collection, sample storage and sample treatment, in: "Trace Element Analytical Chemistry in Medicine and Biology," P. Brätter and P. Schramel, eds., 2:909, W. de Gruyter, Berlin, New York (1983).

90. M. Stoeppler, Processing biological samples for metal analyses, in:

"Chemical Toxicology and Clinical Chemistry of Metals," S. S. Brown and J. Savory, eds., p.31, Academic Press, London - New York (1983).

91. M. Stoeppler, M. Bernhard, F. Backhaus and E. Schulte, Mercury in marine organisms from the Western Italian coast, the Strait of Gibraltar and the North Sea, Sci.Tot.Environm., 13:209 (1979).

92. R. Ahmed, P. Valenta and H. W. Nürnberg, Voltammetric determination of mercury levels in tuna fish, Microchim.Acta, 171 (1981).

93. M. Stoeppler and K. Brandt, Comparative studies on trace metal levels in marine biota. II. Trace metal in Krill, Krillproducts and fish from the Antarctic Scotia Sea, Z.Lebensm.Unters.Forsch., 169:95 (1979).

94. H. D. Narres, P. Valenta and H. W. Nürnberg, Die voltammetrische bestimmung von Schwermetallen in Fleisch und inneren Organen von Schlachtrindern, Z.Lebensm.Unters.Forsch., in press.

95. J. Golimowski, P. Valenta and H. W. Nürnberg, Toxic trace metals in food. I. A new voltammetric procedure for toxic trace metal control of wines, Z.Lebensm.Unters.Forsch., 168:333 (1979).

96. J. Golimowski, P. Valenta, M. Stoeppler and H. W. Nürnberg, Toxic trace metals in food. II. A comparative study of the levels of toxic trace metals in wine by differential pulse anodic stripping voltammetry and electrothermal, Z.Lebensm.Unters.Forsch., 168:439 (1979).

97. J. Golimowski, H. W. Nürnberg and P. Valenta, Die voltammetrische bestimmung toxischer Spurenmetalle in Wein, Lebensmittelchemie u.gerichtl.Chemie., 34:116 (1980).

98. M. Stoeppler, H. W. Dürbeck and H. W. Nürnberg, Environmental specimen banking: a challenge in trace analysis, Talanta, 29:963 (1982).

99. P. Ostapczuk, M. Gödde, M. Stoeppler and H. W. Nürnberg, Kontroll- und routinebestimmung von Zn, Ni und Co mit differentieller pulsvoltam- metrie in materialien der Deutschen Umweltprobenbank, Z.Anal.Chem., 317:226 (1984).

100. M. Stoeppler, Bedeutung von Umweltprobenbanken - anorganisch-analy- tische Aufgabenstellungen und erste Ergebnisse des Deutschen Umweltprobenbankprogramms, Z.Anal.Chem., 317:228 (1984).

101. M. Stoeppler, F. Backhaus, J. D. Schladot and H. W. Nürnberg, Concept and operational experiences of the pilot environmental Specimen Bank Project in the Federal Republic of Germany, in: "Environ- mental Specimen Banking and Monitoring as Related to Banking," R. A. Lewis, N. Stein and C. W. Lewis, eds., p.95, Martinus Nijhoff Publishers, The Hague - Boston - London (1984).

102. H. W. Nürnberg, Realization of specimen banking. Summary and con- clusions, in: "Environmental Specimen Banking and Monitoring as Related to Banking," R. A. Lewis, N. Stein and C. W. Lewis, eds., p.23, Martinus Nijhoff Publishers, The Hague - Boston - London (1984).

103. P. Valenta, H. Rützel, H. W. Nürnberg and M. Stoeppler, Trace chemistry of toxic metals in biomatrices. II. Voltammetric deter- mination of the trace content of Cadmium and other toxic metals in human whole blood, Z.Anal.Chem., 285:25 (1977).

104. H. W. Nürnberg, Potentialities and applications of voltammetry in the analysis of toxic trace metals in body fluids, in: "Analytical Techniques for Heavy Metals in Biological Fluids," S. Facchetti, ed., p.209, Elsevier, Amsterdam (1983).

105. J. Golimowski, P. Valenta, M. Stoeppler and H. W. Nürnberg, A rapid high-performance analytical procedure with simultaneous volt- ammetric determination of toxic trace metals in urine, Talanta, 26:649 (1979).

106. M. Stoeppler, P. Valenta and H. W. Nürnberg, Application of indepen- dent methods and standard materials: an effective approach to

reliable trace and ultratrace analysis of metals and metalloids in environmental and biological matrices, Z.Anal.Chem., 297:22 (1979).

107. M. Stoeppler, C. Mohl, P. Ostapczuk, M. Gödde, M. Roth and E. Waidmann, Rapid and reliable determination of elevated blood lead levels, Z.Anal.Chem., 317:486 (1984).

108. P. Ostapczuk, M. Froning, M. Stoeppler and H. W. Nürnberg, Square wave voltammetry: a new approach for the sensitive determination of Nickel and Cobalt in human samples, in: "Proc. 3rd Int. Conf. Nickel Metabolism and Toxicology," Paris, Sept. 1984, Blackwells, Oxford, in press.

109. H. F. Hildebrand, B. Raumazeille, J. Decouix, M. C. Herlant-Peers, P. Ostapczuk and M. Stoeppler, Biological consequences of long-term exposure to orthopedic implants, in: "Proc. 3rd Int. Conf. Nickel Metabolism and Toxicology," Paris, Sept. 1984, Blackwells, Oxford, in press.

ELECTROCHEMISTRY IN DRUG AND FOOD CONTROL

J. Volke

The J. Heyrovský Institute of Physical Chemistry
and Electrochemistry
Chechoslovak Academy of Sciences, Prague

INTRODUCTION

The present way of life in advanced industrial countries, i.e. in
Europe, North America, some parts of Asia and in Australia, and the complicated social structure there, impose much heavier demands on mankind
than those of just one generation ago. These demands are of both somatic
and psychic nature and are caused by the much closer interaction of individuals in the work process, by life in large communities and in close
contact with other people, by the daily stress in the traffic and often by
some other adverse effects of the environment. In this region of the
world, the quantity of food is no longer the decisive environmental factor
– as it was in the past and it is still the case in developing countries –
but this is now its quality, its properties and its contamination.

In a parallel way the need for drugs and pharmaceuticals has also
largely increased compared to the situation several decades ago or even
only twenty years ago. A great many more pharmaceutical preparations are
produced, prescribed and consumed. This is of course also due to the
factors mentioned in the preceding paragraph. For this reason, drug and
food control becomes more important and new, sensitive, specific, fast and
inexpensive methods are searched for. At the same time the solution of
analytical problems in this field has taken on very difficult features
since very often it is necessary to perform the assay of a pharmacologically active substance at much lower concentrations than in the past: the
newly developed or synthesized pharmaceuticals are administered in very low
doses since they are more efficacious; further, more complicated matrices
have to be analyzed, particularly, if the substances are assayed in biological materials or in mixtures. Similar or even more serious difficulties are encountered in the food control. The aim of the following
lecture by Professor Oelschläger is to demonstrate the applicability of
electrochemistry in solving all these problems, in particular the out-
standing role of voltammetry and polarography, made substantially more
sensitive by recent instrumental improvements.

Electrochemical Methods and Problems of Pharmaceutical Analysis

From the very beginning of the application of DC-polarography as an
analytical method, approx. in the early thirties, one of the main fields,
and perhaps the most important area of analytical polarography, was the

determination of physiologically and pharmacologically active compounds and drugs, sometimes in more complicated samples. Owing to such an orientation toward these prevalently organic compounds, fundamental knowledge of reduction and oxidation mechanisms at mercury electrodes has been collected using classical polarography which up to the present forms the main treasury of information[1,2,3]. This is being further complemented; a deeper insight is presently obtained with the help of more sophisticated electrochemical methods. Even if some authors such as A. M. Bond[4] question the necessity and/or the usefulness of the knowledge of electrode mechanisms in electroanalysis, we incline to the opinion of most electrochemists[2]. Only when the mechanism – often rather complex – of the electrolytic process is understood, is it possible to interpret unexpected changes in electroanalytical measurements and to predict interferences.

The development of electrochemical applications in pharmaceutical analysis (and to a somewhat lesser degree in food control) was chiefly based on potentiometric titrations, but chiefly on DC polarography and, soemtimes, on coulometric procedures. This phase culminated toward the end of the 1950's when it became evident that the DC polarographic procedures are not sensitive and selective enough to be applied to all the following problems of pharmaceutical analysis[5]:

   a) production control and tasks following from legal regulations for the
      registration of new pharmaceuticals (also including, bioavailability
      and uniformity tests),
   b) qualitative and quantitative assay of drugs in preparations of known
      and unknown composition,
   c) detection and determination of impurities and degradation products,
   d) determination of pharmaceuticals and their metabolic products in
      biological fluids and tissues with respect to toxicology, pharmacology
      and pharmacokinetics.

The matrices which are to be analyzed from this point of view comprise pharmaceutical preparations, raw materials, chemical products, plant material, body tissues and fluids (e.g. blood, blood serum, urine, spinal fluid etc.).

Historical Development

   Voltammetric and polarographic methods were, and still are, only exceptionally used for determining the main component of a sample (pure substance determination) since the accuracy of about 2 to 3% is not sufficient here. The other electroanalytical methods, e.g. coulometry and coulometric titrations are to be recommended here because of their higher accuracy.

   In DC polarography the lower sensitivity limit is in the region from 10^{-5} to 10^{-4} mol.1^{-1}, perhaps still lower with higher electron consumptions (e.g. with nitro compounds). An increase in sensitivity was reached only by charging current compensation[5] or, later, by the so-called tast polarography[6].

   A climax in the application of DC polarographic methods in drug analysis was seemingly attained by introducing these methods in pharmacopoeias of some countries. This happened first in Czechoslovakia[7] in 1954, later in the USA[8]. In Czechoslovakia DC polarography served for determining pure substances such as ascorbic acid, nicotinamide, chloramphenicol and a by-product in its synthesis, as well as drug contents in some preparations, e.g. in tablets - Tabuletta hydrocodonii (Hydrocodon tablets) or injections.

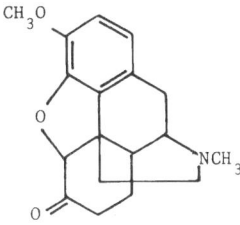

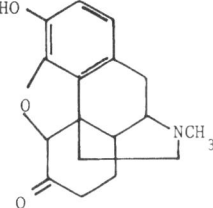

I Hydrocodon   II Hydromorphon

The former group, where about 98-99% of the pure substance are contained in the sample, is rather inconvenient with respect to the 2 to 3% reproducibility of the method. In the following,(3rd Edition of PhBS III 1970 Pharmacopoea Bohemoslovenica, Editio tertia,) the polarographic determination of the content in pure substances disappeared. The most recent edition of the USP XX also contains articles with polarographic assays, e.g. in case of Phenylmercuric Nitrate, Thimerosal, Diluted Isosorbide Dinitrate, Isosorbide Dinitrate Tablets etc. (Moreover the USP XX contains an excellent general chapter on fundamentals of various polarographic methods).

In general, polarographic methods make themselves useful in assays of pharmaceuticals in tablets, injections, various solutions etc. It is convenient that a substance is determined which is present at a not very low concentration, its concentration being usually approximately known, interfering substances are only seldom present here and in contradistinction to optical methods such as spectrophotometry, the turbidity of the solution does not interfere, i.e. the voltammetric analysis can be done even in suspensions containing insoluble particles of the excipient. It is a somewhat surprising fact that no special extraction is required in the analysis, e.g. in the analysis of tablets: the pulverized tablets are only shaken with a suitable solvent for several minutes; the solvent is usually a nonaqueous one with organic analytes; it may be exemplified with dimethlformamide or its mixture with water[9]. The dimethylformamide extract is mixed with the aqueous solution of the supporting electrolyte. The final concentration of dimethylformamide (or acetonitrile) varies from 5 to 50%. Then the solution is deaerated and a voltammetric curve recorded.

In the course of the 1960's voltammetric and polarographic method in pharmaceutical analysis started to be displaced by more sensitive methods[1], e.g. atomic absorption. Why did this happen? DC polarographic techniques were not sensitive enough for the low doses administered with new, highly efficacious deugs and their levels often need to be followed in biological materials, e.g. in metabolism studies. Moreover it is necessary to have methods for the analysis e.g. of single tablets, possibly in presence of degradation products. Such techniques are closely connected with the uniformity tests. Since large quantities of samples are usually to be analyzed, an automation of the procedures has been attempted. One of the first automatic methods has been published for lorazepam tablets[10], a tranquilizer of the 1,4-benzodiazepine group. In general, such automatic determinations may be based on:

a) current measurement at controlled potential,
b) automated coulometric titrations,
c) automated recording of whole current-potential curves combined with an automatic processing of the sample.

The first two techniques are not specific, nevertheless they are more often used than the last one, applied in cases of lorazepam. The necessity of recording complete i-E curves is due to the presence of electroactive decomposition products which must be also detected and assayed. They would interfere in constant potential techniques. The problem was solved by combining a polarographic analyzer with a continuous-flow system. The samples (e.g. powdered tablets) and standards are placed on a sample plate. Then they are introduced into the so-called SOLID-prep unit and dispersed in the solvent with the supporting electrolyte. After dissolution the sample is aspirated into a flow-system and automatically filtered. After decantation it passes the proportioning pump. It is segmented with nitrogen and enters the "steady-state" extension unit. In the continuous deaerator a stream of solvent-saturated nitrogen converts the solution into a turbulent liquid film which enters the polarographic flow-cell. The analyzer and recorder are sychronized to initiate a scan at the moment the sample attains steady state at the flow-cell. The recording is performed in the DC mode. The measurement is only carried out with a constant concentration at the detector. Hence a steady state of long duration must be achieved with the help of the above-mentioned steady-state extension unit.

Pulse Polarographic Methods

A radical change in the applicability of voltammetric methods in food and drug control took place in the 1970's and is closely connected with the introduction of pulse-polarographic methods, in particular with that of DPP (differential pulse polarography), and with the appearance of pulse polarographs on the market. The DPP increased the sensitivity by several orders of magnitude from 10^{-4} to 10^{-6} -10^{-8} mol.L^{-1}, or perhaps 10^{-9} mol.L^{-1}, in practice this allows a determination of a substance with M = 300 g and n = 2 at a concentration in the range of 1-10 mg.mL^{-1}. Very low concentrations may be measured with the help of the stripping method based on preceding adsorption without an electrochemical process at the electrode as introduced by Kalvoda[11] and reported at this Meeting.

However, there is another recently introduced improvement in polarographic experimental techniques, namely the static mercury drop electrode (SMDE)[12,13]. This electrode represents a non-electronic approach to obtaining a constant electrode surface. It follows from recent literature on analytical applications of polarography and from the comparison of pertinent equations for the polarographic current in DC and DP polarography (under assumption of linear diffusion):

$$i_d = nFAc\ D/t_D \qquad (1)$$

(A = electrode area, i_d = diffusion-controlled current, t_D = drop time, c = concentration of the electroactive species, D = diffusion coefficient, F = Faraday constant) for the diffusion-controlled current and,

$$(i_d)_p = nFAc\ D/t_m \qquad (2)$$

for the current in the pulse polarographic method (in normal pulse polarography). Here, t_m is the time at which the current is measured after application of the pulse. The equations demonstrate that a longer decay is associated with a DC measurement; this leads to a smaller faradaic current in DC polarography. It further implies that the faradaic current/charging current ratio is more favorable. With a small noise the DC method can be superior. The following experimental results[13] in Table 1 shows that DC polarography at the SMDE rivals the DPP techniques because of no difference in the detection limit and much simpler instrumentation. Hence, according to Bond[13], a renaissance of the use of DC polarography may be expected if the use of SMDE becomes widespread.

Table 1. DC, Normal Pulse and Differential Pulse Polarography at the
DME and SMDE

			t_D/s	Detection limit $c/mol.L^{-1}$
Reversible:				
Cd(II) + 2e$^-$	DME	DC	5.0	2.10^{-6}
(0.1 M NaNO$_3$)		NP	2.0	4.10^{-7}
		DPP	2.0	8.10^{-8}
	SMDE	DC	0.5*	1.10^{-7}
		NP	0.5*	2.10^{-7}
		DPP	0.5*	1.10^{-7}
Irreversible:				
Ni(II) + 2e$^-$	DME	DC	2.0	5.10^{-6}
(1 M KCl)		NP	2.0	3.10^{-6}
		DPP	2.0	7.10^{-7}
	SMDE	DC	0.2*	3.10^{-7}
		NP	0.5**	5.10^{-7}
		DPP	0.5*	5.10^{-7}

Pulse amplitude E = -50 mV
* electrode growth period 50 ms
**electrode growth period 200 ms.

Interferences and Separation

    With pulse polarographic methods, in particular with the DPP, the
resolution in the analysis of electroactive mixtures is much improved as
compared to DC polarography. The peaks are well resolved even for differences $\Delta E_{1/2}$ or ΔE_p = 30-50 mV. In spite of this, one often has to work
in such a way that the voltammetric determination (or current measurement)
is preceded by a separation procedure where, in addition to the above-
mentioned TLC, gas-liquid chromatography (GLC) and, more recently, high-
performance liquid chromatography[14] (HPLC) proved extremely useful.

    The first published combination[15] of TLC with DC polarography is
based on the separation of a mixture of 2-nitro and 3-nitro-4-acetamino-
phenetole which results after nitration in the indirect determination[15]
of acetphenetidine. Later, the method was used with other adsorbents (i.e.
silica gel and magnesium silicate) and proved successful in the separation
and determination of small quantities (10-100 μg) of psychotropic substances[16] such as nitrazepam and its metabolites containing NH$_2$ and
NHCOCH$_3$ groups instead of NO$_2$. The separation is followed by voltammetric
indication, often also in DC mode at constant potential. Reductions are
mostly carried out with dropping mercury electrodes. However, the experience in electrochemistry with solid electrodes - platinum, gold, glassy
carbon, carbon paste, etc. - enabled the determination of many new substances in the oxidative mode and the construction of electrochemical
detectors.

Narcotic alkaloids of the morphine group, possessing a free 3-hydroxy group, and resembling dopamine in their structure

Morphine

Dopamine

can be easily oxidized[17] in an amperometric detector (TL-3, Bioanalytic Systems Inc., West Lafayette, IND); this detection was successfully applied in HPLC. The following substances are electroactive: morphine, oxymorphone, nalorphine, naltrexone, naloxone. The amperometric response is a linear function of concentration in the nanogram region. The 3-methoxy derivative codeine is allegedly inactive. This behavior corresponds to older findings, but according to a recent paper[18], codeine and hydrocodeine are oxidized at platinum and gold rotating disk electrodes in 0.1 $mol.L^{-1}$ H_2SO_4.

A HPLC procedure also enables a quantitation of serotonine in plasma and spinal fluid at a 5 $pg.mL^{-1}$ level after enrichment[19]; in a flow-through sandwich detector with carbon electrodes adrenaline and l-dopa are determined:

Adrenaline

l-dopa

The lowest quantity determinable is 40 pg and 70 pg, respectively.

The analysis of Lilly Research Laboratories discuss[20] the usefulness of electrochemical detection in liquid chromatography (particularly HPLC) of pharmaceuticals contained in biological media, considering the fact that presently there are over one thousand literature reports on their use. In particular the decision between an ultraviolet and an electrochemical detector is considered. The chief factors are signal-to-noise ratio (i.e. the detection limit), specificity and convenience. The relative merits of these two types of detectors were examined with stilbestrols, an ergot derivative (pergolide), two pairs of antirhinovirals and an octadecylenol. The results are based on a successful assay of about 15,000 examples. For readily oxidazed molecules (stilbestrols or catacholamines) the S/N of electrochemical detection is clearly better than that of UV detection. With increasing anodic peak potentials E_p, S/N decreases for EC detection. If analytes with larger extinction coefficients are investigated, S/N for UV detection increases. With increasing $E_{1/2}$ or extinction coefficient ε or both a point may be reached at which S/N for both types of detectors is equivalent. If ε ($L.mol^{-1}.cm^{-1}$) is plotted versus $E_{1/2}$ (volts) for each compound, a set of experimental points is obtained; a line in the graph represents the points at which S/N is equal for EC and UV. The compounds to the left of the line have lower absolute detection limits by EC detection. The opposite holds for compounds to the right of the line. The advantages of EC and UV detectors relative to each other are summarized in the Table 2. The electrode coating problems are partially or fully removed by lowering the amount of the compound injected and/or decreasing the detector potential. Late eluters may be prevented by the technique of column switching. The influence of temperature is excluded by thermostating both the detector and the column and by working well past $E_{1/2}$.

The most recent advances[21] in LC-EC include pre- and post-column derivatization (mostly with the aim of increasing the number of electrons transferred and hence also the sensitivity). In the LC-EC determination of amino acids it may proceed via oxidative detection of o-pthaldehyde/mercaptoethanol. In this case the sensitivity has been increased so that 5.10^{-13} mol may be determined. A special method is the photolytic derivatization by a broad spectrum mercury discharge lamp (30 to 45 s of irradiation). It is used mainly with nitro compounds but also e.g. in the assay of mannitol in presence of isosorbide dinitrate. Dual-electrode cells with different potentials at the electrodes (the one enabling an oxidation, the other corresponding to reduction) make the distinction of the oxidation states possible of e.g. pteridines in biological materials. In metabolic studies LC-EC is applied e.g. for following the metabolism of benzene to benzoquinone (and its gluathionate) or the hydroxylation of phenol to hydroquinone.

An interference results from excipients in tablets in the DPP assay of some progestogens in combined low-dosage oral contraceptives (CLDOC)[22]. In such a case no polarographic response can be observed. An ultrafiltration device allows rapid prior extraction of the unknown interfering substances. The first extraction of the tablet is performed with water, the slurry is filtered under pressure (Millipore vessel and Millipore FH filter 0.5 μm pore size) and the filtrate is discarded. The extraction proper proceeds with methanol (30 min at 55°C). The interference is caused by competitive adsorption and is minimized at short drop times.

Table 2. HPLC Detection Modes

		EC	UV
Advantages	{	Sensitivity	Ease of Use
		Specificity	Universality
			Isosbestic Points
			Choice of Internal Standards
Disadvantages	{	Coating of the Working Electrode	Drug Interferences
		Reference Electrode Instability	Temperature Sensitivity
		Temperature Sensitivity	
		Late Eluters	

Large-scale preparative electrolysis is only exceptional in pharmaceutical industry. One should perhaps point out that there are intersting laboratory possibilities. This could be examplified with the electrochemical simulation[23] in preparing N-dealkylated metabilites of some drugs the synthesis of which is only accessible with difficulty. (The procedure is more convenient than a micrososomal method and proceeds in the following way:

Diazepam

The yield is 8.9% by micrososomal method and 12.9% anodic oxidation).

In one point the applicability of polarographic measurements in pharmacology is not quite evident: it is the serveral times repeated attempt to find a correlation between electrochemical parameters; for instance reduction potentials (i.e. $E_{1/2}$ or E_p) of substituted sulphonamides[24]

and their bacteriostatic, i.e. physiological, activity. In this series an in essence linear correlation is assumed to result. It will be the aim of future investigations to find whether an actual significance can be ascribed to such a correlation, i.e. whether the activity is controlled by the redox properties of such compounds or by some other properties, for example by the solubility in water or by the rate of permeation through cell membranes.

Food Control

Food control, making use of electrochemical methods possesses many features in common with the drug control and the control of pharmaceutical preparations. One may conclude, however, that the problems are often even more complicated and more difficult. The reason is the more complex matrix (except for the drug analysis in biological samples described in some of the preceding paragraphs) and the presence of high concentration levels of surface-active compounds which usually adversely affect the course of the electrode processes.

One has to consider that the production of raw materials and of final products in the food industry has become completely different as compared with the situation in the past. Because of the large food quantities required by the population, the production approaches the way raw materials are obtained in the industry and, consequently, may be affected by production factors as well as by environmental influences. The analytical food control must therefore ensure a knowledge of the concentration level of useful and important components, e.g of vitamins, but among the un-

158

desirable substances present in food the following compound groups are electroactive:

a) carcinogenic naturally occuring mycotoxins. The most important group are aflatoxins[25] which after a separation on Sephadex LH-20 may be determined by voltammetry. There are similar possibilities with deoxynivalenol[26] and zearalenon[29]. The latter may be present in wheat and in crops in general.

b) substances introduced in food in order to increase the yield.

This group includes pesticides, i.e. insecticides, herbicides, rodent-icides etc. Some of them are determined down to concentrations in the $ng.mL^{-1}$ region. The electrochemically active ones comprise the following groups:

carbamates,
ureas,
nitro compounds,
nitrogen-containing heterocycles (e.g. subst. triazines or the so-called viologens),
organophosphorus compounds,
organotin compounds,
organometallics.

Their electrochemical behavior and the analytical procedures are described in several monographs[27,28,29]. The determinations in contaminated fish meat[30] (shark, tuna, etc.) are characteristic.

A further subgroup comprises the substances protecting the livestock from diseases or stimulating the growth. It includes mainly steroids and antibiotics. The applicability of polarography can be best examplified by the determination of chloramphenicol in milk and meat[31] in the 10 ng to $1 \mu g.mL^{-1}$ region. In this particular case the so-called HPDPP (high-performance differential-pulse polarography) is used. Chloramphenicol is extracted with diethylether which is then evaporated, the residue is redissolved in acetate buffer pH 4.7 and filtered through a membrane filter. Recoveries from spiked samples fortified at levels of 10 to $1000 \ ng.mL^{-1}$ are 60% for milk and 50% for minced meat. The analysis is necessary because of possible dangerous effects in man such as aplastic anemia and Grey's syndrome in infants which both may be fatal.

The last group are the pollutants from the environment, also present in trace concentrations. The following compounds are electroactive: N-nitrosamines, aldehydes, azo dyes, amines, phenols, isocyamates and various surfactants. For most of them electroanalytical methods of determination have been worked out.

A typical modern example for the combination of high performance liquid chromatography and voltammetric detection in the field of food control is the determination in meat of phenolic hormones[32] promoting the growth of livestock. These substances comprise e.g. oesetriol, oestron, diethylstilbestrol, hexoestrol etc. The meat is homogenized and extracted with CH_2Cl_2. A methanolic solution of the extract (100 µL) is separated by HPLC. With hexoestrol $1-2 \ ng.g^{-1}$, with the remaining substances (except for oestrol) $2-3 \ ng.g^{-1}$ are the lower detection limit.

An interesting example of the determination of a substance which is neither toxic nor a pollutant but which determines the quality of butter because of its flavor is that of diacetyl[33]. From a melted butter sample diacetyl is removed by steam-distillation and determined by DPP at pH 5.2. Acetoin must be first oxidated with $FeCl_3$ to diacetyl. According to Dutch standards, first quality butter must contain $1 \ mg.kg^{-1}$ of diacetyl.

CONCLUSION

The lecture which will follow now is based on a vast experimental material and will describe a number of examples of both mechanistic investigations and of practical applications in analysis. Special attention will be paid to analytical procedures in presence of excipients and in biological matrices, and to the functionalization of electrochemically inactive substances.

REFERENCES

1. J. Volke, Polarographic and voltammetric methods in pharmaceutical chemistry and pharmacology, Bioelectrochem.Bioenerget., 10:7 (1983).
2. G. J. Patriarche, M. Chateau-Gosselin, J. L. Vandenbalck, and P. Zuman, Polarography and related electroanalytical techniques in pharmacy and pharmacology, in: "Electroanalytical Chemistry," A. J. Bard, ed., 11:141-289, M. Dekker, New York (1979).
3. M. M. Baizer, ed., "Organic Electrochemistry," p.1072, M. Dekker, New York (1973).
4. A. M. Bond, "Modern Polarographic Methods in Analytical Chemistry," M. Dekker, New York (1980).
5. D. Ilkovič and G. Semerano, Increased sensitivity of micro-analytical estimations by a compensation of current, Collec.Czech.Chem.Comm., 4:176 (1932).
6. A. J. Bard and L. R. Faulkner, "Electrochemical Methods," p.718, J. Wiley and Sons, New York (1980).
7. Československý lékopis - Pharmacopoea Bohemoslovenica, p.1040, Editio Secunda, Prague (1954); Československý lékopis - Pharmacopoea Bohemoslovenica, 3 volumes, Editio Tertia, Prague (1970).
8. US Pharmacopoeia XX/National Formulary XV, US Pharmacopoeia Convention, Rockville MD, A general chapter on polarography and individual determination described in detail, p.801 (1979).
9. H. Oelschläger, J. Volke, and E. Kurek, Polarographische gehaltsbestimmung des diazepams (Valium Roche), Arch.Pharmazie., 297:431 (1964); H. Oelschläger, E. Kurek, F. I. Sengün, and J. Volke, Polarographische gehaltsbestimmung des chlordiazepoxids (Librium) in seinen arzneiformen, Z.Anal.Chem., 282:123 (1976).
10. L. F. Cullen, M. P. Brindle, and G. J. Popariello, Automated polarographic analysis in pharmacy, J.Pharmac.Sci., 62:1708 (1973).
11. R. Kalvoda, Adsorptive accumulation in stripping voltammetry, Anal. Chim.Acta, 138:11 (1982).
12. W. M. Peterson, Static mercury drop electrode, Amer.Lab., 1/2:51 (1980); L. Novotný, Czech. Pat. A.O. 202 772, Prague (1978); L. Novotný, Czech. Pat. A.O. 202 316, Prague (1978); L. Novotný, The multi-purpose dropping mercury electrode, J. Heyrovský Memorial Congress, Proc. II, Prague, p.129 (1980).
13. A. M. Bond, Developments in polarographic (voltammetric) analysis in the 1980's, J.Electroanal.Chem., 118:381 (1981).
14. H. B. Hanekamp, P. Bos, and R. W. Frei, Design and selective application of a dropping mercury electrode amperometric detector in column liquid chromatography, J.Chromatograph., 186:489 (1979); H. B. Hanekamp, W. H. Voogt, P. Bos, and R. W. Frei, A pulse polarographic detector for HPLC: Determination of nitrazepam, J.Liq.Chromatograph., 3:1205 (1982); K. Bratin, C. L. Blank, I. S. Krull, C. E. Lunte, and R. E. Shroup, Recent advances in LCEC and voltammetry, Intern.Lab., 14:24 (1984).
15. H. Oelschläger, J. Volke, and G. T. Lim, Polarographische auswertung von dünnschichtchromatogrammen, Arch.Pharmazie, 298:213 (1965).
16. H. Oelschläger, J. Volke, and G. T. Lim, Polarographische auswertung

von magnesiumsilikat-dünnschichtchromatogrammen, Arzneimittel-forschung, 17:637 (1967); J. Volke and H. Oelschläger, Polarographische auswertung von dünnschichtchromatogrammen, "Scientia Pharmaceutica," 2:105, Butterworths, London (1967).

17. R. G. Peterson, B. H. Rumack, J. B. Sullivan, and A. Makowski, Amperometric high-performance liquid chromatographic method for narcotic alkaloids, J.Chromatography, 188:420 (1980).

18. E. Bishop and W. Hussein, Anodic voltammetry of codeine and dihydrocodeine at rotating disc electrodes of platinum and gold, Analyst, 109:143 (1984).

19. H. P. van Bennekom, PhD Thesis, University Leiden (1981).

20. D. J. Miner, M. J. Skibic, and R. J. Bopp, Practical aspects of LC/EC determinations of pharmaceuticals in biological media, J.Liq. Chromatogr., 6:2209 (1983).

21. H. Oelschläger, Advances in electroanalytical methods, in: "Topics in Pharmaceutical Sciences," D. D. Breimer and P. Speiser, eds., p.357, Biomedical Press, Elsevier, North-Holland (1981).

22. A. M. Bond, I. D. Heritage, and H. M. Briggs, Removal of interference in the differential pulse polarographic determination of progestogens, Anal.Chim.Acta, 127:135 (1981).

23. T. Shono, T. Toda, and N. Oshino, Preparation of N-dealkylated drug metabolites by electrochemical simulation of biotransformation, Drug Metab.Disposition, 9:481 (1981).

24. R. Andreoli, G. Batistuzzi Gavioli, G. Grandi, L. Benedetti, and A. Rastelli, Half-wave potential of sulfa drugs as structural index, J.Electroanal.Chem., 108:77 (1980).

25. M. R. Smyth, P. W. Lawellin, and J. D. Osteryoung, Polarographic study of Aflatoxins B_1, B_2, G_1 and G_2: Application of differential-pulse polarography to the determination of Aflatoxin B_1 in various foodstuffs, Analyst, 104:73 (1979).

26. F. Palmisano, A. Visconti, A. Bottaligo, P. Lerario, and P. G. Zambonni, Analyst, 106:992 (1981).

27. W. F. Smyth, ed., "Electroanalysis in Hygiene, Environmental, Clinical and Pharmaceutical Chemistry," Elsevier, Amsterdam (1979).

28. W. F. Smyth, ed., "Polarography of Molecules of Biological Significance," p.326, Academic Press, London (1979); W. F. Smyth, L. Goold, D. Dadgar, M. R. Jan, and M. R. Smyth, Polarographic and voltammetric methods of environmental analysis, Intern.Anal., 40 (1983).

29. J. Volke and M. Slamnik, Polarography and related methods in the determination of pesticides, in: "Pesticide Analysis," K. G. Das, ed., p.175, M. Dekker, New York (1981).

30. W. A. MacCrehan, R. A. Durst, Measurement of organomercury species in biological samples by liquid chromatography with differential pulse electrochemical detection, Anal.Chem., 50:2108 (1978).

31. J. J. van der Lee, W. P. van Bennekom, and H. J. DeJong, Determination of chloramphenicol at ultra-trace levels by high-performance differential-pulse polarography. Application to milk and meat, Anal.Chim.Acta, 117:171 (1981).

32. M. R. Smyth and C. G. B. Frischkorn, Trace determination of some phenolic growth-promoting hormones in meat by high-performance liquid chromatography with voltammetric detection, Z.Anal.Chem., 301:220 (1980).

33. E. Lechner, I. Kunder, and H. Klostermeyer, Polarographische bestimmung von diacetyl und diacetyl in butter, Z.Lebensmittel-Untersuch.Forsch., 173:372 (1981).

THE ROLE OF ELECTROANALYTICAL METHODS

IN PHARMACY AND FOOD-CONTROL

H. Oelschläger

Institut für Pharmazeutische Chemie
der Johann Wolfgang Goethe-Universität
Frankfurt

INTRODUCTION

The use of electroanalytical methods in the broad field of pharmacy
has to be estimated with respect to the morbidity and mortality rate of the
population of the world. Round about 30,000 diseases including tropical
diseases are known. 4,000-8,000 of these diseases are under control by
drug therapy. This medication is predominantly symptomatic; only therapy
of bacterial infections with the aid of chemotherapeutics and antibiotics
is a curative one. The necessity for drug research is focused by the small
success in the chemotherapy of cancer. Thirty five percent of patients
suffering from cancer may be cured by surgery and/or radiation. With the
aid of chemotherapy only 6% of those patients, especially those with
metastases, may be healed.

Physicians and the public demand drugs of high quality without side
effects and with low toxicity. The pharmacist is obliged to guarantee
their identity, further a minimal amount of impurities due to synthesis and
decomposition during storage, and the content of the active ingredient.
The importance of such analytical procedures may be demonstrated by a
drastic event which occurred in Belgium some years ago. Due to analytical
error many patients had been treated with digitoxin instead of estradiol
benzoate. Thirty four people died. A further problem is formed by the
increasing amount of allergic responses in the population of western
countries. Even small amounts of impurities reacting as haptenes e.g.
aldehydes, sulfides, ketones may cause lethal shock. By analytical control, therefore, the presence of these impurities must be excluded.
Analytical quantifications are also needed for problems in pharmacokin-
etics. Very often plasma levels correspond with the success of drug
therapy. For instance, the plasma levels of antiepileptics of patients
suffering from epilepsy and treated with antiepileptic drugs must be
controlled at short intervals.

Whilst on one hand the quality requirements of drugs are increasing
world-wide, progress in receptor research, on the other hand, enables the
syntheses of more specific active compounds. Smaller doses, therefore are
needed for the same therapeutic effect. This development may be demonstrated by the dose regimen of benzodiazepines used as tranquilizers. Over
a long period since 1960 diazepam and related compounds were given with a
single dose of 10 mg. The second generation of these tranquilizers, e.g.

represented by triazolam, is now administered with a dose of 0.25-0.5 mg. This is a world-wide trend with the consequence that further improvements are required in the development of analytical methods to be used also for biological samples.

ANALYTICAL METHODS

Which methods are available for drug-analysis in vitro and in biological samples? This question has also to consider the fact that there is an increasing tendency for application of ng-analysis in this field. The usual micromethods are gas chromatography, densitometry, high-performance liquid-chromatography with different detectors as well as photo and fluorimetry. Among the routine electroanalytical methods used up to now, only differential pulse polarography is of great importance[1,2,3]. The significance of titrimetry must not be underestimated, since the pure drug and many pharmaceutical preparations can be readily and accurately analyzed by titrimetry in research laboratories in pharmaceutical industry as well as in pharmacy laboratories. That almost all micromethods have certain restrictions in application may be demonstrated by gas chromatography, whose limitation is determined by the volatility of the samples. Derivatization can be used to convert substances of higher molecular weight to more volatile and stable derivatives which make them amenable to gas chromatography analysis.

HPLC, at its current state of development, is clearly not a method for analytical problems with a high repetition rate because the receptive conditioning of the system requires 24-36 h. On the other hand, electroanalysis is a manageable method, which is suitable for various problems (see Table 1).

It is surprising, that the European Pharmacopoeia in contrast to the United States Pharmacopoeia has not yet introduced polarographic methods (see Table 2). Only biamperometric endpoint indications, e.g. in the titration of sulphonamides, are given. One convincing advantage of polarography is the fact, that the current signal $\bar{i}_d$ is calculable, if the reduction or oxidation mechanism is known.

Table 1. Applications of Polarography in
Pharmaceutical Analysis

1) Pure drugs	limited use compared to spectrometric methods
2) Pharmaceutical formulations	┌content ├detection of impurities and byproducts └stability trials and quality controls

Advantages of the polarographic techniques:

- often no or only simple sample preparation
- turbid / coloured solutions are measurable
- the intact molecule is determined

3) Determination of intermediates in production control

4) Determinations of kinetic and physicochemical parameters

- stability constants
- reaction rate constants

164

Table 2. Polarographic Assays in USP XX

Acetazolamide Tablets	Diluted Isosorbide Dinitrate
Azathioprine Tablets	Isosorbide Dinitrate Tablets
Azathioprine Sodium for Injection	Methazolamide Tablets
Chlorothiazide Tablets	Nitrofurantoin
Dichlorphenamide Tablets	Nitrofurantoin Tablets
Ethacrynate Sodium for Injection	Nitrofurantoin Oral Suspension
Ethacrynic Acid Tablets	Procarbazine Hydrochloride Capsules

DIFFERENTIAL PULSE POLAROGRAPHY IN PHARMACEUTICAL ANALYSIS IN VITRO

Some examples may illustrate the advantages of polarography in pharmaceutical analysis[1,2,3,4,5].

1. DPP is often superior to other techniques in the control of chemical production, e.g. the synthesis of steroids with a fluorine atom in C-6 (α) is always linked with the formation of a smaller amount of the β-isomer, which can be detected easily by its more positive peak in DPP (see Figure 1). No other analytical method is known for the determination of this impurity with such accuracy.
2. Figure 2 shows the degradation of the tranquilizer Flurazepam[6] in acid solutions with subsequent formation of the corresponding benzophenone which can be estimated by CRP simultaneously due to its more negative $E_{1/2}$.
3. The XOD-blocker allopurinol (see Figure 3) is determinable in tablets only by DPP due to the reduction of its C = N double bond[6,7]. The assay is not disturbed by the presence of the excipients.
4. Phenylbutazone and prednisone, constituents of Delta-Butazolidin® Dragees are determinable at different electrodes without interference. Phenylbutazone will be oxidized at a glassy-carbon electrode and prednisone will be reduced at the DME (see Figure 4).

FUNCTIONALIZATION OF POLAROGRAPHICALLY INACTIVE COMPOUNDS

Over the past years efforts have been particularly successful in making polarographic alloy inactive molecules accessible for the electrochemical detection by the so-called functionalization reaction. While the conversion of tertiary amines in aminoxides, the nitrosation or nitration are favored methods, our school (compare Tables 3 and 4) attacked this difficult problem by treating polarographic inactive drugs with reactive organic molecules, for example by the Schotten-Baumann-reaction to introduce polarographically-active groups.

Thus, in accordance with our investigations secondary amines could be converted by reaction with nitrobenzoylchlorides into polarographically-active benzamides (see Figure 5).

Polarographically-inactive alcohols or phenols, mercaptans and thiophenols could be transformed by reaction with 3,5-dinitrobenzoic acid into polarographically-active 3,5-dinitrobenzoic acid esters (see Figure 6).

Somewhat of a problem arises in the removal of the excess of the second coupling component, which is polarographically active, if such excess is required. Metamphetamine can be polarographically activated[5] by treatment with picrylfluoride. Twelve electrons are consumed during reduction at the dropping mercury electrode (see Figure 7).

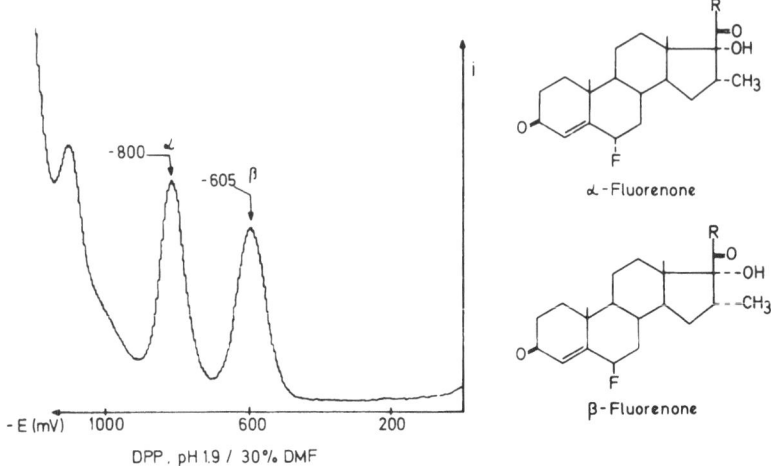

Fig. 1. Determination of intermediates (α- and β-fluorenone) in
pharmaceutical production. Differential pulse polarograph;
buffer solution pH 1.9; 30% (by volume) dimethylformamide.

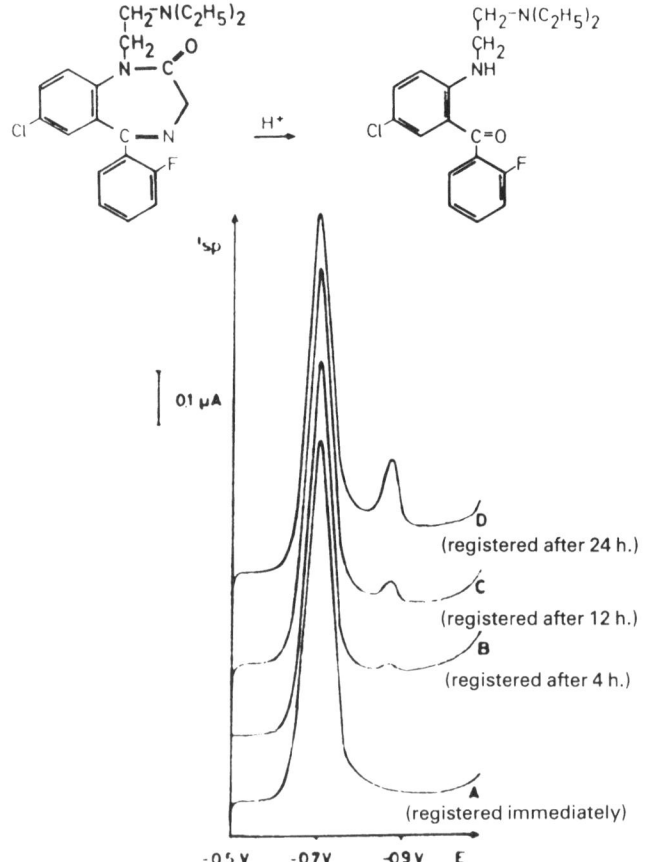

Fig. 2. Cathode-ray-polarographic investigation of flurazepam degradation
in acid media. Curve A – registered immediately; B – registered
after 4 h; C – registered after 12 h; D – registered after 24 h.

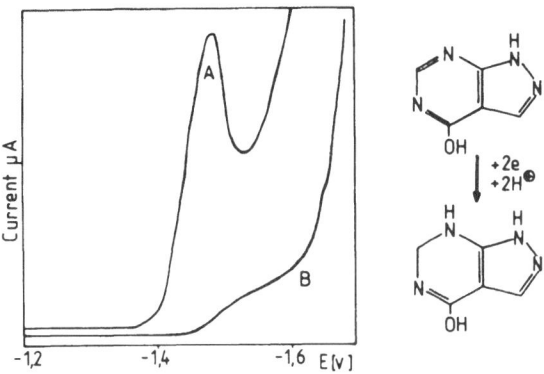

Fig. 3. Differential-pulse polarographic determination of allopurinol in
         tablets. Soerensen citrate buffer pH 4.75 with dimethylformamide.
         A – DPP (current 50 μA); B – DC polarography (current 100 μA).

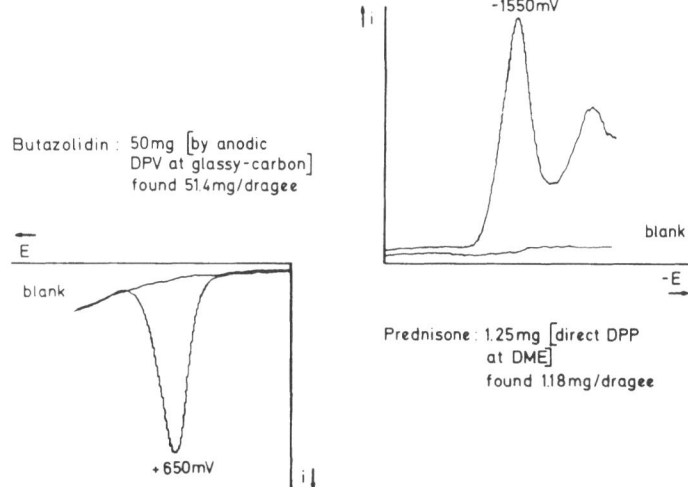

Fig. 4. Polarographic analysis of Delta–Butazolidin (the drug contains two
         active components). Butazolidin: 50 mg (by anodic differential-
         pulse voltammetry at a glassy-carbon electrode 51.4 mg found per
         1 coated table). Prednisone: 1.25 mg (by differential-pulse
         polarography at a DME) 1.18 mg found per 1 coated tablet).

     Quite often activation is successful, even in the presence of pharma-
ceutical excipients, as we were able to demonstrate in the case of the
neuroleptics perazine, prothipendyl, levomepromazine and thioridazine. The
oxidation to the sulphoxides with diluted HNO_3 and their subsequent reduction is carried out with good results in numerous dosage forms (tablets,
drops, ampoules, suppositories) (compare Figure 8 and 9).

SAMPLE PREPARATIONS

     It would be an advantage if drugs and/or their metabolites in bio-
logical samples (blood, urine, feces, bile, tissues) could be analyzed
directly. However, this has only been achieved with radio immuno assay
(RIA) procedures and, in some cases, with voltammetric methods, but with
high detectability limits (20 μg/ml). In all other cases background

Table 3. Functionalization of Polarographic Inactive Drugs (1). [Oelschläger and Coworkers 1963-1982]

Drugs	Derivative	e - Consumption
Lidocaine	N-oxide	2
Sulpiride	"	2
Phenacetin	2-and 3-Nitro-d.	4 (+2)
Phenazone	4-Nitrophenazone	6
Chlorpromazine	Sulfoxide	2
Levomepromazine	"	2
Perazine	"	2
Prothipendyl	"	2
Thioridazine	"	2

Table 4. Functionalization of Polarographic Inactive Drugs (2). [Oelschläger and Coworkers 1963-1982]

Drugs	Derivative	e-Consumption
Thioesters	Sulfones	2
sec. Amines	Nitrobenzamides	4
Alcohols	3,5-Dinitrobenzoic acid esters	4 (+4)
Phenols	"	4 (+4)
Mercaptans	"	4 (+4)
Thiophenols	"	4 (+4)
Methamphetamine	N-Picryl-d.	12
Chlorprothixene	1-Nitro-3-dimethyl-aminopropane	4

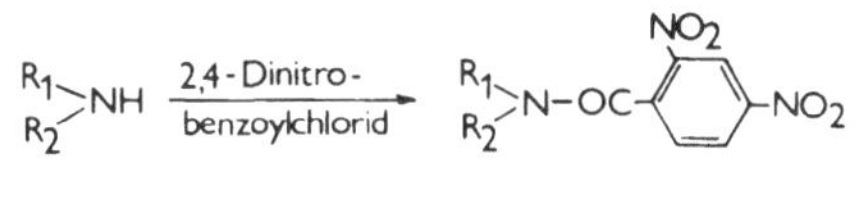

Fig. 5. Functionalization of electroinactive compounds. Exemplified with secondary amines.

material has to be excluded so that the selectivity and specificity of the compound in question can be enhanced and detection limits of the quantification procedure can be lowered. So the final stage of assay has to be preceded by separation and purification steps.

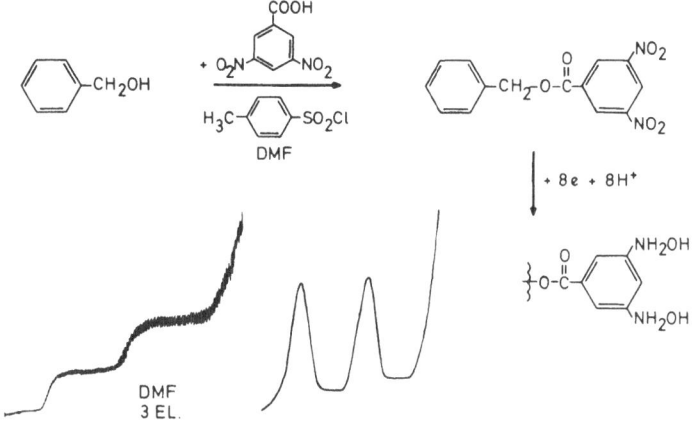

Fig. 6. Functionalization of electroinactive molecules: alcohols, phenols, mercaptans, thiophenols.

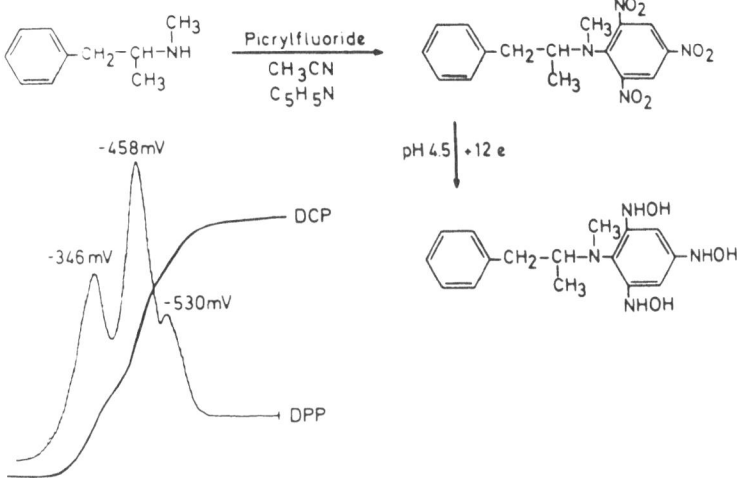

Fig. 7. Functionalization of electroinactive drugs: methanpletanine (as developed by Oelschläger and Müller[5]).

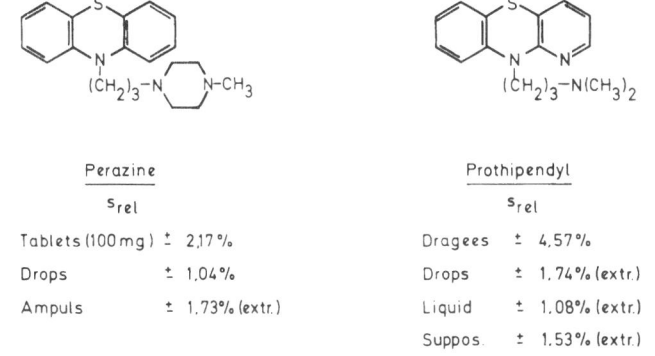

Perazine				Prothipendyl	
s_{rel}				s_{rel}	
Tablets (100 mg)	± 2,17 %		Dragees	± 4,57 %	
Drops	± 1,04 %		Drops	± 1,74 % (extr.)	
Ampuls	± 1,73 % (extr.)		Liquid	± 1,08 % (extr.)	
			Suppos	± 1,53 % (extr.)	

Fig. 8. Polarographic determination of phenothiazines.

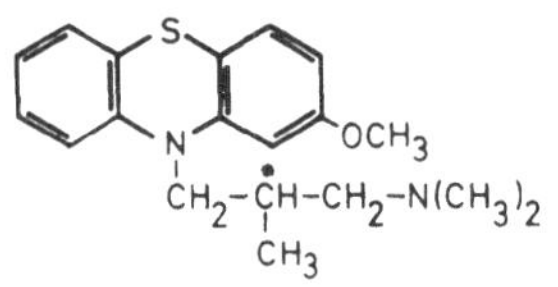

Levomepromazine

[2 Diastereomeric Sulfoxides]

	s_{rel}
Tablets (25mg)	± 2,28%
Tablets (100mg)	± 0,77%
Drops	± 1,55%
Ampuls	± 0,70%

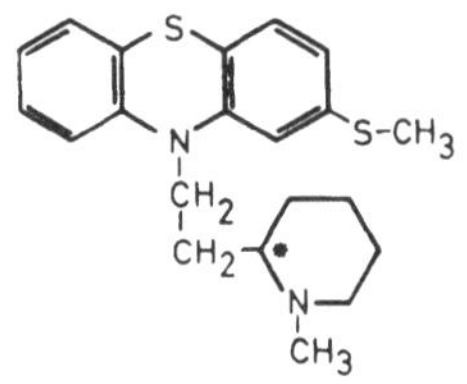

Thioridazine

[2 Diastereomeric Sulfoxides (5) as Racemates]

	s_{rel}
Dragees	± 2,13%
Drops	± 1,03% (extr.)
Liquid	± 3,11% (extr.)

Fig. 9. Phenothiazines investigated.

The repertoire of isolation techniques includes:

1. Solvent extraction (heterogeneous, homogeneous) under pH-control: with and without deproteinization.
2. Selective adsorption onto a solid agent with subsequent elution.
3. Gel chromatographic separations and purifications with controlled pore beads.

A brief description of these methods may be useful since the choice of the final step governs the mode of sample preparation.

ad 1. The most common first step in the treatment of the biological specimen, possibly after homogenization (or lyophilization), is liquid-liquid-partitioning with water-immiscible solvents after adjustment of the pH level. If further purification is required the organic solvent containing the drug can be shaken with an aqueous solvent where the pH is such that the drug is far more soluble in the aqueous phase than in the organic phase. The aqueous solution of the drug can be removed, its pH readjusted to that of the first extraction and shaken with a suitable organic solvent mixture, so taking the drug back to the organic phase, leaving undesired organic materials in the former (back extraction).
However, the use of water miscible solvents is often of an advantage: firstly a suitable solvent (an alcohol, acetone, acetonitrile) is added to the sample. In this case most of the proteins will precipitate and may be removed by centrifugation. The remaining mixed upper phase may be treated either by salts or a low miscible component (e.g. hexane) thus creating a two-phase system in which subtle partitioning processes may help further purification (homogeneous extraction). Removal of proteins may, of course, be achieved by suitable reagents (trichloroacetic acid, perchloric acid, sulphosalicylic acid, tungstic acid, zinc hydroxide etc.).

ad 2. Liquid-solid extraction offers a number of advantages over solvent extraction: high extraction efficiency; no losses due to formation of emulsions; minimal introduction of contaminants; reusable materials reduction of safety hazards; easily miniaturized.
The materials used range from the classical adsorbents silica gel, alumina, charcoal, aluminium silicates, XAD-2 resins to chemically-

modified silica gels, carrying a wide variety of functional groups (alkyl, cyano etc.).

The effectiveness of such agents consists in their ability to remove solutes from aqueous media, such as plasma, in a highly reproducible manner, whilst leaving proteins behind. In this way, e.g. quinidine may be isolated from whole blood with a recovery rate of 100% and detected by DPP.

ad 3. Methods developed during the last 10 years in the field of bio-chemical analysis offer further possibilities of purification and preanalytical separation. Gel-permeation and gel-filtration materials used in these techniques are small, pore-controlled beads of polymeric materials prepared from dextrans, polyacrylamide or polystyrene: The term gel-filtration is generally associated with separations in aqueous media when materials like dextrans are used. Lipophylic dextran derivatives (Sefadex LH 20, LH 60) swell in organic solvents thus allowing very efficient separation procedures even with short columns.

The term gel-permeation is reserved for separations effected with hydrophobic resins like polystyrene in non-aqueous media (non-polar solvents). This latter method is especially suitable for removal of interfering lipids, e.g. diethylstilbestrol can be removed completely from fatty tissues.

All above mentioned methods must work with high recovery rates. The national authorities now demand a recovery rate higher than 90%. Ten years ago recovery rates of 50% were sufficient.

In the lower ng- and pg-range losses often occur by adsorption on glass materials. A very impressive example is the adsorption of selenazines (see Figure 10) and methylene-blue used as dyes for selective monitoring of the parathyroid glands under operation. We observed losses of nearly 50% of the dye. After treating the glassware with silylating agents no adsorption could be observed.

DIFFERENTIAL PULSE POLAROGRAPHY IN PHARMACEUTICAL ANALYSIS IN VIVO

Differential pulse polarography is in some cases suitable for in vivo analysis too. Above all it can be used for the diagnosis of intoxication as a result of overdosage or often in cases of potential suicide. So we were able to detect the new triazolobenzodiazepine triazolam[4,7] in a concentration of up 30-40 ng/ml in serum, after it had been previously separated by an Extrelut® - column (see Figure 11).

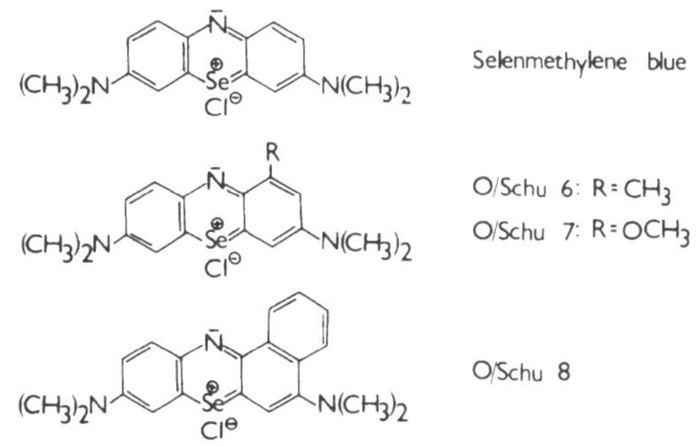

Fig. 10. Selenazines as dyes to stain the parathyroid gland.

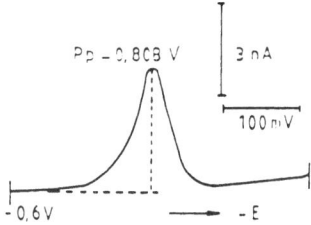

Pp – 0,808 V   3 nA

100 mV

– 0,6 V   — E

3. Mechanism of Reduction at DME

Fig. 11.   Determination of triazolam in human serum (according to
Oelschläger and Sengün[4]).   Procedure: 1. Extraction on
Extrelut® columns) elution with toluene; recovery rate >90%.
2. Determination by DPP or CRP; limit of detectability 30–
40 ug.mL^{-1}.   3. Mechanism of reduction at a DME.

Hoffmann and Kruskopf succeeded in determining phenols and sulphonamides in plasma by DPP after bromination (see Figure 12).   Estimation of
drug blood levels requires very efficient methods.   After oral administration the drug or its metabolites should be detected in serum at least
for more than 6 half-lifetimes.   The consequence of this demand may be
illustrated: If a drug with an elimination half-lifetime of 24 reaches a
plasma level of 100 ng/ml after administration then a plasma level of about
3 ng/ml could be expected after 6 days.   The concentration of those drugs,
which are extremely effective and therefore applied in low dosage, drops to
1 ng/ml and less.   Differential pulse polarography cannot manage this any
more.   Therefore it is desirable, that more sensitive electroanalytical
methods must be developed.   It is hoped, that in this case second-order-
techniques will open the way for success.

The electrochemical detector (ECD), which is still used for anodic
oxidation of phenols and amines offers interesting perspectives.   With the
aid of a commercial detector we were recently able to detect diethylstilbestrol in plasma in the low pg-region by an oxidation reaction (see
Figure 13).   Diethylstilbestrol represents the main metabolite and one of
the active principle of fosfestrol, which has been used for treating
metastasizing prostate carcinomas for about 30 years.

DIFFERENTIAL PULSE POLAROGRAPHY IN FOOD CONTROL

The application of electroanalytical methods in food control is
closely related to the use of these methods for the determination of drugs
and their metabolites in vivo.   The main feature in common is the fact that
the contaminants have to be first detected and then quantitated.   Mainly
pesticide materials used in agriculture, antibiotics necessary in cattle-
breeding and actual environmental pollutants are to be determined mostly in
very low concentrations.   The addition of anabolics like estrogens to
fodder is strictly forbidden in most countries but the farmers very often

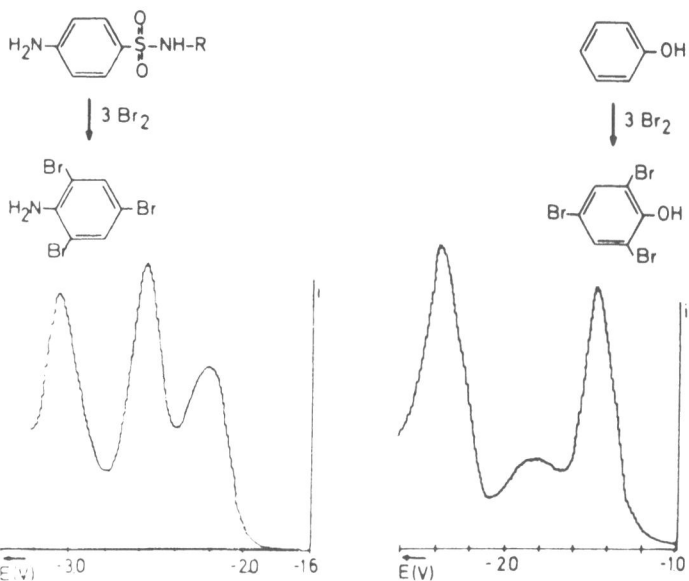

Fig. 12. Functionalization of electroinactive drugs: Determination of sulfonamides and phenols in plasma. 1. bromination in the deproteinized plasma. 2. extraction. 3. polarography in $CH_3CN/Et_4N+ClO_3^-$ (DPP).

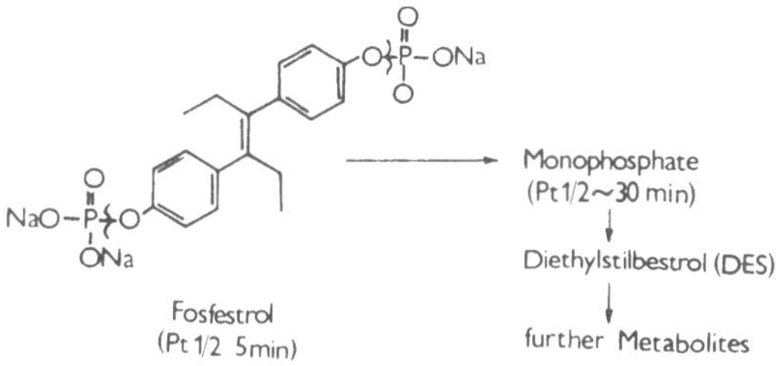

Fig. 13. Reactions of the neoplastic agent Fosfestrol (Houvan®).

ignore this inhibition. The complexity of analyses is increased by the fact that many of these substances are very unstable and are easily decomposed or metabolized into various degradation and metabolic products under the influence of the environment.

The sampling itself and preservation of the sample prior to the analyses is similar to the sampling of drugs in biological fluids and tissues. The sample must be immediately wrapped properly and stored as soon as possible at a temperature of -20° to -30°C. The low temperature prevents the loss and changes of volatile residues of pesticides, decreases the rate of enzymatic and non enzymatic processes and enables a higher recovery due to the damage of membranes during freezing and defrosting.

The isolation of the above mentioned compounds, especially of pesticides, is mainly carried out by the extraction. The following clean-up is governed by the physical and chemical properties of the compounds to be analyzed. A very annoying factor in the analytical procedure is fat, which is extracted additionally by the solvent used. For this no uniform procedure exists. Contaminants in repurified extracts can be determined by different methods. GLC, HPLC, and DPP have found widespread utilization.

S-triazines, like prometryne, chlorinated pesticides, like DOT, methoxy-chlor and hexachlor-cyclohexane and nitro compounds like parathione and related substances have been successfully determined polarographically (see Table 5). Prometryne is reduced in acid solution giving rise to a single wave at about - 1,0 V (vs.s.c.e.). The γ-isomer of hexachlorocyclohexane, known as γ-BHC, may be determined in several formulations by polarography according to the mechanism

$$C_6H_6Cl_6 + 6e^- \longrightarrow C_6H_6 + 6 \ Cl^-$$

The other active stereoisomers of γ-BHC are reduced at a more negative potential than γ-BHC, using $[(C_2H_5)_4N]$ as supporting electrolyte. Parathione can be determined by DPP in the presence of its metabolite p-nitrophenol in BR buffer pH 3. The polarographically active group is the nitro group which is reduced with consumption of $6e^-$ to the corresponding amine. Polarographic methods are also suitable for the determination of organometallic pesticides. Ethylenbisdithiocarbonate fungicides containing metals (Zn, Ziram) have often been determined by means of this method. Ziram gives an anodic wave in 0,5 M NH_4OH/NH_4Cl due to the oxidation of dimethyldithiocarbamic acid, whereas in 0.2 M-NaOH a cathodic wave is observed due to the reduction of Zn^{2+} ions.

Furthermore indirect polarographic methods for pesticides have been reported, e.g. carbaryl, dimetilan, isolan etc., may be functionalized by nitrosation procedures after hydrolysis.

Table 5. Electroanalysis of Agrochemicals

1. s-Triazines — Prometryne

2. Organochlorine Compounds — γ-BHC

3. Nitro Compounds — Parathion

4. Organometallic Compounds — Ziram

5. Carbamates — Carbaryl
$R_1 = CH_3$
$R_2 = 1$-Naphthyl

Intensive studies have been done in the group of antibiotics. Polarography is suitable for the determination of chloramphenicol, tetracyclines, penicillins and cephalosporins.

The combination of HPLC with flow-through detectors, using a glassy carbon electrode as working electrode, made possible the separation and quantitation of the forbidden phenolic growth-promoting hormones like estrone, hexestrol, diethylstilbestrol and dienestrol in meat.

In some cases environmental pollutants in food may be assayed electrochemically. These contaminants are formed mostly by industrial activities and some of them appear in food at trace level. They include N-nitrosamines, phenols, amines, isocyanates, aromatic hydrocarbons, azodyes etc. Due to its carcinogenity the determination of N-nitrosamines at trace level is very important. Fortunately not all nitrosamines which may be found in food and beverages are dangerous. They come into contact with the body firstly from the uptake of contaminated food and beverages and secondly by formation in the body. Secondary amines e.g. need nitrites as precursors for this in vivo reaction occurring in the stomach. The amount of N-nitrosamines formed depends on the chemical structure of the secondary amine. N-nitrosamines with small substituents for instance N-nitroso dimethylamine are the most potent carcinogens. Furthermore N-nitrosomethylphenylamine is a carcinogen but not N-nitrosodiphenylamine. But the latter compound transfers its nitroso group to other secondary amines e.g. dibutylamine. These N-nitroso compounds are determinable by DPP. Optimum conditions for analysis are obtained in acid media where the detection limits are of the order of 5×10^{-8}M. The reduction product was confirmed to be the corresponding hydrazine:

$$\begin{array}{c} R_1 \\ \diagdown \\ N - NOH^+ + 4e^- + 4H^+ \longrightarrow \\ \diagup \\ R_2 \end{array} \quad \begin{array}{c} R_1 \\ \diagdown \\ N - NH_3^+ + H_2O \\ \diagup \\ R_2 \end{array}$$

CONCLUSION

The examples given of successful application of polarography in food control may produce the impression of a dominating analytical technique. This is not true. In most laboratories densitometry and GLC are the common methods of analysis and polarography is restricted to special determinations. In my opinion the education at the Universities is responsible for this situation. If we train our students in electrochemistry in future then they will apply these advantageous methods as routine methods in daily use.

REFERENCES

1. H. Oelschläger, Advances in electroanalytical methods, in: "Topics in Pharmaceutical Chemistry," D. D. Breimer and P. Speiser, eds., p.357, Biomedical Press, Elsevier, North-Holland (1981).
2. H. Oelschläger, Polarographic analysis of psychotropic drugs, Bioelectrochem.Bioenerget., 10:24 (1983).
3. H. Oelschläger, Drug biotransformation as a source of drug development, in: "Strategy in Drug Research," J. A. Keverling Buisman, ed., p.203, Elsevier Publ. Co., Amsterdam (1982).
4. H. Oelschläger and F. I. Sengün, Polarographische bestimmung (DPP und CRP) des triazolam im plasma und serum, Archiv.Pharmazie, 317:69 (1984).

5. H. Oelschläger and M. Müller, Vergleichende densitometrische und polarographische (DPP) bestimmungen des methamphetamins, <u>Pharmazie</u>, 36:808 (1981).

6. H. Oelschläger, F. Druckrey, and F. I. Sengün, Polarographische bestimmungen des euhypnicums flurazepam in seinen arzneiformen, <u>Pharma.Acta Helv.</u>, 51:353 (1976).

7. H. Oelschläger, F. I. Sengün, and J. Kruskopf, Über das elektro-chemische verhalten (DCP, DPP, CRP) des triazolem (halcion) und gehaltsbestimmung seiner arzneiformen, <u>Z.Anal.Chem.</u>, 315:53 (1983).

DISCUSSION

J. Volke

The J. Heyrovský Institute of Physical Chemistry
and Electrochemistry
Chechoslovak Academy of Sciences, Prague

CHAIRMAN'S SUMMARY

Professor Kalvoda commented that with very low concentrations of drugs or
of their metabolites, where - according to the lecture - the sensitivity of
DPP does not suffice, the so-called adsorptive stripping polarography (vide
supra in the introduction) could be very advantageous. With respect to
this new, promising method it will be necessary, however, to solve the
problem of presence of other, interfering surface active substances by
developing simple separation processes, e.g. by means of molecular exclu-
sion chromatography. Only then it may be used in practical pharmaceutical
or environmental analysis (e.g. urine or serum analysis). Further he asked
whether a combination of TLC separation with DPP determination would be
possible in a similar manner as that of TLC with DCP in some papers published by Oelschläger and co-workers.

Professor Oelschläger considers the TLC + DPP combination to be very
convenient. In separation preceding the polarographic determination step
he often uses the Extralut cartridges. The legal requirements regarding
the detection of drugs in biological matrices are still increasing since,
at present, the drug levels after 6 half-life times are to be determined.
With the starting concentration in the 5-10 $ng.mL^{-1}$ region, the final
concentration level may be in the region 100-300 $pg.mL^{-1}$. In such a case,
DPP would probably still not be sensitive enough.

Professor Berg emphasized the possibility of using polarography or
voltammetry in modelling the course of processes taking place in a living
organism. This is the case e.g. with the splitting reactions in anthra-
cyclins where different mechanisms could be found depending on the sugar
residue.

Professor Kalvoda asked, in connection with the polarographic behavior of
N-nitroso derivatives, whether there is a correlation between electrochemical activity and physiological activity of substances, e.g. carcinogenic properties.

Professor Oelschläger did not agree with the authors advocating such a
dependence and thought that other properties of drugs or of pollutants were
responsible for the activity, such as the adsorption on the cell membranes,
permeation through the wall membrane or the distribution of the drug in the

tissues or in the body fluids. These properties may even change in the course of the day.

Dr. Volke pointed out that the results of plotting $E_{1/2}$ of differently substituted sulphonamides versus their activity parameters are highly doubtful.

This criticism was confirmed by Professor Berg. According to his view these sulphonamides differ in their penetration rate into the cell and, consequently, in their concentration in it. Professor Oelschläger: It has to be borne in mind that there are two groups of drugs, one with a special affinity to receptors (e.g. cholinergic drugs), the other causing a change in the action potential (e.g. marcotics). According to his view the primary factors influencing the activity of drugs are their configuration and conformation. It is because of this that legal regulations in the Federal Republic of Germany prescribe a separation of enantiomers and their independent pharmacological testing. Thus, hexabarbital is a hypnotic in its l-form and an anticonvulsant in the d-form. In the case of methadon the racemate first appeared on the market, but at present the l-isomer is produced.

Professor Berg pointed out that the separation of racemates often encumbers the pharmaceutical with too high a price which does not correspond to the increased activity.

SECTION III
ELECTROCHEMISTRY IN BIOSCIENCES

ANALYTICAL BIOELECTROCHEMISTRY

D. R. Thévenot

Laboratoire de Bioélectrochimie
Université Paris-Val de Marne
94010 Créteil Cedex, France

INTRODUCTION

Among the various analytical methods using biochemistry or physio-chemistry bioelectrochemistry is one of the most active and rapidly developing fields. It is indeed rarely necessary to add any chemical in the sample, thus allowing direct in vivo application of these methods, and their selectivity and sensitivity is often sufficient to avoid any separative or concentrative step. Besides the direct monitoring of a given species and of its reacting properties analytical bioelectrochemistry is also able to follow the heterogeous reactions occurring at a metal-solution inter-face with biopolymers such as nucleic acids or proteins and evaluate their conformation[1,2].

This introduction to the session devoted to electrochemistry in biosciences is intended to present briefly the various methods used in analytical bioelectrochemistry with a particular emphasis on in vivo sensors (Table 1).

METAL-SOLUTION INTERFACE SENSORS

Non-destructive Methods: Use of Microelectrodes

The use of carbon or mercury micro-cathodes and platinum or carbon micro-anodes enables the identification of a compound by the potential necessary for its reduction or oxidation. It enables also the quantitative determination of such compounds using either limiting currents - in steady-state methods - or peak currents - in transient method. Detection limits reach ca 1 nM when background or capacitive currents are filtered or compensated, in transient methods such as differential pulse polarography (DPP) or voltammetry. Numerous redox molecules, either metabolites or drugs, may be determined in biological samples either directly or after an extraction procedure[3,4]. Particularly miniaturized electrodes using carbon fibers or paste may be implanted and allow the continuous monitoring of some metabolites and drugs in neuronal tissues or arteries[5-8].

Table 1. Main Characteristics of Bioelectrochemical Sensors

Sensor type	Metal electrode	Ion selective electrode (ISE)	Gas selective electrode (GSE)	Enzyme of whole cell electrode
Detected species	Electroactive (oxidant or reducer)	Inorganic ions	Electroactive or acid-base gases	Metabolites inhibitors
Electro-chemical detector	Metal + ref. electrode	2 ref. electr. IFSET + ref.	2 ref. electr. metal + ref.	Metal + ref. ISE, GSE
Membrane	Without	Glass crystal organic	Hydrophobic polymer permeable to gas	Hydrophylic polymer with immobilized enzyme or cells
Properties of the analyte	Electrochem. reduction or oxidation	Specific ionic conductivity of membrane	Gas diffusion and redox or acid-base	Enzyme or cell catalyzed reaction
Calibration curve	I $\underline{vs}$ c q $\underline{vs}$ c	E $\underline{vs}$ log c	E $\underline{vs}$ log c I $\underline{vs}$ c	I $\underline{vs}$ c (dI/dt)max $\underline{vs}$ c E $\underline{vs}$ log c
Response type	Steady state transient	Equilibrium	Equilibrium steady state	Steady state transient

Destructive Methods: Coulometry

These methods, which to our knowledge are only used in vitro, enables the determination of a compound by its complete oxidation or reduction on a macroelectrode at a fixed potential or current: current is integrated till the electrochemical detection of the final point. Coulometry may be either direct if the analyte is electroactive, such as seric iron, or indirect: the electrogenerated reagent may then be:

- a complexing agent, such as Ag^+, Hg^+ or Hg^{++}
- a strong oxidant, such as bromine, Ag^{++}, Mn^{+++} or Au^{+++},
- a strong reducer, such as Ti^{+++} or viologen radical cations[9].

Because of their coulometric origin, if the end point is well determined, all these methods are absolute and do not need any calibration. Some clinical instruments, such as blood chloride meters, use such Ag^+ generated coulometries.

An alternative to these method was originally proposed by Bruckenstein and Albery[10-12] and used with biological species or samples by Rauwell and Thévenot[13-15]: with a rotating ring-disk electrode a reactant is electrogenerated on the disk, reacts with the analyte in the vicinity of the electrode and its excess is eventually detected on the ring. When titration reaction is rapid enough, detection limit of this non-destructive alternative to indirect coulometry reaches 100 nM.

Another alternative to, or more precisely, application of in vivo coulometry has recently been proposed by Albery for long term implanted carbon electrodes: a controlled electrolysis on a brain microelectrode surrounded by living cells allows the oxidation of ascorbate and thus its elimination in a further anodic voltammogram: it is therefore possible to detect much lower concentration of neurotransmitters or of their metabolites and study the effect of chronic drug treatment[5,8,16].

ION SELECTIVE ELECTRODES (ISE)

Glass, crystalline and organic membranes which present specific ionic conductivity have been intensively used for direct determination of ion activity. Measurement of membrane potentials using either 2 reference electrodes such as Ag/AgCl, Cl- or Hg/Hg2Cl2, Cl- [17-19] or a field effect transistor (ISFET) and a reference electrode[20] allows the determination of ion activities. Most of the present studies involve the preparation and characterization of organic membranes impregnated with water insoluble compounds specifically chelating some ions.

Such ISE are being progressively introduced into clinical laboratories for hydrogen, sodium, potassium and calcium ion determination[21,22]. Miniaturized models have been developed for in vivo intra-tissue or intra-cellular ion determinations[23-27] and for the detection of foetal suffering during delivery by detecting a slight pH decrease in the scale tissue[28].

GAS SELECTIVE ELECTRODES (GSE)

Some gases may be selectively detected: they need to be either electroactive (oxygen, bromine) or acido-basic (carbon dioxide, sulfhydric acid, ammonia) and diffuse through an hydrophobic membrane (teflon, polyethylene). Thus such GSE are generally built with a metal or a pH electrode covered with a gas permeable membrane[17,18,29,30].

Very frequently used in analytical laboratories, GSE have been introduced into clinical laboratories in the last few years for the direct determination of blood gases on very small samples[31] and for intravascular monitoring of oxygen, tension of blood[32,34]. Non-invasive transcutaneous oxygen or carbon dioxide electrodes may alternatively be placed on the surface of the heated skin of neonates or adults: the electrode response is directly proportional to arterial oxygen and carbon dioxide tensions[34-38].

ENZYME OR CELL ELECTRODES

This last type of specific electrode is certainly the most versatile but also, and this is often ignored, the most complex one. One or several enzymes are immobilized on a membrane by entrapment, adsorption, coreticulation or covalent coupling[39-41]. Alternatively bacterial or vegetal whole cells are entrapped or fixed on a membrane[41,42]. When such membrane are placed in class vicinity to an appropriate electrochemical detector, it is possible to associate the extremely selective properties of enzymes and the versatility of the previously mentioned electrochemical sensors. Indeed several tens of such electrodes have been described in the literature for the determination of either metabolites (enzyme or cell substrates) or toxic compounds (enzyme inhibitors); more recently, immuno-enzymatic electrodes have been described using immobilized enzymes coupled to antibodies.

Since the pioneer work of Clark in 1962[43], several hundred publications and several international meetings have been devoted to enzyme electrodes and especially to glucose sensors[39,40]. However very few instruments are commercially available besides the glucose sensor commercialized for several years by Yellow Spring Instrument. One may find in France glucose sensors made by Solea Tacussel and Seres and a lactate sensor made by Setric; these enzyme electrodes are respectively based upon the previous research activity of Thévenot, Coulet and Gautheron[44,45] and Romette and Thomas[46] and of Durliat and Comtat[47]. Such a small number of commercially available sensors and instruments is mainly related to the complexity of the heterogeneous kinetics and thermodynamics involved in these sensors, but also to the usually poor characterization of the membrane and sensing head and to the absence of theory taking into account the numerous experimental parameters.

The major interest and possibilities of such sensors for clinical, fermentation control and food industry control and on-line analysis have led several research groups to spend time and effort for a better understanding of enzyme electrodes. This effort is even more important for in vivo devices and especially for the glucose sensing part of an artificial pancreas[40]: extra-corporal closed-loop devices have for example been developed and tested on animals by Abel in Karlsburg [48,49], whereas needle-type glucose sensors have been presented by Shichiri[50,51]. Both devices include multimembrane systems for enzyme entrapment, glucose and oxygen flux regulation, and protection from blood clotting. For such in vivo glucose sensors it is indeed necessary to solve several specific problems simultaneously.

- glucose concentration is usually much larger in blood than the Michaelis constant for glucose oxidase,
- oxygen concentration is different in blood and in air-saturated solution and may vary upon ventilation,
- hydrodynamical conditions, i.e. blood velocity in the sensor vicinity, may vary or be poorly controlled especially if the glucose sensor is implanted subcutaneously,
- response time should be smaller than 2-5 min for a good insulin regulation,
- glucose oxidase being normally absent from blood should not be released by the sensor.

In collaboration with Reach from the Faculté Villemin of the Paris University, we have studied these problems using an extra-corporal blood circulation in a conscious freely-moving rat (Figure 1). Preliminary results obtained during intravenous glucose tolerance test (IVGTT) have given interesting results when an in vivo calibration was performed and when the delay period of our device, i.e. response time + circulation time, was taken into account (Figure 2 and 3). Studies are in progress in order to compare results obtained with different enzymatic membranes and to characterize the most important parameters (permeabilities, partition coefficients, enzyme load and activity) of such membranes.

CONCLUSION

This rapid survey of present possibilities of analytical bioelectro-chemistry shows that the following research fields are particularly active:

a) ultra-microelectrodes implanted in neuronal tissues, intravascularly or subcutaneously in order to monitor continuously in vivo electro-active species, ions or metabolites,
b) development of ISE or ISFET using organic ion conducting membranes and studying alternatives to the external reference electrode,
c) specific electrodes using either immobilized enzymes or whole cells in which the vessel and electrode design is optimized for a given application,

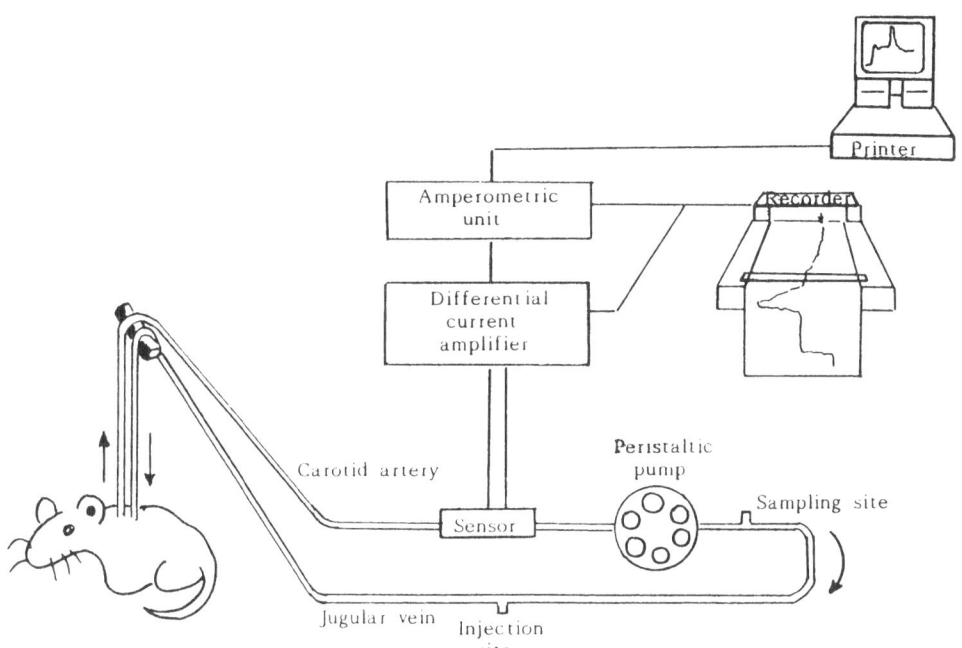

Fig. 1. Schematic diagram of in vivo evaluation of glucose sensor on conscious freely-moving rat.

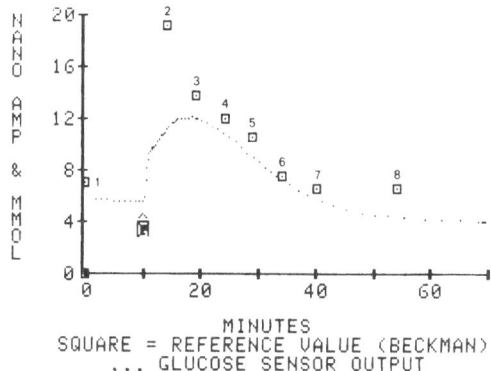

SQUARE = REFERENCE VALUE (BECKMAN)
... GLUCOSE SENSOR OUTPUT

Fig. 2. Intravenous glucose tolerance test (IVGTT) on a sensor connected to a conscious freely-moving rat. (...) direct signal from glucose sensor using a glucose oxidase cellulose acetate membrane, (o) reference values using a Beckman glucometer.

d) and finally, development of modern instrumentation using autocalibration and control procedures associated with signal processing and interpretation. This latter field is of great importance in clinical applications.

The following paper, presented by Wightman, gives a more detailed review on implanted ultramicroelectrodes especially developed for neuronal tissue. The continuous monitoring of neurotransmitters and their metabolites is indeed one of the present challenges of analytical bioelectrochemistry.

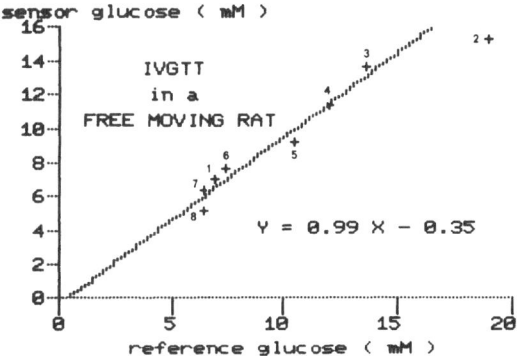

Fig. 3. Comparison of glucose values obtained with <u>in vivo</u> calibration before IVGTT and reference values using Beckman glucometer. A 4.5 min time-lag of the sensor, including 3.0 min response time and 1.5 min external circulation time has been taken into account for this comparison. Figures refer to blood samples identification numbers plotted on Figure 2.

Acknowledgements

    The work from my laboratory summarized in this review was supported by the Centre National de la Recherche Scientifique (Laboratoire associé LA 329). These results would not have been possible without the efforts of R. Sternberg, T. Tallagrand and G. Reach MD.

REFERENCES

1. H. W. Nürnberg, "Bioelectrochemistry I," G. Milazzo and M. Blank, eds., Plenum Press, London (1983).
2. C. Jacubowitz, L. T. Yu, and J. A. Reynaud, <u>Electrochem.Acta</u>, 28(1):57 (1983).
3. W. F. Smyth, "Polarography of Molecules of Biological Significance," p.326, Academic Press, London (1979).
4. W. F. Smyth, Electroanalysis in hygiene, environmental, clinical and pharmaceutical chemistry, <u>Anal.Chem.Symp.Ser.</u>, 2:473 (1980).
5. W. J. Albery, M. Fillenz, N. J. Goddard, M. E. McIntyre, and R. D. O'Neill, <u>J.Physiol.</u>, 132:107 (1982).
6. R. N. Adams and C. A. Marsden, "Handbook of Psychopharmacology," Plenum Press, 15:1-74 (1982).
7. W. N. Brooks, C. E. W. Hahn, P. Foex, P. Maynard, and W. J. Albery, <u>Br.J.Anaesth.</u>, 52:715-722 (1980).
8. R. M. Wightman, this volume p.   (1985).
9. G. J. Patriarche and J. C. Vire, <u>Anal.Chem.Symp.Ser.</u>, 2:209-225 (1980).
10. S. Bruckenstein and D. C. Johnson, <u>Anal.Chem.</u>, 36:2186 (1964).
11. W. J. Albery, S. Bruckenstein, and D. C. Johson, <u>Trans.Faraday Soc.</u>, 62:1938 (1966).
12. W. J. Albery and M. L. Hitchman, "Ring-disc Electrodes," Oxford Science Research Papers, Clarendon Press, Oxford (1971).
13. F. Rauwel and D. R. Thévenot, <u>J.Applied Electrochem.</u>, 6:119-126 (1976).
14. F. Rauwel and D. R. Thévenot, <u>Bioelectrochem.Bioenergetics</u>, 3:284-301 (1976).
15. F. Rauwel and D. R. Thévenot, <u>J.Electroanal.Chem.</u>, 75:579-593 (1977).
16. W. J. Albery, N. J. Godelard, T. W. Beck, M. Fillenz, and R. D.

O'Neill, J.Electroanal.Chem., 161:221-233 (1984).

17. N. Lakshiminarayanaiah, "Membrane Electrodes," p.368, Academic Press, New York (1976).

18. K. Camman, "Das Arbeiten mit Ionenselektiven Elektroden," p.227, Springer Verlag, Berlin (1977).

19. P. C. Meier, D. Ammann, H. F. Osswald, and W. Simon, Med.Prog.Technol., 5:1-12 (1977).

20. P. Bergveld and N. F. de Rooij, "Proc. Int. Conf. on Monitoring Vital Parameters," p.113, Nijmegen, Karger (1981).

21. J. D. Czaban, A. D. Comier, and K. D. Legg, Clin.Chem., 28:1936-45 (1982).

22. J. H. Ladenson, F. S. Apple, J. J. Aguanno, and D. D. Koch, Clin.Chem., 28:2383 (1983).

23. W. McD. Armstrong, W. McD. Wojtkowski, and W. R. Bixenman, Biochim. Biophys.Acta, 465:165-170 (1977).

24. R. A. Steiner, M. Ochme, D. Amman, and W. Simon, Anal.Chem., 51(3):351 (1979).

25. W. Crowe, A. Mayevsky, and L. Mela, Am.J.Physiol., 233(1):C56-C60 (1977).

26. K. Shimada, Y. Yano, K. Shibatani, Y. Komoto, M. Esashi, and T. Matsuo, Med.Biol.Eng.Comp., 18:741 (1981).

27. G. Koning and S. J. Schepel, in: "Proc. Int. Meeting on Chemical Sensors," T. Seiyama, K. Fueki, J. Shiokawa, and S. Suzuki, eds., p.597-602, Elsevier, Amsterdam (1983).

28. M. R. Neuman, "Theory Design and Biomedical Applications of Solid-state Chemical Sensors," C.R.C. Press, p.277-287.

29. M. A. Jensen and G. A. Rechnitz, Anal.Chem., 51(12):1972 (1979).

30. M. E. Lopez and G. A. Rechnitz, Anal.Chem., 54(12):2085 (1982).

31. C. E. W. Hahn, J.Phys.Sci.Instrum., 13:470 (1982).

32. H. P. Kimmich, F. Kreuzer, J. G. Spaan, K. Jank, de J. Hemptinne, and M. Demeester, Adv.Exp.Med.Biol., 75:33-40 (1976).

33. B. Hagihara, K. Kurosawa, S. Hashimoto, H. Sugimoto, and T. Sugimoto, in: "Proc. Int. Meeting on Chemical Sensors," T. Seiyama, K. Fueki, J. Shiokawa, and S. Suzuki, eds., p.591-596, Elsevier, Amsterdam (1983).

34. R. Huch, D. W. Lubbers, and A. Huch, Arch.Disease in Childhood, 49:213 (1974).

35. J. L. Peabody, G. A. Gregory, M. M. Willis, and W. H. Tooley, Am.Rev. Respir.Disease, 118:83-87 (1978).

36. O. Löfgren and L. Jacobson, Acta.Paediatr.Scand., 68:789 (1979).

37. B. Hagihara, K. Kogo, K. Nakayama, S. Shiraishi, M. McCabe, and S. Ohkawa, Jap.J.Med.Electr.Biol.Eng., 18:262 (1980).

38. P. Eberhard and R. Schäffer, J.Clin.Eng., 6:36 (1981).

39. G. G. Guilbault, in: "Immobilized Enzymes, Antibodies and Peptides," H. H. Weetall, ed., p.293-417, Dekker, New York (1975).

40. D. R. Thévenot, Diabetes Care, 5(30):184-189 (1982).

41. G. Rechnitz, Science, 214:287-291 (1981).

42. S. Suzuki and I. Karube, Applied Biochem.Bioeng., 3:145-174 (1981).

43. L. C. Clark and C. Lyons, Ann.NY.Acad.Sci., 102:29-45 (1962).

44. D. R. Thévenot, P. R. Coulet, R. Sternberg, and D. C. Gautheron, Bioelectrochem.Bioenerg., 5:548-553 (1978).

45. D. R. Thévenot, R. Sternberg, P. R. Coulet, J. Laurent, and D. C. Gautheron, Anal.Chem., 51:96-100 (1979).

46. J. L. Romette, B. Froment, and D. Thomas, Clin.Chim.Acta, 95:249 (1979).

47. G, Durliat, M. Comtat, J. Mehenc, and A. Baudras, Anal.Chim.Acta, 85:31-40 (1976).

48. P. Abel, U. Fischer, and E. J. Freyse, Life support systems, J.Europ.Soc.Artif.Organs, suppl.1, 94-97 (1982).

49. P. Abel, U. Fischer, A. Muller, and E. J. Freyse, Life support systems, J.Europ.Soc.Artif.Organs, suppl.1, 45-48 (1983).

50. M. Shichiri, R. Kawamori, Y. Yamasaki, N. Hakui, and H. Abe, The Lancet, pp.1129-1131 (1982).

51. M. Shichiri and R. Kawamori, Diabetologia, 24:179-184 (1983).

IN VIVO ELECTROCHEMISTRY

R. M. Wightman

Department of Chemistry
Indiana University
Bloomington, Indiana

INTRODUCTION

Electrochemical methods of analysis have seen wide application as in vivo biosensors. Several features of electroanalytical methods have enhanced their utility in these applications. Electrochemical techniques show sufficient sensitivity and selectivity that in many instances they can be used directly to measure chemical activity. In addition, the probes can be miniaturized so the measurements cause minimum trauma to the tissue.

Although there are many types of in vivo electrochemical sensors. All of these must meet certain criteria to provide interpretable results. The electrodes must exhibit a known selectivity for the ions or molecules of interest, and the response of the electrode should allow predictable calibration of the experiments. For situations where the electrode will be implanted for chronic use, sometimes for periods of time as long as a year, these requirements also exist, but require more stringent consideration of the stability of the electrode response. Nevertheless, these criteria have been met in many different types of applications. Thus, electrodes can be used as chemical sensors in a particular organ, or with ultramicroelectrodes, in a single cell.

In this review, many of the different types of electrodes that have been used in in vivo applications will be described, along with an explanation of the types of information that can be obtained, and a special emphasis will be placed on sensors which have been used in the central nervous system. As will be shown, the use of many different types of electrodes has expanded our knowledge of biochemical processes at the cellular level. This type of research can lead to probes that can be used routinely in diagnosis of the health and well-being of human patients[1].

TYPES OF IN VIVO SENSORS

Potentiometric Sensors

Electrochemical methods for chemical analysis can be divided into two types: potentiometric and voltammetric. Potentiometric measurements are made under conditions where no current flows; i.e. the potential of the test electrode is measured against that of a suitable reference electrode.

To provide a degree of specificity in these measurements, a membrane is usually placed between the test electrode and the reference electrode. The specificity of the measurement, and the resulting relationship between concentration of analyte and the measured potential difference, arises from the design of the chemically-selective material at the tip of the electrode. An alteration of the activity of an ion in solution results in a potential change across the membrane, and this change is related in a logerithmic sensor to the activity change. Thus, the key to the successful design and use of this type of electrode is to select membranes which exhibit a specific and reproducible change for the desired species. Many types of membrane materials are available, some from commercial sources[2].

The most widely used ion selective electrode is that which is sensitive to the pH of a solution. This electrode is fabricated with a glass membrane which shows a selective permeation for hydronium ions. Glass electrodes are also available for many other cations, such as lithium, potassium, sodium, and calcium. Membranes which employ liquid ion exchangers or neutral carriers, crystal membranes, and gas sensors, have been developed and used successfully for the determination of cations[3]. However, to be useful for _in vivo_ applications, these electrodes must be constructed with micrometer dimensions. The liquid ion-exchange membranes are particularly useful in this regard[4]. These electrodes are constructed from a glass pipette with a very small tip (1-2 μm diameter), and the liquid ion exchanger is placed in the tip.

An example of a potentiometric electrode suitable for intercellular work is the Ca^{2+} selective electrode. This electrode can be constructed with a tip diameter of 0.5 μm[5,6]. They are fabricated from bevelled glass micropipettes filled with o-nitrophenyloctyl ether, which serves as the immobile lipid, polyvinylchloride which stabilizes the matrix, and a calcium chelating agent. Typical response times for a change in Ca^{2+} concentrations are 1 s. However, optimization of the electrode design has resulted in reported response times of 7 ms[7]. Methods for building other types of ion selective electrodes have recently been reviewed[8].

Amperometric and Voltammetric Electrodes

An alternative method for electrochemical analysis is to measure the current that passes at a test electrode as a function of the applied potential between it and a reference electrode. This technique takes advantage of the fact that many molecules can be easily oxidized or reduced. At a fixed applied potential, this is referred to as an amperometric experiment. If the current is measured as a function of voltage (termed a voltammogram) it provides a curve that is characteristic of a particular molecule. The sign of the current indicates whether the molecule can be oxidized or reduced, the potential at which current starts to flow is indicative of the formal potential of the compound, and the amplitude of the current is proportional to the concentration of the particular species of interest. Because in the voltammetric experiment one has the added capability of determining the formal potential of the compound of interest. The use of chemically specific membranes over the electrode surface is not as great a requirement.

Electrodes of micrometer dimensions can be fabricated[10]. Electrodes of this diameter cause minimal damage and are useful as _in vivo_ sensors. The electrode materials that have been most useful are platinum and carbon[12]. Electron transfer processes at the electrode surface from biological important molecules can be obscured by the electrochemical response originating from surface oxides on platinum. For this reason, the most successful results have been obtained with carbon electrodes. However, even at carbon electrodes the inhibition of electron transfer caused by the adsorption of large biomolecules onto the electrode surface tends to cause a deterioration of the electrode response in vivo.

Amperometric techniques have been widely used for the determination of oxygen[12]. Oxygen is easily reduced at a platinum electrode, and thus miniaturized platinum electrodes have been successfully employed in vivo. The specificity of this type of measurement can be improved by placing a membrane which is only permeable to oxygen over the platinum surface. Teflon is ideal in this regard. Oxygen electrodes prepared in this manner are operated at a fixed potential and the current measured is directly proportional to the oxygen concentration. If the potential is pulsed to a reducing potential at periodic intervals the sensitivity of the measurement increases since the membrane is not depleted of oxygen[13]. Sensitivity can also be optimized by proper selection of the membrane thickness and the use of small platinum wires as cathodes[14]. Since oxygen is required for tissue metabolism, this electrode has been extremely important in biochemical and physiological studies.

The membrane-covered, amperometric electrode can also be used for the detection of hydrogen peroxide. For this type of measurement, the potential of the electrode is simply reversed from a reducing to an oxidizing potential. The oxygen electrode can also be coupled with immobilized enzymes[15]. In this manner, the concentration of a species which is a substrate for the enzyme can be monitored by measuring the changes in oxygen or hydrogen peroxide induced by the enzymatic reaction. This indirect form of amperometric detection can be used for the determination of glucose, using glucose oxidase as the immobilized enzymes.

APPLICATIONS

Potentiometric Measurements in Neuronal Tissue

Both the voltammetric and potentiometric sensors described previously in this review have been used extensively to understand processes that occur in neuronal tissue. Neurons are essentially specialized cells that collect and array information (Figure 1). In the resting state, the neurons contain a low concentration of sodium ions relative to the concentration in extracellular fluid and contain an enhanced concentration of potassium ions. Because of the concentration gradients, a potential difference exists across the cell membrane. To send a message from the cell body of the neuron down the axons to the terminal field, the permeability of the membrane to sodium ions increases. This causes an influx of sodium and a subsequent egress of potassium ions. The change ion ionic permeability propagates down the neuron resulting in the movement of the membrane potential down the axon. This is referred to as an action potential and has been recognized for many years by electrophysiologists. Electrophysiologists have used microelectrodes in the potentiometric mode to detect these potential changes and to characterize their response[16]. The potential changes happen on a millisecond time scale and thus the amplifiers that are used in this work are adjusted to have the proper band pass to observe these changes. The advent of ion selective electrodes has enabled confirmation of the chemical characterization of these responses[11,17,18].

An example of the new type of information that can be obtained from the use of ion-selective electrodes in vivo is given by the work of Nicholson[19]. In these experiments, the apparent diffusion of anions and cations through brain tissue was examined. Ions were injected into the brain from a micropipette placed approximately 100 μm away from the ion-selective electrode. The ions were ejected from the pipette by means of an

191

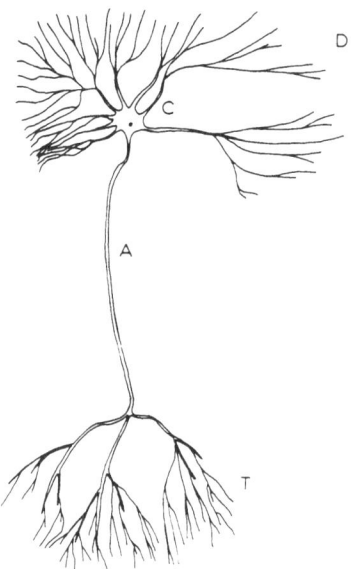

Fig. 1.   Schematic diagram of a neuron. D - dendrites or input processes
of cell. C - cell body. A - axon or output process of cell.
T - terminal field; region from which neurotransmitters are
secreted.

electric current. A technique known as iontophoresis. By observing the
response of the ion-selective electrode as a function of time following
injection of an ion, the rate of diffusion in the extracellular fluid of
brain tissues can be determined by determining the concentration of the
injected ion which reached the electrode. In addition, an estimate of the
volume of the extracellular fluid can be made. Using this approach,
Nicholson was able to show that tetraethylammonium ion and sodium ion
diffusion through brain tissue with a diffusion coefficient that is approximately one third of that obtained in aqueous solution. The extracellular volume was found to be 20% of the total volume of brain tissue.
For ions such as potassium which interact strongly with the neuronal and
glial uptake processes, the transport appears to have a much larger rate.

Determination of Neurotransmitters In Vivo

     Neurotransmitters are compounds which are stirred in nerve terminals
and which are secreted by these neurons to relate information from one
neuron to another neuron. The structures of several well identified
neurotransmitters are given in Figure 2. As recognized by Ralph Adams in
the early 1970's, dopamine, serotonin, norepinephrine, and epinephrine are
of particular interest from an electrochemical point of view because they
are easily oxidized and, thus, can be detected by the amperometric and
voltammetric techniques[20]. It also turns out that these chemicals are of
particular importance in normal neuronal function. For example, serotonin
is thought to be important in the regulation of sleep mechanisms. Dopamine
is thought to be important in the regulation of motor function as well as
the control of mood and emotion[21]. Therefore, it would be desirable to
measure the concentration of these species in extracellular fluid and
relate this to neuronal activity.

     Neurotransmitters are stored in vesicles at the nerve terminals as
shown in Figure 3. The secretion of the neurotransmitters most commonly
occurs as a result of an action potential. The close proximity of the

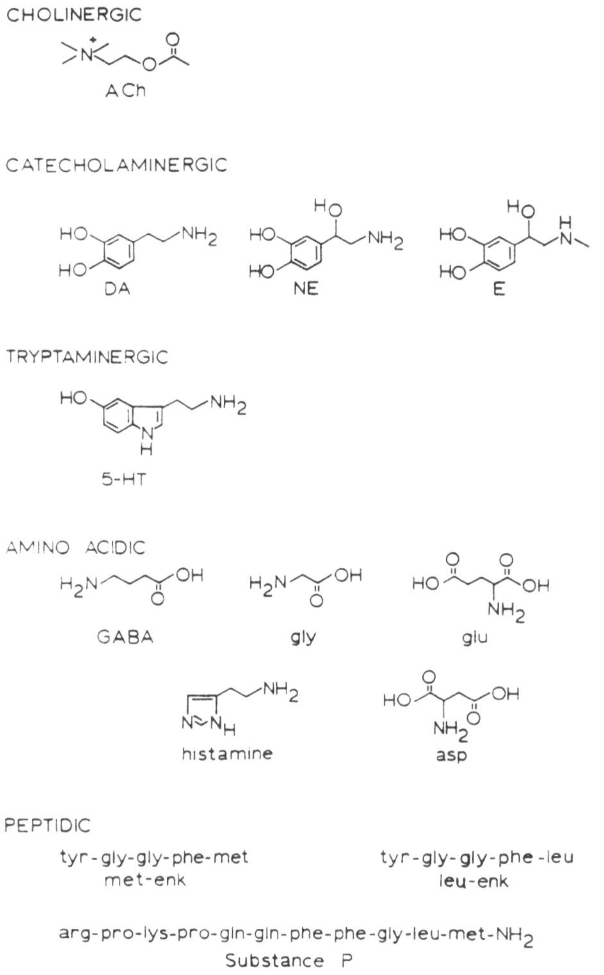

Fig. 2. Structure of several recognized neurotransmitters. Abbreviations used: Ach - acetylcholine; DA - dopamine; NE - norepinephrine; E - epinephrine; 5HT - 5-hydroxytryptamine; GABA - γ-aminobutyric acid; gly - glycine; glu - glutamate; asp - aspartate; met-enk - met-enkaphalin; leu-enk - leu-enkaphalin.

nerve terminals to the input processes of other neurons permits the rapid diffusion of neutrotranmitters from the point of secretion in the opposing neuron. On this cell are located specific receptors which alter the permeability to ions in the presence of the proper neurotransmitter, which either increases or decreases the probability of an action potential from the opposing neuron. In this way, electrical information is conveyed from one neuron to the next. This junction of neurons where all of this chemistry occurs is referred to as the synapse.

Neuroscientists have been interested for many years in being able to make measurements of the concentration of neurotransmitters in extra-cellular fluid. Two non-electrochemical methods have seen widespread use for these measurements. One is the use of the push-pull cannula. In this experiment, two concentric tubes are placed in the brain region of interest. Fluid is pumped into the brain through the inner tube and removed by the outer tube (Figure 4). This sampling technique, coupled with the use

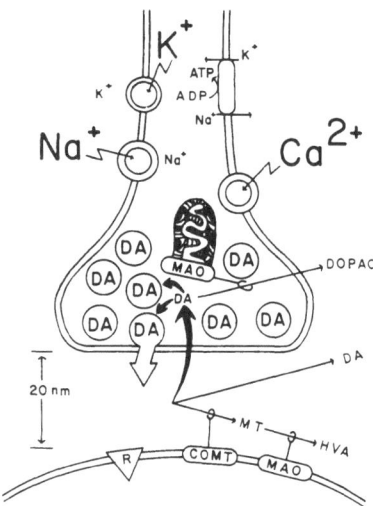

Fig. 3. Schematic representation of the nerve terminal region of a
dopamine neuron and synapse. Ion channels are represented by
circles, with the size of the ionic symbol representing its
concentration in the normal, resting state. DA represents
dopamine stored in the nerve terminal. R represents a dopamine
receptor on the post-synaptic side.

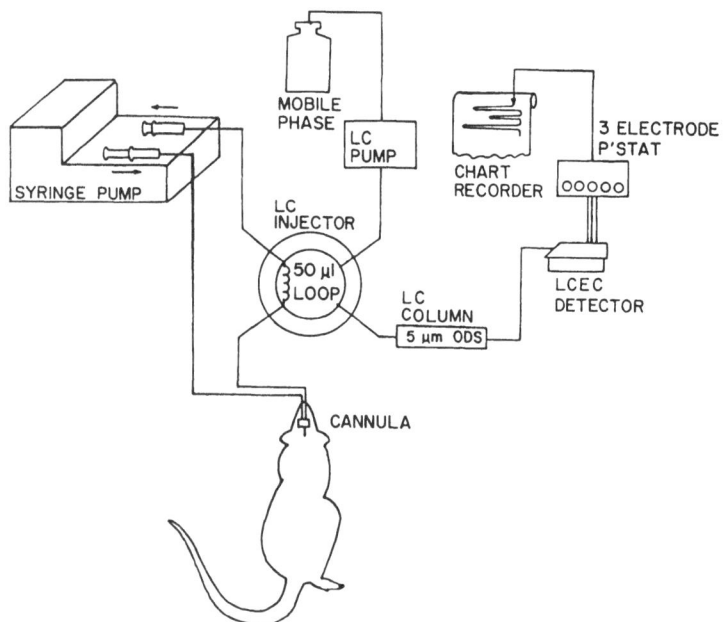

Fig. 4. Schematic diagram of the push-pull cannula experiment.
Physiological fluid is passed from the syringe pump, through the
cannula in the rat brain, to a loop injector in the chromatograpic
system. In the example shown, amperometric detection is employed.

of subsequent chemical analysis, permits the determination of the relative
concentration of the species in extracellular fluid[21]. More recently
this technique has been modified by placing dialysis tubing over the end of

the cannula[23,24]. This technique reduces damage caused by the flow of liquid through the brain tissue, but allows small molecules such as the neurotransmitters and their metabolites to pass into the perfusion fluid. These types of measurements provide important information which is useful in planning strategies to make in vivo electrochemical measurements. For example, perfusion experiments have shown that the concentration of neuro- transmitters is much below the concentration of their metabolites in extracellular fluid (Figure 5). Furthermore, these studies have shown that ascorbic acid and uric acid, two very easily oxidized species, are present in relatively high concentrations, and thus would serve as potent inter- ference to the electrochemical technique. In many of these experiments, the technique of liquid chromatography with electrochemical detection has been used to quantitate the species in the perfusion fluid. This technique is particularly useful in this application because of the very high sensi- tivity obtained with electrochemical detection.

Despite the exciting results that have been obtained with these perfusion techniques, they do have some restriction. The most obvious restriction is that these probes necessarily must be of a relatively large size. Because of their size, which may exceed 500 μm, they tend to cause damage to the brain region where measurements are being made. The second limitation is the time scale over which measurements can be made. In general, very low perfusion rates are used so that sampling times less than

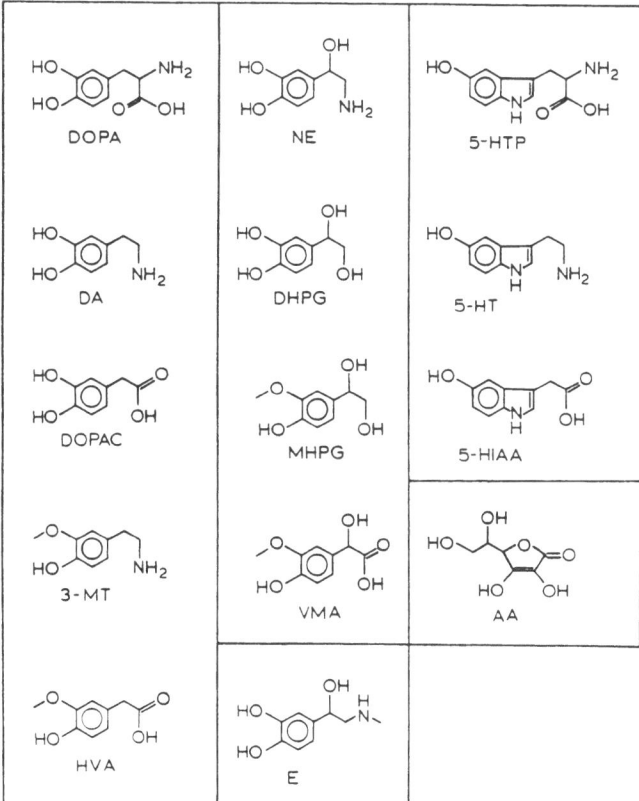

Fig. 5. Structures of metabolites of dopamine (DA), norepinephrine (NE), and 5-hydroxytryptamine. The primary compounds found in per- fusates from the caudate nucleus are dihydroxyphenylacetic acid (DOPAC), homovalillic acid (HVA) and ascorbic acid (AA).

10 minutes are quite difficult. Thus, _in vivo_ electrochemistry can make significant contributions by employing smaller probes and by making measurements on a shorter scale. As noted previously, neurons tend to fire on a millisecond time scale, so this would be the ultimate goal for a sensor of _in vivo_ chemical neurotransmission.

Factors Which Affect In Vivo Electrochemical Measurements

Several factors must be considered when interpreting data from _in vivo_ measurements. The dimensions of the synapse are submicrometer, and, thus, it is unlikely neurotransmitters can be measured directly as synaptic transmission is being carried out with electrodes of micrometer dimensions. Rather, it is likely that measurements that are _in vivo_ will be of those neurotransmitter molecules that have spilled out of the synaptic region. Furthermore, since synapses are often found in close proximity to one another, measurements may well involve secretion from several synapses. Neurotransmitters are actively metabolized by brain tissues, so the probe that is employed must be able to distinguish metabolites from the authentic neurotransmitter as well as other compounds that are found in brain tissue. In addition, the amounts of neurotransmitter in the brain are of quite low levels, as seen from the push-pull and dialysis measurements. Therefore, a technique capable of micromolar concentration measurements is required.

Microvoltammetric electrodes based on carbon fibers have greatly enhanced the feasibility of making interpretable measurements in brain tissue[25]. Their size is sufficiently small (Figure 6) so that there is no visible tissue damage in the region of electrode implantation when examined by post-mortem histology. Characterization of the factors that

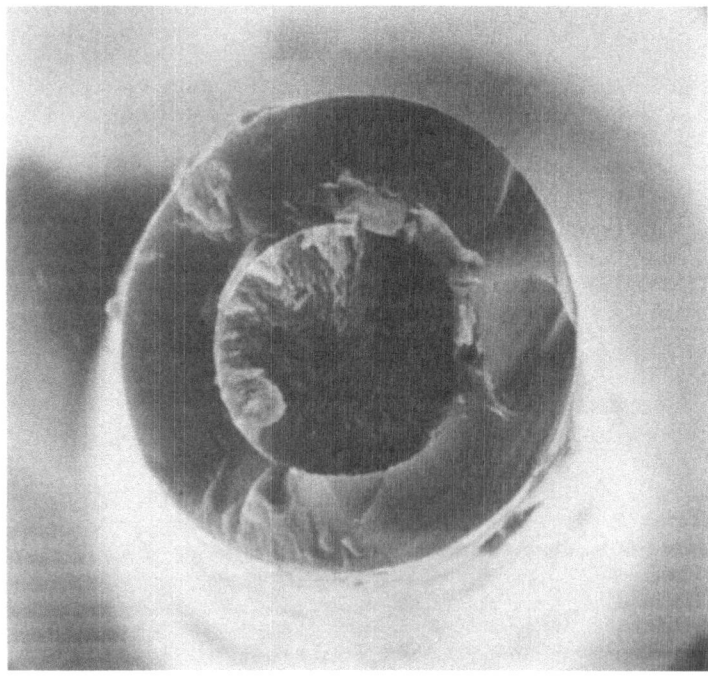

Fig. 6. Tip of a carbon fiber microvoltammetric electrode as viewed with a scanning electron microscope. The carbon fiber, viewed in the center of the assembly, has a radius of 10 μm.

196

effect mass transport to the electrode in the brain has led to optimization of the applied potential waveform. Potential pulses of short (92 ms) duration insure that measurements have a restricted special and temporal basis[26] while minimizing the generation of reactive chemical products at the tip of the electrode. The small currents that are generated at micro-voltammetric electrodes (less than nanoamperes minimize the possibility of inducing neuronal excitation while making electrochemical measurements[27]. These electrodes also have a considerable degree of chemical specificity. The catecholamines give relatively well-defined voltammograms while their metabolites, and other easily oxidized components that are found in brain tissue, yield voltammograms that are poorly defined or at different potentials from the catecholamines[28].

The factors which effect mass transport is extracellular fluid as demonstrated in the potentiometric measurements of sodium and potassium transport also effect transport to amperometric electrodes in vivo. However, in amperometric measurements this problem becomes more difficult if long electrolysis times are used. The electrolysis itself causes a depletion of the substance being determined[29]. One model that has been employed to explain the experimentally observed results is shown in Figure 7. In this model, there is a region next to the electrode that is comprised primarily of extracellular fluid. Molecules which are at a distance remote from the electrode can only be determined when they diffuse through the tortuous environment of the brain cells. If electrolysis times are kept short so that the concentration of electroactive species in the solution volumes next to the electrode is minimally perturbed, then the current observed is controlled by the diffusion coefficient of species in the extracellular fluid. This hypothesis has been tested by injecting electroactive substances into the brain at a distance remote from the electrode (Figure 7). Diffusion coefficients in excellent agreement with those obtained potentiometrically have been obtained for compounds such as dihydroxyphenylacetic acid and ascorbic acid[26]. However, neurotransmitters which are electroactive tend to show reduced values of the diffusion coefficient, implying that their diffusion is restricted in brain tissue. This may be a result of interaction with brain tissue by an uptake process or by an interaction with charged cell surfaces[30].

$$C = \frac{N}{8(\pi Dt)^{1/2}} \exp\left(\frac{-x^2}{4Dt}\right)$$

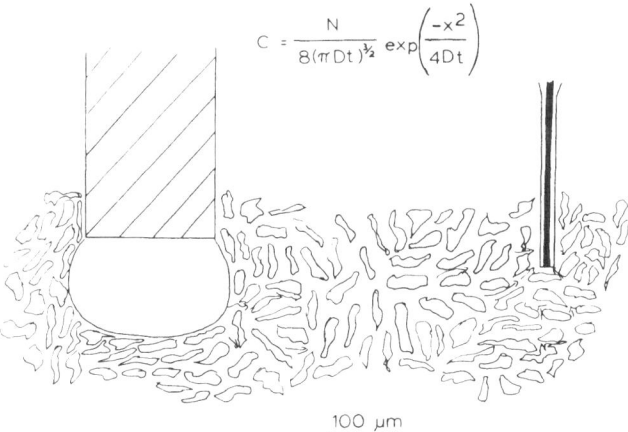

100 µm

Fig. 7. Diagramatic view of an electrode (right) and injection syringe (left) implanted in brain tissue. Diffusion through brain tissue involves the tortuosity induced by the tissue; at the tip of the electrode diffusion is unimpeded for a short distance (∼ 50 µm) into tissue. The equation is for the concentration-time response for an injected substance.

As expected from the push-pull and dialysis data, measurement of electroactive substances in extracellular fluid has indicated that the composition of the primary detectable species are ascorbate and dihydroxy-phenylacetic acid[31,32]. These measurements were made in the caudate nucleus, a region of the brain known to contain large amounts of dopamine, in the hope that this substance would be determined. The first report of the release of dopamine which was confirmed by voltammetry came as a result of injecting potassium at a distance remote from the electrode into this caudate nucleus region[33]. The basis of this experiment is that the potassium would cause the depolarization of many neurons resulting in an increased extracellular concentration of dopamine. This is indeed the observed result.

In a more controlled type of experiment, the neurons were electrically stimulated in order to cause secretion of neurotransmitter, while the amperometric sensor was placed in the nerve terminal region. The measurements were made in the caudate nucleus of male, Sprague-Dawley rats anesthetized with chloral hydrate[34]. When the electrode is inserted into this brain region, voltammograms obtained without stimulation resemble those for ascorbate, the easily oxidized compound which is of highest concentration in this brain region. However, when the neurons were forced to fire because of an electrical stimulation of the axons with tungsten electrodes using a 60 Hz sinusoidal signal with an rms value of 100 μA, dopamine can be observed. The concentration of easily oxidized substances detected by the electrode at a potential of 0.5 increases with the stimulation. When voltammograms are obtained before and during the stimulation and are subtracted, the voltammograms obtained is identical to the voltammograms obtained in dopamine solutions using the same electrode (Figure 8). As

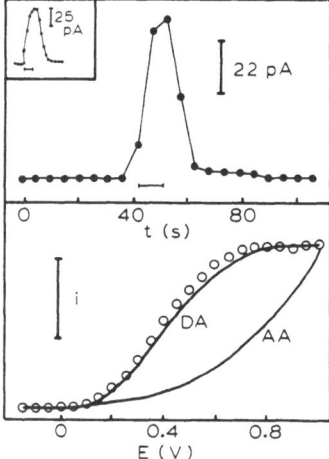

Fig. 8. Oxidation current from a microvoltammetric electrode placed in the caudate nucleus of an anesthetized rat during an electrical stimulation of the medial forebrain bundle. Top: chronoampero-metric current vs. time at 0.5 V vs. SCE. Bottom: difference of voltammogram obtained at the peak (circles), compared to voltammograms obtained for dopamine (DA) and ascorbic acid (AA) in vitro. Current scales for voltammograms: i = 16 pA, 25 μM dopamine; i = 28 pA, 200 μM ascorbic acid; i = 32 pA, stimulation. Insert: chronoamperometric current vs. time for an identical experiment with 2-s intervals between measured points. Reprinted with permission from reference 34 (copyright 1983 by the AAAS).

shown in Figure 5, the voltammograms for dopamine and ascorbate are quite distinct, facilitating the identification of the released substrate. It should be noted that there is no known substrate except dopamine in the caudate nucleus, which gives voltammograms identical to that substance released by electrical stimulation[28].

Examining Ascorbate With In Vivo Electrochemistry

One of the goals of this line of research is to measure the effects of pharmacological agents on neurotransmitter release. For example, the behavioral actions of amphetamine are thought to arise because this drug can increase the amount of dopamine in the synaptic region. Early experiments with in vivo electrochemistry demonstrated that this drug caused a change in the concentration of an easily oxidized species in the brain of anesthetized rate. However, refined voltammetric techniques, as well as the dialysis technique, have shown that this compound is not dopamine, but rather is ascorbic acid[23,24,32,33]. Although it has long been recognized that ascorbic acid is present in the brain at relatively high concentrations, a role in neurotransmission had not previously been proposed for this molecule. Although the data is still incomplete, the results of these in vivo electrochemical experiments do suggest that ascorbic acid plays an important role in overall brain function[35].

FUTURE TRENDS FOR IN VIVO ELECTROCHEMISTRY

The electrochemistry described so far in this report have primarily been with bare carbon electrodes. Recently, other voltammetric probes have been employed. It has been found that passing a large electrical current through a carbon fiber electrode greatly alters its voltammetric properties and allows the resolution of a greater number of the compounds that are present in vivo[32]. This electrode was used to demonstrate that ascorbate is in fact the compound which increases with systemic injections of amphetamine (Figure 9). In addition, this electrode shows very good sensitivity for monitoring the concentration of dihydroxyphenylacetic acid in extracellular fluid.

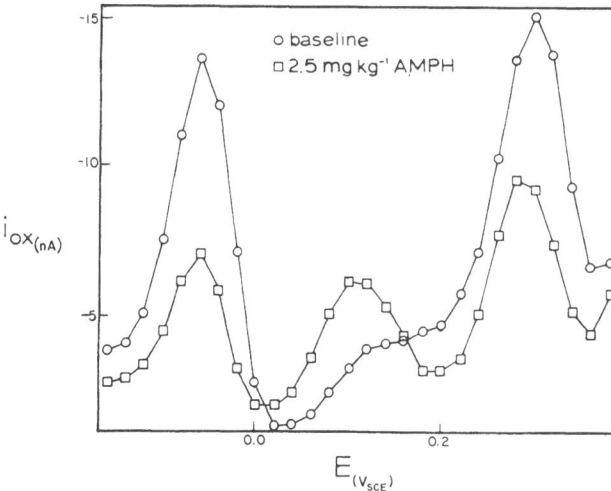

Fig. 9. Differential pulse voltammogram obtained in the caudate nucleus of an anesthetized rat with an electrochemically oxidized carbon fiber electrode. □ - before drug; ○ - after intraperitoneal administration of amphetamine (AMPH).

The ultimate goal of this area of _in vivo_ electrochemistry is to have _in vivo_ probes that can measure chemical concentrations in alert, freely moving animals. Neuroscientists are particularly excited about correlating a particular neurotransmitter with various different forms of behavior. Relatively new evidence suggests that this is in fact a reality. Electrodes have been implanted in the rat brain for up to three months and have continued to provide interpretable voltammetric data[36]. In these experiments, metabolites of the neurotransmitter dopamine have been observed rather than the neurotransmitter itself. The use of electrodes in rat brains for such a long period of time enables the study of the effect of chronic drug treatments. This is especially important in examining the properties of the drugs used to control the symptoms of schizophrenia. These are given to patients for very long periods of time, and in many cases cause disquieting side effects. Since these drugs are known to interact strongly with the dopamine receptors on neurons in the caudate nucleus, this technique may show immediate benefits in the design and evaluation of drugs with fewer side effects.

Future research using _in vivo_ probes will be directed at further understanding these types of drug interactions with neurotransmitters. In addition, future research will be directed at the development of more specific probes so that dopamine and other neurotransmitters can be measured on time scales that are closer to those that occur during normal brain function. An area of research which is receiving widespread attention is developing more specific electrodes. One approach to this problem is to employ polymer films on the electrode surface which restrict access to the electrode surface to molecules which are of interest. For example, Nafion is a polysulfonated polymer which acts as a cation exchange membrane. Since the neurotransmitters are amines which are protonated at physiological pH, and their metabolites are neutral or negatively charged molecules, coating the electrode with this polymer greatly enhances the selectivity[23]. In addition, these coatings may also increase stability.

CONTROL OF CELLULAR FUNCTION WITH ELECTRODES

The existence of electrical potentials across cell membranes suggests that electrodes could be used to control _in vivo_ processes as well as to monitor their activity. Indeed, in order to measure dopamine release, it was shown earlier in this report that electrodes can be used to stimulate the activity of neurons. This has been recognized by biologists for many years, and the use of local electrical stimulation has been used to probe the anatomy and function of neuronal circuits.

An area of research that is currently being explored is the use of electrical stimulation for the control of chronic pain. Many patients suffer acute pain, and several methods of control are being explored[38]. It is believed that the technique of acupuncture, which has been claimed to reduce pain, is thought to have its effect by electrical stimulation of peripheral nerves. However, understanding and evaluation of acupuncture and related techniques has been slow. Stimulation of discrete brain regions relieves pain in some cases, but can also result in large changes in behavior.

A major problem with the chronic implantation of electrodes, especially for control of cellular function, is the likelihood that the electrode will poison or cause trauma to the tissue[16]. When large currents are passed through implanted electrodes, the dissolution of ions and the generation of gases can cause lesions. However, the use of transcutaneous stimulation avoids this problem. For example, an area that has received particular attention is the use of electromagnetically induced currents, and their beneficial effect on the time it takes for bone repair. Isolated human red blood cells[39] and bone tissue[40] have been investigated in the presence of these fields to examine the optimum parameters and the basic underlying mechanism. It has been proposed that one of the mechanisms involves the effects of calcium uptake into cells.

CONCLUSIONS

The results summarized in this report demonstrate that electro-chemistry can indeed provide new information concerning the chemistry of physiological functions. While many of the developments described here are at a preliminary stage and only suitable for use with laboratory animals, the information that is being learned from these experiments should be useful in understanding more about various different biological mechanisms. This knowledge will then lead to improved pharmacological aspects and perhaps the understanding of the origin of various different diseases.

Acknowledgements

The work from my laboratory summarized in this review was supported by the National Science Foundation (BNS 81-00044) and the National Institute of Health (RO1 NS-15841). These results would not have been possible without the efforts of M. A. Dayton, A. G. Ewing, R. L. Wilson, D. S. Brown, J. C. Bigelow and W. G. Kuhr.

REFERENCES

1. J. D. Moirell, Science, 210:263-267 (1980).
2. R. A. Durst, "Ion Selective Electrodes," National Bureau of Standards Special Publication, 314 (1969).
3. G. A. Rechritz, Anal.Chem., 54:1194A-1200A (1982).
4. J. L. Walker, Anal.Chem., 43:83A-93A (1971).
5. R. Y. Tsien and T. J. Rink, J.Neurosci.Methods, 4:75 (1981).
6. M. V. Thomas, "Techniques in Calcium Research," Academic Press, New York (1982).
7. E. Ujec, O. Keller, N. Kriz, V. Pavlik, and J. Machek, Bioelectrochem. Bioenerg., 7:363-369 (1980).
8. R. C. Thomas, "Ion-sensitive Intracellular Microelectrodes," Academic Press, New York (1978).
9. L. R. Faulkner and A. J. Bard, "Electrochemical Methods, Fundamentals and Applications," J. Wiley and Sons, New York (1980).
10. R. M. Wighton, Anal.Chem., 53:1125A-1130A (1981).
11. J. Koryta, M. Brezina, J. Pradac, and J. Pradacova, in: "Electro-analytical Chemistry," A. J. Bard, ed., Vol.11, M. Dekker, New York (1979).
12. L. C. Clark and G. Sachs, Ann.NY.Acad.Sci., 148:133-153 (1968).
13. K. D. Wise, R. B. Smart, and K. H. Mancy, Anal.Chim.Acta, 116:297-305 (1980).
14. J. S. Ultman, E. Firouztale, and M. J. Skerpon, J.Electroanal.Chem., 127:29-66 (1981).
15. P. W. Carr and L. D. Bowers, "Immobilized Enzymes in Analytical and Clinical Chemistry," J. Wiley and Sons, New York (1980).
16. R. Plonsey, "Bioelectric Phenomena," McGraw Hill, New York (1969).
17. H. O'Doherty, J. F. Garcia-Diaz, W. McD. Armstrong, Science, 203:1349-1351 (1979).
18. E. Sykova, J.Electroanal.Chem., 116:231-246 (1980).
19. C. Nicholson, Neurosci.Res.Progr.Bull., 18:183-322 (1980).
20. R. N. Adams, Anal.Chem., 48:1126A-1138A (1976).
21. J. R. Cooper, F. E. Bloom, and R. H. Roth, "The Biochemical Basis of

Neuropharmacology," Oxford Univ. Press, New York (1982).

22. R. D. Meyers, ed., in: "Methods in Psychology," 1:169-211, Academic Press, London (1972).

23. T. Zetterstrom, T. Sharp, C. A. Marsden, and U. Ungerstedt, J.Neurochem., 41:1769-1773 (1983).

24. J. B. Justice, S. A. Wages, A. C. Michael, K. D. Blakely, and D. B. Neill, J.Liquid Chrom., 5:1873-1896 (1983).

25. A. G. Ewing, M. A. Dayton, and R. M. Wightman, Anal.Chem., 53:1842-1847 (1981).

26. M. A. Dayton, A. G. Ewing, and R. M. Wightman, J.Electroanal.Chem., 146:189-200 (1982).

27. F. Hefti and D. Felin, J.Neurosci.Meth., 7:151-156 (1983).

28. P. M. Kovach, A. G. Ewing, R. L. Wilson, and R. M. Wightman, J.Neursci.Methods, 10:215-227 (1984).

29. H. -Y. Cheng, J. Schenk, R. Huff, and R. N. Adams, J.Electroanal. Chem., 100:23-31 (1979).

30. M. E. Rice, G. A. Gerhardt, P. M. Hierl, G. Nagy, and R. N. Adams, Society for Neuroscience 13th Annual Meeting (1983), Abstract No.998.

31. J. O. Schenk, E. Miller, M. E. Rice, and R. N. Adams, Brain Res., 277:1-8 (1983).

32. F. Gonon, M. Buda, R. Cespuglio, M. Jouvet, and J. -F. Pujol, Nature, 286:902-904 (1980).

33. A. G. Ewing, R. M. Wightman, and M. A. Dayton, Brain Res., 249:361-370 (1982).

34. A. G. Ewing, J. C. Bigelow, and R. M. Wightman, Science, 221:169-170 (1983).

35. J. A. Clemens and L. A. Phebus, Brain Res., 267:183-186 (1983).

36. W. J. Albery, N. J. Godelard, T. W. Beck, M. Fillenz, and R. D. O'Neill, J.Electroanal.Chem., 161:221-233 (1984).

37. G. A. Gerhardt, A. F. Oke, G. Nagy, B. Moghaddam, and R. N. Adams, Brain Res., 290:390-395 (1984).

38. F. W. L. Kerr and K. L. Casey, Neurosci.Res.Prog.Bull., 16:6-174 (1978).

39. R. Schmukler and A. A. Pilla, J.Electrochem.Soc., 129:526 (1982).

40. H, Assailly, J. -D. Mouet, Y. Goureau, P. Christel, and A. A. Pilla, Bioelectrochem.Bioenerg., 8:515 (1981).

DISCUSSION

D. R. Thévenot

Université Paris
Val de Marne
Paris

CHAIRMAN'S SUMMARY

R. Kalvoda: On the microelectrode that you present, currents may be very
small, in the range of picoamperes. I want to know whether some precautions have to be taken in signal processing because of the noise level.

R. M. Wightman: When you go to very small electrodes, you indeed have very
small currents to measure. It turns out that with modern electronics it is
not a particularly difficult problem. Only two precautions were taken:
first, experiments were done in a Faraday cage, second, all our equipment
is home-made so that we can build into it the noise-free level current
measurements that I showed you here. It is important to have very low
currents, because if we pass high currents we may stimulate the neurons
themselves and perturb the system that we are trying to make a measurement
on. So the fact that we use very small electrodes has two advantages: we
have very small currents and we cause minimum perturbation to our tissues.

H. Oelschläger: Dr. Wightman do you see any chance in the future for brain
surgery in human beings after measuring the dopamine concentration?

R. M. Wightman: Well, I think that it is certainly too premature at the
present time to suggest making such measurements. I firmly believe that
even though dopamine is present in very low concentration in extracellular
fluid, this problem of being able to measure it will be surmounted by one
of the different things that are being developed today. Perhaps then it
will be a useful type of technique for clinical purposes.

N. Hackerman: On one of the stimulation experiments you said that it
returns to the base line. Actually, it does not return to the base line
before for a period of 60 to 100 s.

R. W. Wightman: In the potassium stimulation, it does not return to base
line very quickly because there is lots of potassium around. In the
electrical stimulation, it gets to within 90% of base line within 10 to
12 s and by 8 min it is absolutely indistinguishable from base line. If
the dopamine was metabolized we would see a very small contribution to that
base line change and I think that might be part of the problem. But the
important fact is that it is decaying in a very nice exponential fashion,
following first order kinetics out to about 15 s.

203

M. Fleischmann: What actually limits the frequency response of your microelectrode?

R. M. Wightman: The technique that we use is normal pulse voltammetry with pulses that are 100 ms long. But in the last 6 months, working with some people at the London hospital we have been using very fast cyclic voltammetry, at 300 V/s and we can get a complete record in 10 ms: it gives exactly the same quantitative numbers as I showed here. So that is the direction in which certainly all of us will be going.

M. Fleischmann: At that point you should be able to locate from time history where the source was. You should be able to map fairly exactly the source if you could implant several of these things.

R. M. Wightman: As you go faster and faster your diffusion layer gets closer and closer to the electrode. Then you are sampling from neurons that are next to your electrode which is actually a good thing: we would like to be able to sample from the smallest region possible. We would like the measurements to be fast because as I pointed out neurons act on a millisecond time scale and we would like to correlate the chemical with the physiological measurements.

R. M. Dell: Could you tell us something more about the importance of the control of the sodium and potassium ratio in the body and the consequence of this loss of control during surgery?

R. M. Wightman: I am not a physician and I prefer not to comment on this point. The control is there, it is well-built in; the brain uses something like a third of the oxygen that comes in your body for energy to keep these pumps going in your head that are keeping the sodium and potassium levels exactly right in the brain. So it is all important to normal functioning that this level of sodium and potassium be maintained. If you do not have this maintenance, you are going to go into convulsions and the whole central nervous system is going to be in an extremely bad condition.

H. Hurwitz: Usually there is a problem when one tries to measure potentials in site in cells because frequently one is not sure about the bulk potential and there may be a colloidal effect. So I wonder if there could not be such a shift of the potential scale in the case where you use the potassium method.

R. M. Wightman: The voltammetric resolution that we have is not very good, even for the catecholamines: if you look carefully at those voltammograms, the rising part of the wave went over a couple of hundred millivolts. So, precise potential accuracy is not necessary for these experiments. We really base our identity in part on the position of the wave but also on the shape of the wave: it is very different for the other compounds that we see.

H. Hurwitz: If you are changing the potassium concentration, you can just have a shift of potential due to the colloidal effect and that might be important because you might be in a situation where you have no bulk potential.

R. M. Wightman: Certainly! All I can say is that our experimental results suggest that there is no measurable shift: it could well be a shift of 25-50 mV and I am sure that there probably is! Yet this effect is not big enough for us to be able to see it or to disturb our measurements.

H. Berg: I wish to mention that there are some new methods in competition to yours: for instance, for pH measurements in tissue, the phosphorus

NMR. On the other hand what do you think of the Pilla currents, i.e. the electromagnetic induced currents pulsed by Helmholtz coils, which can also stimulate nerve dendrite growth? In the future they might be in competition with your methods.

R. M. Wightman: Right. What electrochemistry offers, because it is one of the few analytical techniques that you can make into a very small sized probe - 10 micrometers or less - is that you can get a localization of effects. Another method in addition to those mentioned is the 2-deoxy-glucose method which allows people to see where energy changes are occurring in the brain. But that gives you a global picture of what is going on, electrochemistry gives you a local picture. There is probably no best method and there may be certain methods that are better than others in certain applications.

A. Despic: Can you give us some technical details such as the nature and position of your reference or auxiliary electrode, where is the ohmic drop?

R. M. Wightman: When you use microelectrodes of this size, you get rid of lots of your electrochemical problems. You are passing such very small currents that, as Professor Fleischmann pointed out, you do not have to worry about ohmic drop. We thought for a while that we might want to have our reference electrode very close to our microelectrode; however, since we have no ohmic drop, we do not need that to be the case. So, the reference electrode simply rests on the skull. It is a conventional normal size SCE electrode which rest on the skull and has contact with sealing. Because you do not have ohmic drop, you do not need to use a 3 electrode potentiostat, you can go back to a 2 electrode potentiostat. So, the other parts of the treatment are not so tricky! In addition, physiologists for years have been using micro reference electrodes. So eventually when more and more people get to use freely moving animals and make electrochemical measurements, the idea of making very small reference electrodes and implanting them under the skin is not going to be a problem either.

D. R. Thévenot: Can you comment on the type of electrode pretreatment that you use and on the behavior of these microelectrodes after long periods of operation?

R. M. Wightman: We do not do any treatment to the electrode that I describe. Electrochemical treatment can greatly change the properties of the electrode and make it very different. What we find is that the electrode does not respond as well as after we take it out of the brain as it did when we put it in: electrode kinetics are much too much slower. But this lowering of the rate of electrode kinetics occurs in the first 15 min after implantation. For hours after that it stays exactly the same. What it is going to do for days is a different problem, but electrodes have been implanted in the brain for 3 months are reported to give very well defined voltammetric curves.

D. R. Thévenot: Professor Albery describes some dramatic damage of carbon fibers after several months implantation.

R. M. Wightman: Right. It is a hostile environment!

E. Gileadi: The use of implanted microelectrodes is presently limited by the gradual change in their sensitivity over periods of days or weeks. This can probably not be overcome by improved designs and ways of in-situ periodic calibration will have to be developed. This could be achieved in one of two ways: (i) injection of known amount of the analyzed material and recording the response of the electrode (ii) implantation of an additional microelectrode which can trigger the production of a known amount of the material to be analyzed (e.g. a neurotransmitter) and calibration of the detecting electrode at set intervals of time.

R. M. Wightman: This has been addressed in part: people have found thylanol, an aspirin substitute, is electroactive and if you give an IP injection of that, it gets into the brain very rapidly and has well characterized voltammetry and it can be used as an external standard as you suggested. In addition, it is interesting because electrochemistry is monitoring what this drug is doing as it comes into the brain and suggest a whole different area into which in vivo electrochemistry could go.

A. Bard: In extending what you have been saying to other types of tissues, such as blood stream or heart muscle, one starts to find rather serious problems about adsorption of proteins onto the electrode and deactivation: is the cerebro-spinal fluid a particularly good medium?

R. M. Wightman: Cerebro-spinal fluid and the extracellular fluid, both are very protein free from a biological point of view and, so, they help a great deal in making such measurements. The fact that we can make these measurements does not mean that you can make direct measurements in blood: that is a lot more complicated and a lot more hostile environment!

A. Bard: Maybe polymer films of the right type may be important in those media.

R. M. Wightman: Polymer films may indeed increase longevity and help in a wide variety of other problems.

H. Hurwitz: There is still a problem concerning the use of ISE where you need at least a stationary state and where you have to be sure about the bulk potential.

R. M. Wightman: The remarks that I made so far are to voltammetry. It certainly is going to be a very different sort of story with ISE. But, again, electrophysiologists are very adept in pulling multi-barrel electrodes, i.e. several glass capillaries pulled together in one of which they can put the liquid (on exchanger, in the other they can make a reference electrode, so you have the two electrodes right there together.

H. Hurwitz: It is still a complex situation: you may certainly have some problems of activity in cytoplasm and have difficulties in translating your results easily.

R. M. Wightman: Absolutely.

H. Oelschläger: A very severe problem of public health is the increasing number of patients who are suffering from diabetes. Do you see any chance of realizing the construction of an electrode which would be able, after implantation, to measure the glucose concentration in blood over a long period? If you should succeed, then you would have solved the problem of the insulin injector. And I think that is more important than to measure dopamine concentration in rat brains!

R. M. Wightman: To my knowledge there is no such sensor, but I should turn over to the chairman who has been looking at that.

D. R. Thévenot: This is a problem which has been addressed for several years, if not several decades. Besides the direct electrochemical detection of glucose which is very poorly sensitive, selective and reproducible, glucose is generally detected by its oxidation with oxygen in the presence of glucose oxidase. Enzymatically generated hydrogen peroxide may be detected by anodic oxidation; alternatively enzymatically reacting oxygen may be detected by a Clark type electrode. Both systems have their advantage and drawbacks but none of them is presently commercially available for non diluted blood glucose determination or for implant. As presented in my introduction, my feeling is that the complexity of such enzyme electrodes, the poor understanding of their behavior and the difficulties arising when they are all involved in this restricted progress. As you pointed out, I think that much research activity should be done in the field of the implantable glucose sensor since the development or an artificial pancreas is one of the important challenges of today and one of the domains where, as stated in the title of this forum, electrochemistry could be of the service of mankind.

H. Oelschläger: May I add a further remark. We have had a Symposium on diabetes in Austria. I have learned from my physical collegues, who have long term experience, that the insulin pump is only to be handled by intelligent patients. The personal behavior of the patient, influences the success of the insulin injector. If diabetes could not be solved by an injector coupled to a glucose sensor, only intelligent people could benefit of such insulin injectors.

D. R. Thévenot: I agree completely. Such miniaturized insulin pumps are not given to every insulin-dependent diabetic patient. Their use will be restricted as long as no in vivo glucose electrode is available. My research group is working in the field: the goal is not yet reached, even if significant progress has been obtained.

MEMBRANES

H. D. Hurwitz

Université Libre de Bruxelles
Belgium

Many reports have been presented in this UNESCO forum, which concern the technological achievements of electrochemical science. Fundamental scientific questions have been but little raised. We have no doubts that chemistry is pulled by technology and is pursuing technological results. Though, let us stress that scientists involved in research are not primarily concerned with technological goals. Science, as a human practice, produces a clearly defined discourse about the universe. This discourse leads us to perceive progress as being, fundamentally, an improvement in our way to question nature and in our logical understanding of the experimental processes.

The intrusion of electrochemistry upon the field of biological membranes illustrates perfectly our conception of progress. Bioelectrochemistry has become the pattern of more consistent approaches of many aspects of life. We wish to show in this forum what part of membrane science is most pertinently covered by electrochemistry and how the interdisciplinarity between biology and electrochemistry comes into action. The work of Prof. Chismadjev has already given adequate answers to these questions. Prof. Chismadjev has had an education as physicist and started his scientific career by performing theoretical research on charge transfer mechanism at electrodes. Later on, he was concerned with the ionic transport in excitable membranes. He gave a proof that thermodynamic and kinetical electrochemical concepts are suitable for the solution of many problems in the field of stability and conduction of biomembranes and that these concepts are absolutely necessary too for animal and vegetal physiologists.

We all know that the number and diversity of biochemical functions of a living cell membrane imply a complex architecture of the membrane. Accordingly, an understanding, at the molecular level, of the modes of action of each separate membrane component is extremely difficult and sometimes even meaningless because of the synergic nature of the living process. In order to gain some insight of the membrane mechanisms, one direction of research consists in isolating the membrane components and assembling them afterward in a model membrane. Such a bilayer lipid membrane permits the investigation of specific functions of the membrane by selecting carefully its constituents and their environment.

The study of ionic conduction, pore formers, defects, electric breakdown in membranes, their interaction and fusion, the influence of an external field are some of the many problems which can be handled in this way. The work of Prof. Chismadjev contributes also largely to our fundamental understanding of these problems.

MEMBRANES

Yu. A. Chizmadjev

Frumkin Institute of Electrochemistry
USSR Academy of Sciences
Moscow

INTRODUCTION

It should be said in advance that the title of the paper does not
reflect its contents accurately. I shall not discuss any of the problems
connected with the use of membranes in applied electrochemistry. Above
all, this is because I am not expert in that field and for that reason
probably see no new scientific problems. Even a cursory glance at the
immense literature in that field confirms the view that the task now is
that of seeking optimal technological solutions. At the same time the
advances made by modern biology have pushed to the forefront a new object -
cell membranes, the role of which has been underestimated in the past. It
was believed that their main function was to serve as a barrier, i.e. their
job was to maintain a constant and different content inside and outside the
cell. Gradually it became clear that the membranes are responsible for
many of the principal functions of the living cell. Membranes introduced
in modern biology the concept of vectorial reactions, they compelled
scientists to revise their views in bioenergetics, so that now there is
every reason to regard the mitochondria - the power stations of the cell
- as a fuel cell. The triumph of the membrane theory compelled scientists
to turn to electrochemistry, and, at the same time, stimulated the interest
of electrochemists in biology. That is why UNESCO added to the long list
of traditional sciences a new one which was designated as biological
electrochemistry.

There were at least two reasons for its emergence. First, the attractions of modern biology which has superseded physics as the Number 1
science, and secondly, that objective fact that most life processes are
founded on electrochemical phenomena. Consequently, electrochemistry armed
with its own methods and theories has certain chances of success, although
the road to it is, as usual, by no means an easy one. Additional difficulties are caused by the fact that bioelectrochemistry is developing in
fierce competition with the long established sciences, such as biophysics,
biochemistry, electrophysiology. And it is, above all, a struggle for
territory, for the object of investigation, since a combination of methods,
even first class methods, cannot be regarded as a new independent branch of
science. Meanwhile the territory has long been divided up between the
great predecessors. However, this analogy, happily, does not run deep.
Take, for example, biophysics. To this day arguments continue on what
actually should come under the umbrella of biophysics. The source of the

argument is similar. It stems from the fact that biophysics developed on another science's territory, although it used its own ideas, approaches and methods, going deep down into the subject and building a new infra-structure. That, as we see it, is also the future awaiting biological electrochemistry, although with certain differences that we shall examine later.

THE SUBSTANCE OF THE SUBJECT

Bioelectrochemistry studies the electrochemical phenomena on which biological processes are based. These include photosynthesis, reception, the interaction and fusion of cells, etc. Bioelectrochemistry also studies the effect of external electric fields on various biological systems. The common factor in all these biological processes is the existence of a stage of separation of charges, electrons or ions which takes place either in the course of a redox reaction or, as a result of the active or passive trans-port of ions through the membrane. This leads to the formation of a membrane potential and gradients of concentration of the different ions between the internal part of the cell and the surrounding media. The free energy, accumulated as the membrane potential or concentration gradients, ensures the generation and transfer of nerve impulses, the synthesis of ATP, certain types of mechanical movement, etc. So the main task of bioelectrochemistry is to study the thermodynamics and the kinetics of the separation of charges, which occurs first of all in the membrane systems.

Bioelectrochemistry as a field of science lies at the point of inter-section of electrochemistry and biology. It studies the phenomena and objects that belong to the fields of electrophysiology, biochemistry and biophysics, but it uses electrochemical approaches and methods. Picking up problems from various fields of biology, bioelectrochemistry seeks to resolve them using various models as similar as possible to the object and reconstructing separate cellular systems and functions.

HISTORY

It would be naive to attempt in a short report to reconstruct the entire history of bioelectrochemistry. It is possible only to name the sources and sketch the highlights. [See 1 for greater detail].

The term bioelectrochemistry was first officially recognized in 1971 at the first International Symposium on Bioelectrochemistry in Rome organ-ized by Professor Milazzo. However, the birth of bioelectrochemistry goes back to 1791, when Galvani published his famous treatise "De Viribus Electricitatis in Motu Musculari Commentarius," which led to extended polemics with Volta and resulted in the discovery of the so-called galvanic cell. Full recognition was given to animal electricity following the works of Dubois Raymond (1843) who measured the current of muscles and nerves at rest and after excitation. Helmholtz at about the same time measured the rate of propagation of nerve impulses. The physical views on the nature of bioelectricity developed by Bernstein were based on the classical works on the thermodynamics of a galvanic element by Gibbs, Helmholtz, Nernst and Ostwald. In 1961 Mitchell formulated the chemiosmotic principle of the transformation of energy in biological systems. His theory was confirmed by numerous experiments, and it was thereby established that bioenergetical processes, including photosynthesis are based on electrochemical phenomena. Prigogine evolved the thermodynamics of irreversible processes, occurring far from equilibrium, and thereby laid the foundation for a quantitative description of various biological phenomena, including those studies by bioelectrochemistry.

Bioelectrochemical research is conducted at three levels - at the molecular, cellular and organism levels, and these are represented by the following trends:

1. The thermodynamics and kinetics of redox reactions with the participation of biologically active compounds.
2. Equilibrium electrochemical properties of membranes.
3. Mechano-electrical phenomena.
4. Membrane transport.
5. Bioenergetics.
6. Electrochemically active (excitable) media.
7. Applied bioelectrochemistry.

In effect membranes are present in all these subjects, even in the first, since the possibility of using electrochemical methods for studies of biological reactions are founded on the fact that both electrochemical and many biological redox reactions are heterogeneous - the first take place on the electrode surface, and the second on the enzyme-solution interface with the enzyme often attached to the membrane. Each uses its own methods and models. The thermodynamics and kinetics of redox reactions are studied by polarographic and voltamperometric methods on mercury electrodes. Using these methods it is possible to determine the number of electrons involved in the reaction of each value of the potential, as well as to detect unstable intermediate compounds, including short-lived radicals, which cannot be registered by other methods.

Compared with other methods, electrochemical ones have a wider range of application, which makes it possible to study the details of the reaction's mechanism. They are suitable for unique syntheses and for the solution of analytical problems. The use of electrochemical methods made it possible to obtain detailed information about the thermodynamics (redox potentials), kinetics (number of electrons, etc.) and mechanism of reactions with the participation of heterocyclic nitrogen compounds (purines, pyrimidines, porphyrines, etc.). [For more details see 2]. Capacity measurements provided important information [see, for example 3] on the adsorption properties of low-molecular and high-molecular biologically-active compounds (proteins, DNA, RNA).

The following model systems are used in the bioelectrochemistry of membranes[4]: planar lipid bilayers, liposomes, monolayers on the water-air boundary, interface of immiscible liquids (for example water-octane). Each of the model systems has its own sphere of application and requires the uses of certain definite methods. Bilayer lipid membranes are used to reconstruct transport cellular systems: ionic channels of excitable biomembranes, the proteins performing active transport (ATPases, bacteriorhodopsin etc.). They are convenient for studies of ionic transport by liposoluble anions (dipicrylamine, tetraphenylborate, etc.) and membrane-active complexons (valinomycin, gramicidin, etc.) The surface properties of membranes, such as the structure of the electric double layer, the adsorption of ions and various surface-active substances are studied on lipid bilayers and monolayers. Finally, lipid bilayers are used to study the mechanical properties of membranes, their stability in an electric field and the mechanism of their fusion. The methods used are: registration of the conduction current and capacitive current when an electric voltage that changes according to a certain law, is imposed, measurements of surface tension or pressure (in the case of monolayers), registration of the Volta-potential (in the case of water-air interface, water-octane); and various optical and spectral measurements.

The experimental and theoretical studies of ionic transport on lipid
bilayers in the presence of ionophors revealed two main mechanisms of
transport - through mobile carriers (valinomycin) and through channels
(gramicidin A). It was shown that the transport systems of excitable
biological membranes operate as selective ionic channels. A study of the
mechano-electrical phenomena, such as the movement and orientation of cells
in external electric fields, the restructuring of membranes during electrical breakdown and electro-stimulated fusion of cells laid the foundation
for medico-biological and biotechnological applications (the development of
artificial carriers of medical drugs, membrane diagnostics, and the development of hybrid cells). A major achievement of bioelectrochemistry was
the provision of proof of Mitchell's chemiosmotic hypothesis obtained in
experiments with the reconstruction of bioenergetic membrane systems in
various model systems, including liposomes.

Although the propagation of excitation along nerve fibers and the
neuron network traditionally belongs to the sphere of electrophysiology and
biophysics, studies in electrochemical systems, such as with passivating
electrodes (Lilley Bonhöffer model) and charged porous membranes (Teorell-
Frank electro-osmotic model), contributed a great deal to the understanding
of the mechanisms of these processes.

Applied bioelectrochemistry includes such things as the development of
ion-selective microelectrodes for intra-cellular use, microelectrodes for
intra-cellular injections of electroactive substances, enzyme electrodes,
electrochemical biosensors (bacterial and tissue electrodes) and ion-
selective electrodes using ionophores. The medico-biological applications
cover studies of extracellular electric fields and studies of the mechanism
of the action of external fields and currents on various physiological
processes, including the regeneration of tissues.

After this general survey of the subject, it would be useful to
examine a number of concrete examples.

EQUILIBRIUM ELECTROCHEMICAL PROPERTIES OF LIPID BILAYERS

A bilayer lipid membrane usually carries an electric charge on each
of its surfaces, which is ensured by the polar heads of the phospholipid
molecules. When the membrane is polarized by an external field or when
non-symmetrical conditions in the concentration of the electrolyte in the
surrounding solutions are created, there arises a transmembrane potential
ϕ_m. Its structure is as follows: two surface potentials ϕ_s, two dipole
potentials ϕ_d localized in the region of the polar head groups and the
intra-membrane potential ϕ_{in}. The different membrane functions are governed by different components of the membrane potential. Thus, for instance, the mobility of the cell in the external field is determined
exclusively by the value of ϕ_s, while the transport of ions through the
membrane, if the limiting stage of the transfer is the passage through the
hydrophobic zone, depends only on ϕ_{in}. It is therefore highly important to
be able to measure and calculate all the components of ϕ_m, which is a
sufficiently daunting task. Taking an equivalent scheme, the membranes
could be regarded as three condensers connected in series, with the
capacity of the membrane itself less than that of the diffuse double layers
by about two orders of magnitude. It should be underlined that the membrane capacity depends on ϕ_{in} because of the deformability of the membrane.
It was this that made it possible to develop a number of potentiodynamic
methods for measuring the components of ϕ_m [5-8]. These methods are founded
on the membrane capacities non-linear dependence on the intensity and
duration of the electric field and the principle of the symmetricization of
the capacity current by compensating the intramembrane field[7,8]. As a

result it is possible to measure directly the difference between the boundary potentials $\phi_s + \phi_d$ and at the same time study the elastic and viscosity properties of lipid bilayers. These methods augment considerably the earlier existing ones, one of which, applicable to liposomes and cells, is based on measurements of the electrophoretic mobility and another on measurements of the conductivity of flat bilayers in the presence of ionophores. The data obtained by the new methods indicates that the double layer on lipid membrane is much more complex than believed in the past when the potential was measured by electrophoretic methods and a good agreement with the Gouy-Chapman theory was obtained[9]. It turned out that it is necessary to take into consideration the shift in the proton-adsorption equilibrium on the surface of the bilayers when conditions in the bulk of the electrolyte are changed[10]. In addition, an increasing number of facts indicates that when the divalent ions are adsorbed and the protons bonded, the ϕ_d changes[11]. It is hard to say so far whether this is due to a rotation in the dipole moments of the polar head-groups, the segregation of the lipids or changes in their phase state, or to re-structuring of the neighboring water. The situation in this respect resembles that which has developed in classical electrochemistry, where it is becoming increasingly urgent to study the physics and chemistry of the electrode surface. The membrane surface is much more labile than the metal-solution interface, which makes it much more essential to study its properties at the molecular level. To this it should be added that the fact that the greater capacity of the diffuse double layer is shunted by the smaller capacity of the membrane makes it extremely difficult to study the former directly. One more problem that should be resolved in the near future is that of the adsorption on bilayers of single electrolytes, as the values given in literature for the adsorbtion constants vary so greatly.

MEMBRANE TRANSPORT

Bioelectrochemistry makes a study of ionic transport through planar lipid bilayers on the following systems, listed in order of complexity and closeness to the biological situation:

a) transport of hydrophobic ions

b) transport of ions induced by ionophores

c) transport through protein systems obtained from cellular membranes and built-in bilayers

The transfer of hydrophobic ions is interpreted as ordinary electrodiffusion and presents nothing new compared with the approach usually accepted in electrochemistry. The only simplifying feature is that the thickness of the membrane is many times less than the Debye radius of screening, so one can safely use the approximation of a constant field.

In the studies of induced transport it is customary to differentiate between two mechanisms - mobile carriers and channels[14]. The theory of mobile carriers was advanced comparatively recently[13,14], but is already well-established. But the task that remains is to measure the rate of the different stages of transport for different carriers. The most promising results seem likely to come from different variants of the relaxation methods - the temperature jump and potential jump methods[15].

Things are more difficult with the theoretical description of transport through channels. Yet it is this transport mechanism that is of the greatest interest from the biological point of view, since the transport systems of biomembranes apparently operate on the channel principle. What distinguishes this mechanism is that it is a single-file transport, and classical electrochemistry has never come up against such a case before. Another of the system is the small thickness of the membrane. Because of this both possible approaches to an analytical solution - the discrete and the continuous present their difficulties and limitation. In the discrete approach transport is regarded as a chemical reaction passing through a series of consecutive stages. But since, unfortunately, the energy profile of the ion in the channel - both in the model and the biological one - is unknown, a large number of unknown parameters have to be introduced[14,16]. A way out is offered by the use of the method of molecular dynamics[17,18]. The idea behind the method is to choose the structure of a known channel, as for instance the gramicidin channel, prescribe all the charges and dipoles on its frame, describe a molecule of water, according say to Rahman and Stillinger[19], and then, using a computer, solve the system of equations for the movement of a water molecule and ions. So far this program has not been realized, but the first results for the trajectory of the movement of water molecules and binary correlative functions have already been obtained[16]. Despite the large amount of work involved, the method of molecular dynamics should be regarded as a most effective way for resolving the problem. Perhaps it could be instrumental in breaching the gap that has arisen between detailed information on the structure of proteins provided by bioorganic chemistry and the primitive descriptions which are unable to use all the available knowledge.

The last, and probably the most pressing aspect of the studies of ionic transport through reconstructed protein systems of biological membranes has, so far, reached only the preliminary stages. This is due to the fact that in order to start a study of the process it is first necessary to learn how to produce pure membrane proteins and then manage to implant them into a planar bilayer. In recent years bioorganic chemistry has made great progress in obtaining pure substances, but so far it cannot manage to implant them. Imbedding integral proteins into bilayers is today more of an art than a science. Yet the assessment of the success of the procedure of obtaining the protein rests largely on the success in deconstruction, since the sole criteria is the normal functioning of the protein in the model system.

The difficulties stem from the fact that transport proteins belong to the category of integral proteins that spear the hydrophobic core of the biological membrane. They are so hydrophobic that they can be separated only in the form of vesicles in which the proteins are surrounded by lipids. Consequently one way out is to learn to induce the fusion of such vesicles with planar bilayers. The mechanism of fusion will be examined below. At this point we only note that some scientists have in recent years succeeded in fusing vesicles with a planar bilayer and reconstructing certain membrane proteins[20,21]. At the same time other approaches to the problem of reconstruction have also been developed. Thus[22] use was made of the view that liposomes containing proteins open up and form a monolayer when they reach the water-air surface. By bringing two monolayers together, one can obtain a bilayer with an implanted protein. The patch-clamp method[23] was applied in[24] to monolayers. As a result a bilayer with a reconstructed protein was obtained in a micropipette. So far there have been no systematic studies of reconstructed channels, but that is already a matter for the near future.

Studies of single reconstructed channels will have an immense significance for biology. That is the only way we can learn the role of chemical regulation in dynamics of electrocontrolled channels, understand the molecular mechanism of their action, and decide what influence the cytoskeleton has on the work of the channels.

ELECTRIC BREAKDOWN, LYSIS, FUSION

The phenomena of breakdown, lysis and fusion have elementary mechanisms in common and the process of the electric breakdown of membranes and lysis are stages in the complicated and highly important process of cellular fusion. All three phenomena play an important role in the life of the cell and in pathology. Recently it has become obvious that they present promising ground for biotechnology. That is why interest in this branch of membranology is very great, the number of published works is increasing rapidly, special symposiums are being held, and books and surveys are being published (see, for instance, the excellent survey by Zimmerman[25]).

Electric Breakdown of Membranes

It was established a comparatively long time ago that the appearance of a sufficiently strong electric field on the membrane of a cell increases the conductivity and permeability sharply. There can be two outcomes - the mechanical destruction of the membrane (irreversible breakdown) and healing (reversible breakdown). On most cellular membranes there exists a permanent (or periodic) potential difference and tens of milli-volt, so the field voltage is 10^4 - 10^5 V/cm. So there always exists the possibility that the barrier function of the membrane could be violated, which in certain cases could play an important regulatory role, while in others result in pathology. Cell membranes can be made highly permeable by means of an electric field. And that is a very convenient way of stuffing the cell with molecules that cannot pass into it through the native membrane. Gene engineering and pharmacology (the development of carriers of medical drugs) are in dire need of such methods.

In order to resolve all these problems, it was essential to find the answer to the main questions: what does an electric field do to the membrane? After a detailed study of this matter the following was established[26,29]:

1) The membrane is ruptured not as a result of supercritical pressure, as believed in the past, but as a result of the emergence and growth of inverted pores.
2) The electric field sharply accelerates the breakdown process, since energetically it is advantageous to form hydrophilic pores.
3) The quantitative theory of irreversible breakdown which we developed provides a satisfactory description for all experimental data.
4) The stability of membrane is determined not only by macroscopic parameters, such as surface and linear tension, but also by the molecular geometry of lipids: conic molecules of the "broad head-narrow tail" type are prone to forming inverted pores, and membranes made up of them have a short life-time.
5) Membranes with reversible breakdown are prone from the point of view of energetics to the formation of a large number of pores. Under the action of the field their number and average radius increases in time. Suitable molecular geometry of lipids probably makes a membrane reversible in its behavior.
6) All these facts have been established on planar lipid bilayers. Experiments conducted on biological membranes using the patch method showed that all regularities in both systems were identical.
7) The breakdown of vesicular membrane differs radically from that of a planar one bordering on a meniscus. It is reversible, but in this case because the field is shunted by the pore and the membrane is restored.

All these results provide the foundation for resolving a number of important practical medico-biological and biotechnological problems.

Osmotic Lysis

After a through pore has developed in a membrane, say a vesicle membrane, under the influence of an electric field osmolytic processes take over. Osmotic lysis can occur also without the inducing action of a field. It was shown[30] that it begins when under the action of osmotic pressure the stretching of the membrane exceeds a certain critical value. When it reaches that value the membrane rupture and pores are formed in it.

The dependence of the critical stretching on the radius of the vesicles was calculated and the corresponding values for osmotic pressure were determined. An osmotic pressure greater than 30 atm was needed to tear small (10 nm) vesicles, and for big ones (100 nm) a pressure close to 10^{-3} atm. For vesicles larger than 1000 nm the value of the comparative stretch of the membranes at which it ruptures within a reasonable time depend little on the radius of the vesicle and equals about 3%. And the value of the tension at which the membrane ruptures for large vesicles is about 8 mN m^{-1}, which is in agreement with experimental data available in literature.

After rupturing and the formation of pores, the tension in the membrane drops, but does not disappear altogether. It maintains a residual pressure in the vesicle, which is the case of the ejection of the contents of the vesicle (cell). At the same time there begins an osmotic flow of water through the membrane, directed inside. The outcome is either a complete rupture of the membrane with the ejection of the entire contents, or a gradual washing out of the contents which can take place in two regimes – either the quasi-stationary or the pulsating. In the first case the washing out takes place through the stationary open pore, and in the second case the pore opens and closes periodically, each time ejecting 5% of the solution inside. The kinetics of lysis have been calculated; it was shown that pulsating lysis was the most probable regime.

Fusion

During the lifetime of the cell its membranes undergo various transformations that accompany the processes of pynocytosis, secretion, building up, etc. The problem of the mechanism of membrane fusion has recently acquired special urgency in connection with the formation of hybrid cells (specifically, hybridom) necessary for implementing various biotechnological tasks. The most effective model for studying the mechanism of fusion of membranes seems to be two planar membranes that can be brought into contact by increasing the hydrostatic pressure[31-33]. Such an apparatus in our laboratory was equipped with potentiodynamics instruments to measure the surface potentials and surface charges of each monolayer of lipid bilayers brought together. These experiments in combination with theoretical developments make it possible to elucidate the mechanism of fusion and describe all its stages. We begin with a theoretical analysis of the fusion process[34,35].

At large distances, commensurate with the Debye distance, the interaction of membranes is determined by the balance of molecular and electrostatic forces. If the electrostatic repulsion is not very great, the membranes could come into close contact. The question is – could fusion occur through the mutual penetration of two bilayers? Estimates show that the excess pressure in this case should amount to 100 km of a water column. Consequently for fusion to take place structural changes are needed with the formation of energetically advantageous intermediary forms. Two such mechanisms of fusion have been suggested. One of them introduces stalks which form bridges between the membranes. Given a sufficiently large spontaneous curvature of the initial bilayers, the unrestrained growth of

218

the stalks proves energetically advantageous with the formation of a trilaminary structure, i.e. a bilayer zone bordering on two bilayers at the periphery. The second mechanism, adhesion-condensation, is based on the special properties of the zone of close contact between the two membranes, which are manifested in the presence of calcium ions in the solution. The theory of phase transitions developed shows[36] that the critical temperature in such conditions increases to 160 °C, and that means that in the contact area the nearlying monolayers crystallize. The mechanical stresses that arise can in certain conditions lead to the rupture of the monolayers and the formation of a trilaminary structure.

Experimental studies on a two planar bilayer model made it possible to describe in detail all the phases of fusion and confirm the main premises of the stalk mechanism (for so-called "dry" membranes without a solvent). After the membranes are brought into close contact a trilaminary structure is formed spontaneously, and its lifetime depends on the field inside the membrane. By sending a voltage impulse, an electric breakdown can be caused, which brings about complete fusion.

On our experimental model that signified the formation of a cylindrical membrane tube. By increasing the pressure in the outside section it could be made to collapse with the formation of two planar bilayers. Thus the cycle is completed and can be repeated endlessly. The theory of the collapse of the membrane tube made it possible to develop a new method for measuring surface tension, which is particularly promising for "dry" membranes. The performance of the entire cycle of fusion with membranes formed from the lipids with a different molecular geometry (i.e. different spontaneous curvature of bilayers) make it possible to prove[37] the justice of the stalk mechanism.

These results are not only of fundamental significances. They also provide the scientific basis for the development of effective methods of the electrostimulation of cell fusion which finds the application in biotechnology.

CONCLUSION

For biology, medicine and biotechnology electrochemistry is the fundamental science. But for its views and methods to find wide application there, electrochemical should advance more confidently to meet the requirements of those sciences, they should move to the studies of model systems which resemble as closely as possible the real biological objects. We have given a few examples of this kind. Their number could be multiplied. But it remains a fact that in the whole world there are probably fewer than a dozen electrochemical laboratories specializing on bioelectrochemistry. Nor are such courses available in the universities, and that means we are not training young specialists for such jobs, not attracting young people to this fascinating branch of science.

Electrochemistry can contribute a great deal to medicine and biology, but in the process it will acquire even more for itself. Modern biology is advancing at fantastic rates and each day sets new problems, including those in electrochemistry. I believe everyone engaged in allied subjects feels its stimulating effect. What biology promises to electrochemists converted to its faith is a stream of young enthusiasts, genuine actuality and high rate of research and a wide and rapt audience.

The concrete examples in this paper touched on the membrane level of bioelectrochemical research. The fields of tissues, and organisms were not touched upon, although that is where the most significant events can be expected. The practical achievements obtained from weak currents on the regeneration of tissues and the observed effect of external electric fields on the functioning of an organism requires the development of the scientific electrochemical foundations of these important phenomena.

It is nice to realize the mechanisms of the highest nervous activity is by nature an electrochemical phenomena. Realization of this fact must stimulate the development of bioelectrochemistry.

REFERENCES

1. G. Milazzo, Bioelectrochemistry and bioenergetics. An interdisciplinary survey, in: "Bioelectrochemistry," Vol.1, G. Milazzo and M. Blank, eds., Plenum Publishing Co. (1983).
2. G. Dryhurst, "Electrochemistry of Biological Molecules," Academic Press, New York - London (1977).
3. H. W. Nürnberg, Applications of advanced voltammetric methods in bioelectrochemistry, in: "Bioelectrochemistry," Vol.1, G. Milazzo and M. Blank, ed., Plenum Publishing Co. (1983).
4. J. Koryta, "Ions, Electrodes and Membranes," J. Wiley & Sons, Chichester (1982).
5. O. A. Alvarez and R. Latorre, Voltage dependent capacitance in lipid bilayer made from monolayers, Biophys.J., 21:1 (1978).
6. P. Schoch, D. Sargent, and R. Scwyzer, Capacitance and conductance as tools for the measurement of asymmetric surface potentials and energy barriers of lipid bilayer membranes, J.Membrane Biol., 46:71 (1979).
7. Yu. A. Chizmadzhev and I. G. Abidor, Bilayer lipid membranes in strong electric fields, Bioelectrochem.Bioenerget., 7:83 (1980).
8. V. S. Sokolov and V. G. Kuzmin, Measurement of surface potential difference of bilayers by second harmonigue of capacitive current, Biofisika, 25:170 (1980) (in Russian).
9. S. McLaughlin, N. Mubrine, T. Gresalfi, G. Vaio, and A. McLaughlin, Adsorption of divalent cations to bilayer membrane containing PS, J.Gen.Physiol., 77(4):445 (1981).
10. N. S. Matinyan, I. A. Ershler, and I. G. Abidor, Proton equilibrium on the surface of lipid bilayer, Biologicheskie Membrany, 1:254 (1984) (in Russian).
11. N. S. Matinyan and I. G. Abidor, Proton equilibrium on lipid bilayer, Doklady Acad.Nauk,USSR, 274:1226 (1984) (in Russian).
12. Yu. A. Ovchinnikov, V. T. Ivanov, and A. M. Shkrob, "Membrane Active Complexions," Elsevier, Amsterdam (1974).
13. S. Ciani, G. Laprade, G. Eisenman, R. Laprade, and G. Szabo, Theoretical analysis of carrier-mediated electrical properties of bilayer membranes, in: "Membranes - A Series of Advances," Vol.2, G. Eisenman, ed., Marcel Dekker, New York (1972).
14. V. S. Markin and Yu. A. Chismadjev, "Induced Ion Transport," Nauka, Moscow (1974) (in Russian).
15. P. Läuger, R. Benz, G. Stark, E. Bamberg, P. C. Iordan, A. Fahr, and W. Brock, Relaxation studies of ion transport systems in lipid bilayer membranes, Rev.Biophys., 14:513 (1981).
16. Yu. A. Chismadjev and S. Kh. Aityan, Ion transport across sodium channels in biological membranes, J.theor.Biol., 64:429 (1977).
17. W. Fischer, I. Brickmann, and P. Läuger, Molecular dynamics study of ion transport in transmembrane protein channels, Biophys.Chem., 13:105 (1981).
18. S. Kh. Aityan and Yu. A. Chismadjev, Molecular dynamics of water movement across gramicidin channel, Gen.Physiol.Biophys. (in press).
19. F. H. Stillinger and A. Rahman, Improved simulation of liquid water by

molecular dynamics, <u>J.Chem.Phys.</u>, 60:1545 (1974).

20. B. K. Krueger, J. F. Worley, and R. J. French, Single sodium channels from rat brain incorporated into planar lipid bilayer membranes, <u>Nature</u>, 303:172 (1983).

21. R. Latorre, C. Vergara, and C. Hidalgo, Reconstitution in planar lipid bilayers of a Ca^{2+}-dependent K^{+} channel from transverse tubule membranes isolated from rabbit skeletal muscle, <u>PNAS</u>, 79:805 (1982).

22. H. Shindler and J. P. Rosenbusch, Matrix protein in planar membranes, <u>PNAS</u>, 78:2302 (1981).

23. O. P. Hamill, A. Marty, E. Neher, B. Sakmann, and F. J. Sigworth, Improved patch-clamp techniques for high-resolution current recording from cells and cell-free membrane patches, <u>Pfluegers Arch.Eur.J.Physiol.</u>, 391:85 (1981).

24. U. Wilmsen, C. Methfessel, W. Hanke, and G. Boheim, Channel current fluctuation studies with solvent-free lipid bilayers using Neher-Sakmann pipettes, <u>in</u>: "Physical Chemistry of Transmembrane Ion Motions," G. Spach, ed., Elsevier Science Publishers, Amsterdam (1983).

25. U. Zimmermann, Electric field-mediated fusion and related electrical phenomena, <u>Biochim.Biophys.Acta</u>, 694:227 (1982).

26. R. Benz, F. Beckers, and U. Zimmermann, Reversible electrical breakdown of lipid bilayer membranes: a charge-pulse relaxation study, <u>J.Membrane Biol.</u>, 48:181 (1979).

27. I. G. Abidor, V. B. Arakelyan, L. V. Chernomordik, Yu. A. Chizmadzhev, V. F. Pastushenko, and M. R. Tarasevich, Electric breakdown of bilayer lipid membranes, <u>Bioelectrochem.Bioenerget.</u>, 6:37 (1979).

28. V. F. Pastushenko, Yu. A. Chizmadzhev, and V. B. Arakelyan, Electric breakdown of bilayer lipid membranes. Calculation of the membrane lifetime in the steady-state diffusion approximation, <u>Bioelectrochem.Bioenerget.</u>, 6:53 (1979).

29. L. V. Chernomordik, S. I. Sukharev, I. G. Abidor, and Yu. A. Chizmadzhev, Breakdown of lipid bilayer membranes in an electric field, <u>Biochem.Biophys.Acta</u>, 736:203 (1983).

30. V. S. Markin and M. M. Kozlov, Osmotic lysis of lipid vesicles, <u>Biologicheskie Membrany</u>, 1 (1984) (in Russian).

31. E. A. Liberman and V. A. Nenashev, Model of cell junction of lipid bilayers, <u>Biofizika</u>, 17:1017 (1972).

32. E. Neher, Asymmetric membranes resulting from the fusion of two black lipid bilayers, <u>Biochim.Biophys.Acta</u>, 373:328 (1974).

33. G. B. Melikyan, I. G. Abidor, L. V. Chernomordik, and L. M. Chailakhyan, Electrostimulated fusion and fission of bilayer lipid membranes, <u>Biochim.Biophys.Acta</u>, 730:395 (1983).

34. V. S. Markin and M. M. Kozlov, On the first stage of membrane fusion, <u>Biofizika</u>, 28:72 (1983).

35. M. M. Kozlov and V. S. Markin, Probable mechanism of membrane fusion, <u>Biofizika</u>, 28:242 (1983) (in Russian).

36. V. S. Markin and M. M. Kozlov, <u>Gen.Physiol.Biophys.</u>, 2:201 (1983).

37. L. V. Chernomordik, M. M. Kozlov, G. B. Melikyan, I. G. Abidor, V. S. Markin, and Yu. A. Chismadjev, The shape of lipid molecules and monolayer fusion of the membranes, <u>Biologicheskie Membrany</u>, 1:411 (1984).

DISCUSSION

H. D. Hurwitz

Université Libre au Bruxelles
Belgium

CHAIRMAN'S SUMMARY

R. Parsons: I would like to ask you about the mechanism of defect
formation and of fusion which you discussed. Can you tell us on what level
we understand them? You seem to talk about macroscopic concepts following
Gibbs. Do you understand the mechanism at the level of interfacial tension
or do you understand it at the level of molecular motion?

Y. Chismadjev: Some years ago, J. M. Crowley[1] suggested a theory about
the mechanism of breakdown. His idea was to calculate the balance between
the elastic forces in the planar bilayer, considered as an elastic body and
the electric field. It resulted from his treatment that the planar homogeneous membrane broke when the electric field increased and reached some
critical value, which corresponded to a decrease of the membrane thickness
of about 30%. We tried to measure this phenomenon of breakdown as did
Professor Benz and Professor Zimmerman. We studied the capacity changes
which, of course, are proportional to the changes of the membrane thickness
and observed that just before the breakdown the membrane thickness had
decreased to about 1%, hence much less than 30%. So it was clear that the
breakdown mechanism had some other origin. We know that the usual stability of physical bodies is not affected as much by the existence of homogeneous deformation as by point defects. Those defects are critical for the
breakdown of bilayer membranes which are already metastable. An inverted
pore corresponds to such a defect. If during the process of pore formation, lipids can turn their polar heads towards the inside of the pore,
we obtain an hydrophilic inside. Our idea was to check if such inverted
pores are responsible for the process of breakdown. Therefore, we took
different lipids, differently charged and detected the existence of the
inverted pores by measuring the rectification of the current. Small pores
with charges on their inside wall, created by negatively charged lipidic
head groups, increase the selectivity constant for positive ions by a
factor of about 10, and thus make the negative current vanish.

The lifetime of the membrane depends on the evolution of these pores.
The magnitude of their radius is subjected to thermal fluctuations and if
it reaches a critical value the system breaks. If we apply an electric
field and form such pores with given types of lipids, we can increase the
curvature of the pores particularly in the case of lipids of suitable
molecular geometry. In order to verify this model, we produced bilayers of

different lipids, varying in charge and molecular geometry, and studied their influence on the membrane lifetime. Hence, the approach remains at the macroscopic level with respect to the use of surface tension and linear tension but it involves also some molecular understanding. If we turn now from the treatment of membrane breakdown to membrane fusion, we know that we must obtain bridges or stalks between planar bilayers. The molecular geometry which allows the easiest formation of stalks is not the same as the molecular geometry of lipids appropriate to form the inverted pores since the curvature in both cases is quite different and even opposite. Therefore, we made membranes of different lipids in order to relate both the lifetime for breakdown and for fusion to the lipid molecular geometry. We have also calculated the lifetime of breakdown and have predicted the lifetime of fusion for different lipids. I did not believe that the determining mechanism of fusion corresponds to a stalk formation, since there is another possibility which is also very obvious from the physical view point. In the region of tight contact of two membranes, we have a very low dielectric constant, much less than 80, say of 3 or 5. For this reason, in this region of junction, the interaction between charged head groups of lipids belonging to the membrane becomes very different, and since the interaction energy has changed, the temperature of phase transition will also be changed. In the region of tight contact the lipids take a solid state configuration. Hence, whereas the lipids in the outer monolayer of the bilayer membrane of a liposome are in a crystalline state, the inner monolayer remains in the liquid state and the resulting mechanical stress in this region may produce the condition of a breakdown. I believed that this mechanism gave the most plausible explanation of fusion. However, our experiments with lipids of various molecular geometry allowed us to obtain all necessary parameters from breakdown measurements and to use them in the theory of the stalk formation for the determination of the life time of fusion. The agreement between the theoretical predictions and experimental results is excellent.

H. D. Hurwitz: I would like you to add some comments on your answer to Professor Parsons and to tell us how you introduce the chemistry of living cells in your treatment. Membrane fusion and breakdown must depend on the nature of lipids, but in the living cell membrane there are also proteins and the living membrane fluidity will be quite different from that of the model membrane. To what extent can the mechanism, which you observe on bilayers, be transposed to real living systems of the type, for instance, which Professor Berg, Professor Zimmerman and others have investigated?

Y. Chismadjev: When we investigated breakdown in the case of artificial bilayer membranes, we asked ourselves, of course, whether the mechanism is the same in the case of cell membranes. These include proteins and defects which might be formed in regions like channels, contacts between lipids and proteins, etc. We tried to study these phenomena with real cells using a method suggested by Professor De Maeyer. The idea consists of taking a microelectrode made out of a very thin glass capillary, with a hole of radius less than a few μm. By a special technique, it is possible to obtain a junction with very small leakage such that its resistance is about some Gigaohms. We performed two types of experiments. We worked with the cell in contact with the capillary and destroyed the pouch of the membrane in the capillary region. Another possible experiment is to destroy all the cell and have a small pitch of the membrane in the capillary. By these methods, we can study the breakdown and fusion of biological membranes (e.g. erythrocytes) in function of the composition of the solution, the electric field strength, etc. We compared such results with those obtained with artificial bilayers. The results are absolutely similar. There is only one difference. In the case of bilayers, the breakdown is irreversible. In the case of biological membranes, the breakdown may be reversible. These membranes reseal once the electric field is switched off. It might be suggested that the hole, which is formed in the membrane, while growing, reaches a membrane region containing proteins which should have a stabilizing effect on the hole. Then, the system is resealing once the field is switched off. The same effect can be observed in the case of fusion where we did some preliminary experiments with two micropipettes and two cells. By comparing our results with those obtained with two bilayers, we have the impression that the fusion proceeds in the region of contact between two lipidic domains of the cell membrane.

C. Buess-Herman: The mechanism of breakdown seems to be very similar to the rearrangements of monolayers or Hg electrode, for example, which also present very interesting aspects. Maybe you can give some more information on the time range in which this breakdown occurs and some precision about the experimental method you have used. Do you only measure capacities?

Y. Chismadjev: The lifetime of our membranes depends on the electric field. With changes of the applied voltage from 100 mV to 1 V, the lifetime is affected by nearly five orders of magnitude, from 1 s to parts of μs. In the first method which we used, we applied the electric field and observed first the capacitive current charging the membrane, afterwards the base-line current remains and is affected by some fluctuations followed by a very rapid increase. This last current is flowing through the defects. This experimental behavior is explained by the fact that the membrane passes through two types of conformation. In the first type there are only a small number of pores in the membrane which cause the current fluctuations. The current increase corresponds to the irreversible stage, where, if you remember the parabolic curve which I showed in my lecture the pores become larger than the critical value and the system proceeds spontaneously to the formation of very large pores. The second method is a charge relaxation method which has been introduced by Benz and Zimmerman, and which we have also used in our laboratory. We apply, during a very short time, a high voltage on the system and we measure the current after this very short pulse.

H. D. Hurwitz: Professor Chismadjev did not limit himself to the problem of membrane breakdown and membrane fusion which is also tackled by Professor Berg's contribution. Other important aspects of bioelectrochemistry have been treated by Professor Chimadjev, like, for example, the determination of membrane potential, also considered in Professor Adams' contribution, the electric field across membrane, the influence of antibiotics and of other vectors for the ion transport. I wish to ask the audience if there are any comments which concern such aspects of the selectivity of cell membranes to ions and the role of peptides.

M. Fleischmann: I wish to make a bridge to the influence of peptides in my comment. I was very interested in energetics which Professor Chismadjev put forward, because of course you can also get energetics of a pore by doing experiments on fluctuating systems, say with the peptide alamethicin. Then you find that you have a positive surface energy but a negative edge energy for the formation of a pore, which I always explain, as being totally unsurprising because if you had it the other way round, all the pore formers would blow our membranes to pieces and life couldn't exist. My point is that in the case of pore formers, when you have proteins in the membrane, you must have a positive surface energy and a negative edge energy so that you will never find the normal bilayer or the living system in a supercritical condition. It may be that in the case of fusion in the lipid region only, you have it the other way round. Have you looked at the stability criteria and the phenomena in the presence of high fields if you have the sign of your energy terms reversed, which is certainly the situation in presence of pore formers?

Y. Chismadjev: We intend to make experiments in Moscow on breakdown and fusion of bilayers in which are included molecules like gramicidin. This will allow us to investigate the influence of pore formers.

ELECTROMAGNETIC FIELD EFFECTS ON

CELL MEMBRANES AND CELL METABOLISM

H. Berg

Academy of Sciences of the GDR, Central Institute of
Microbiology and Experimental Therapy
Department of Biophysical Chemistry, Jena (GDR)

INTRODUCTION

After the discovery of galvanism at the end of the 18th century by
Galvani and the foundation of electrochemistry by Ritter, since 1796[1]
electric magnetic effects on cells, tissues and organisms have been registered leading to the expansion of electrophysiology by Du Bois-Reymond[2]
and others during the 19th century[3] parallel to the development of
fundamental electrochemistry[4].

However, the effects of single and periodical electric field pulses on
cell - as well as magnetic influences on tissues - have been studied
intensively only recently, because of its importance for good engineering,
biotechnology, and medicine. The reason this this, are the new possibilities of pulses causing:

- electroporation (PEP): opening or enlargement of membrane pores, for
  incorporation of drugs or genetic material,
- electrofusion (PEF): breakdown of adjacent membranes of cells starting
  subsequent fusion,
- electrostimulation (PEMIC): activation of transport processes and
  reactions of cell metabolism.

The main targets are cell membranes and membrane bound enzymes and effec-
tors. High electric field pulse change the specific capacity of the membrane by increasing the transmembrane potential from 0.1 V to 1.0 V causing
the more or less reversible breakdown of membrane structure. Low pulsating induced currents may enlarge pores or disturb double layers, dipole
orientation, mediator dissociation etc. thus changing energetic levels in
cells. At the 5th Intern. Symp. of the Bioelectrochemical Society, 1979 at
Weimar the first morphological evidence for electrofusion of plant protoplasts[5,9] and the first genetic evidence for electrofusion of auxotrophic
yeast protoplasts[7,8,21] were presented, followed by the discovery of a
lot of surprising phenomena from other laboratories for single high electric field pulses generated in three equipments as shown in Table 1:

The purpose of this paper is - besides a description of these new
methods - to present a survey of main results in pulse electroporation
(PEP), pulse electrofusion (PEF), and electrostimulation (PEMIC) illustrating the future possibilities and trends, too.

METHODS AND MATERIALS

Concerning the methods for electric field applications partially mentioned in Table 1 the electronic equipments[6] (Figure 1) consist of a pulse generators combined with many fold electrode vessels (chambers) for different purposes (compare Table 1).

a) the macrotechnique with disc electrodes and agglutinating substances,
b) the microtechnique with wire electrodes connected by dielectrophoresis for tight cell collection,
c) the micromanipulator technique with needle electrodes suitable for compression of two big cells as well as for dielectrophoresis,
d) the large scale technique using Helmholtz coils for generation a pulsating electromagnetic induced current (PEMIC) in cell suspensions, tissues and living beings.

The principle of the technique a), b), c) is to establish a higher trans-membrane potential difference (ΔU) in the order of 1 V according to:

$$\Delta U = 1.5 \ r \ E_o \ \cos \theta$$

with r : radius of the cell
  E_o : applied electric field strength
  θ^o : angle between the field lines and the normal of a surface element.

for a thin membrane without conductivity.

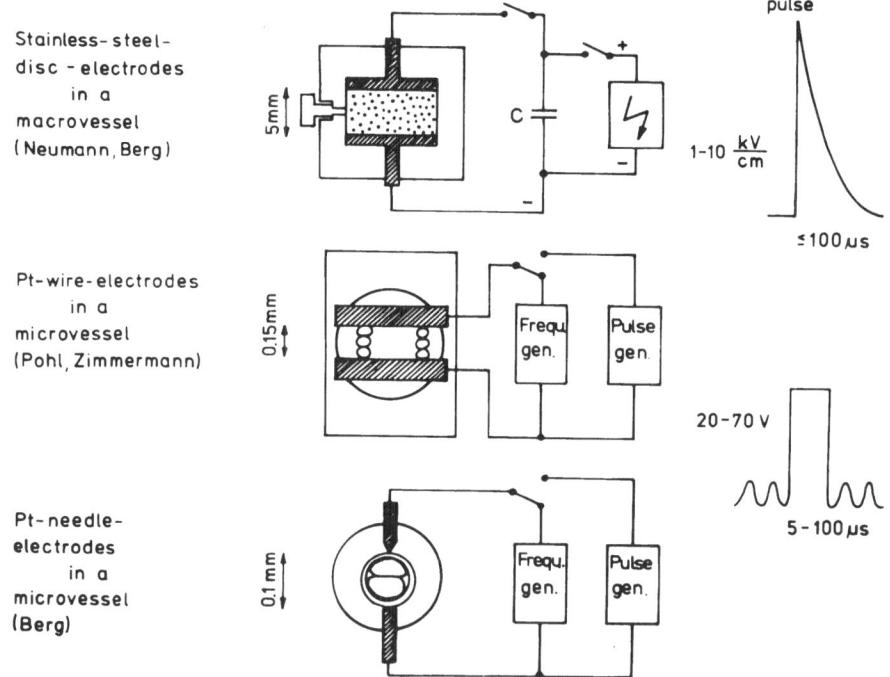

Fig. 1. Electrostimulated fusion techniques shown schematically, vessels and electric circuits: Technique A with stainless steel disc electrodes; technique B with platinum wire electrodes; technique C with platinum needle electrodes; pulse types on the right-hand side.

Table 1. Electro-Fusion and -Transformation

Pulsing in:	Object	Evidence	Group	
Macroscopic Chamber	Yeast Protoplast Mutants	genetic	Berg	1979
	Dictyostelium discoid.	morphol.	Neumann	1980
	Mouse Fibroblasts	physiol.	Tsong	1982
	Bac.thuringiensis	genetic	Berg	1983
	Plasmid DNA → Mouse Cells	genetic	Neumann	1982
	Plasmid DNA → Bac.cereus protopl.	genetic	Berg	1983
Microscopic Chamber after Dielectro-phoresis	Plant Protoplasts, Petunia	morphol.	Zimmermann	1980
	Human Erythrocytes	morphol.	Zimmermann	1980
	Sea Urchin Eggs	physiol.	Zimmermann	1981
	Yeast Protoplast Mutants	genetic	Zimmermann	1982
	Lymphocytes ⎫ Human + Myeloma ⎭ Hybridoma	physiol.	Zimmermann	1982
	Mouse Hybridoma	morphol.	Berg	1982
	Plant Protoplasts, Barley	morphol.	Berg	1983
Microma-nipulator Chamber	Plant Protoplasts, Rauwolfia	morphol.	Senda	1979
	Mouse Blastomers	morphol.	Berg	1982
	L 1210 Ascites Cells	morphol.	Berg	1982

For $\Delta U = 1$ V the breakdown of membrane in field direction occurs more easily the larger r at a given E_o, or in other words, for electrofusion, a higher E_o is needed for small cells[6,7].

Technique A

Disc electrodes of some centimeters in diameter and distance fixed in a cylindrical macrovessel of plexiglass are connected with a high voltage condenser ($U_o \approx 30$ kV) producing an exponential time characteristic of field strength discharge (Figure 1):

$E = E_o \cdot \exp(-t/\tau)$ with

$\tau = RC \sim 5$–100 μs

R – resistance, C – capacity

besides a temperature increase of about 0.5 – 5°C in the cell suspension of 0.5 – 5 ml. For rectangular time characteristics cable discharges can be applied as for a modern temperature-jump apparatus. The macrovessels should be sterilized easily and filled or emptied quickly[6].

Technique B

The microvessel consists of two thin platinum wires fixed on a slide in a distance of about 0.2 mm which are connected with a high frequency alternating current generator (0.5 – 2 MHz, up to 20 V) and the rectangular pulse generator (pulse duration 1 – 300 μs, 1 – 50 V. This combination enables the dielectrophoretic collection of cells like a "pearl necklace" bringing the wires and the microscopic observations of cell motion and sometimes cell rotation. Dielectrophoresis[7] means the movement of particles in a nonuniform alternating electric field to higher field strength by dipole induction (Figure 2). Overcoming the repulsion between membrane surfaces in solutions of low conductivity the cells move into close contact prior to the breakdown effect by pulses of $E = 500$ V/cm. The phenomenon of rotation has been observed for certain resonance frequencies

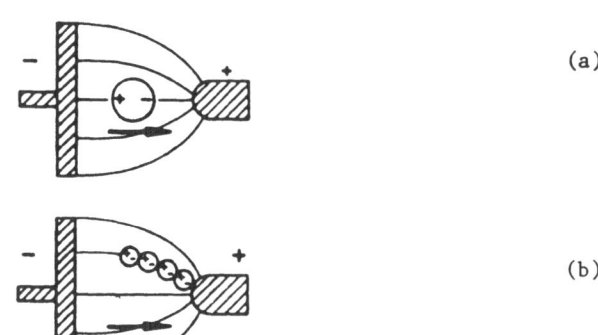

Fig. 2. Scheme of dielectrophoresis of a neutral particle. (a) Induction
of a dipole moving along the field lines to the electrode of
higher field strength. (b) Dipole attraction and formation of
"pearl necklaces" (according to U. Zimmermann, J. Vienken,
J.Membrane Biol., 67:165 (1982).

and therefore it has become a new tool in cell characterization. Unfortunately it disturbs the fusion conditions. In order to get sphere formation
quickly a nutrition medium has to be added during the fusion process.

Technique C

Using a micromanipulator for moving two platinum needle electrodes
(1 - 100 μm in diameter at the end) two larger cells can be contacted by
a slight mechanical pressure[9,11] or by dielectrophoresis[10] which works
for large particles in higher salt medium, too. Moreover suction capillaries are helpful for cell fixation if one of the electrodes should
penetrate the membrane (Figure 3).

Technique D

The equipment for PEMIC was fabricated by Electro Biology Inc.
(Garfield, New Jersey). It consists of a quasi-square wave generator which
proves circular shaped epoxy-embedded Helmholtz coils surrounding a plastic
box or a glass cylinder etc. The current pulses induced in tissue or cell
culture are asymmetrical (Figure 4). The peak current density is about
10 μA cm^2 and consequently the heating effect is only in the order of
10^{-2} °C for which the inner and outer cell membranes are targets of non-
linear behavior. Therefore the most effective wave form must be found out
experimentally for each biological object.

THE ELECTROPORATION OF BIOLOGICAL ACTIVITY SUBSTANCES

The aim of electroporation is to facilate substance penetration
through cell membranes by enlargement of pores, formation of new pores or
reversible lesions. Some of these processes need several minutes after
pulse application for resealing especially at low temperatures (4°C).
Other disturbances are gone within some μs. The kinetics of resealing of
reversible lesions can be followed by conductometry in the μs-range. For
electroporation of pharmaca and genetic material, however, it is more
effective to work in the range of seconds[12,15].

The Electroporation of Pharmaca

It is well known that the toxicity of certain pharmaca can be lowered
by their incorporation into liposomes and also their distribution in the

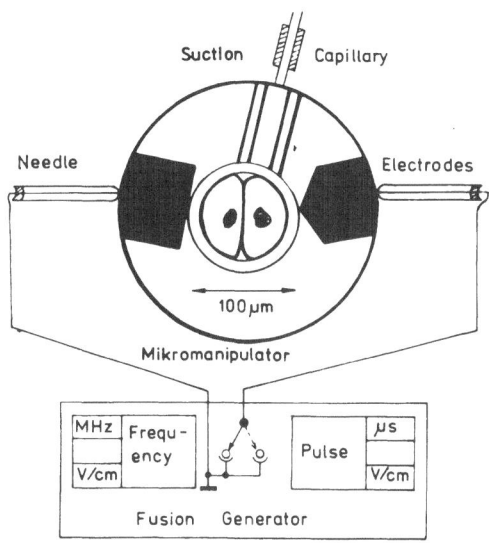

Fig. 3. Details of the electrofusion technique C with the fusion
generator of our Institute. Between the needle electrodes a two
cell blastomer inside the zona pellucida (enlarged) is shown.

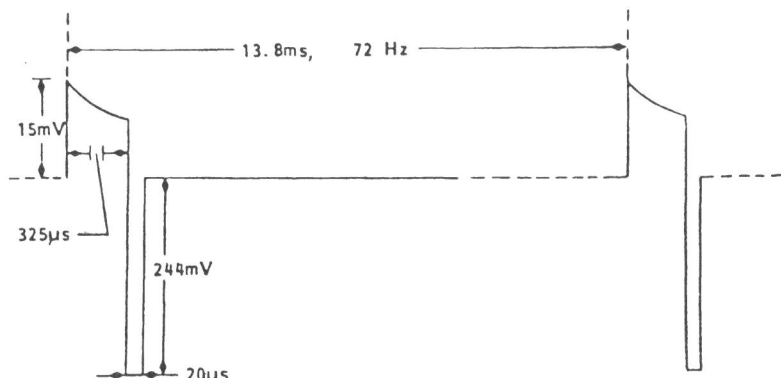

Fig. 4. Current-pulse pattern of PEMIC according to [45].

organism will be changed. The same is possible for electroporation into
appropriate cells, e.g. red blood cells, which was shown with methotrexate,
a cytostatic drug. The erythrocytes have been saturated with this drug
enabled by a 12 kV/cm pulse at 4°C[7]. After injection to mice the concentration of methotrexate in the liver increases about 5 times. Furthermore
attempts are going on to include additionally magnetic particles in order
to bring the drug leaded erythrocytes to distinct regions of the body by
convergent magnetic field forces. Under hemolysis conditions only the
erythrocyte membrane (ghost) may be used, in which case the determination
of drug concentration inside is possible by microspectroscopy. In cultures
of cancer cells it has been found for cytotoxic drugs, e.g. an anthracycline derivate, that the strong membrane barrier for it will be permeable
after a high field pulse[10].

The Transformation with Genetic Material

The transfer of carriers of genetic information of a donor organism
into a receptor organism is nowadays an important manipulation in gene

technology, called transformation of an organism. First of all the isolation of suitable DNA (plasmid DNA) carrying special genetic information (markers) is the prerequisite for this technique. Protoplasts or an animal cell population must be incubated with the genetic material before the field pulse treatment. Afterwards the incubation proceeds on agar containing a substance, e.g. an antibiotic, for selection of such cells transformed showing the new genetic information. An example is the electrotransformation of protoplasts of Bacillus cereus by plasmid DNA of Bacillus thuringiensis carrying the resistance against kanamycine. Applying a pulse of 14 kV/cm (three times) the transformation frequency increases by one order of magnitude ($1.1 \times 10^{-4} \longrightarrow 1.1 \times 10^{-3}$ resistant colonies to protoplasts used). These first electrostimulated transformants in prokaryotes were stable after 30 passages[17]. It should be mentioned that the first electrotransformation for eukaryotes has been performed recently[18].

THE ELECTROFUSION OF MEMBRANES

Fusion of Protoplasts of Yeast

The electric field pulse treatment of a suspension of mixed protoplasts of auxotrophic yeast strains in the presence of PEG and Ca^{2+}-ions strongly enhances the fusion of protoplasts as detected by selection of prototrophic colonies on solid minimal medium. In our experiments the number of prototrophic colonies formed from protoplast suspensions to which an electric field pulse of proper strength had been applied was much higher than from unaffected suspensions (see Table 2). These colonies are the result of complementation of auxotrophic deficiency of the parental strains. The latter are not able to grow and to form colonies on unsupplemented minimal medium. The optimal field strength in the case of intraspecific fusion is about 3 to 5 kV/cm for Sm.lipolytica (Table 2).

An enhancement of the number of colonies on minimal medium by a factor of 20 to 35 was found. For intergeneric fusion high field strengths were necessary for optimal action (Table 2). Here with 10 kV/cm an 80 fold increase of the number of colonies was observed. Individual colonies, isolated from plates of fusion experiments, were maintained over a long period. Hybrids resulting from interspecific fusion have been shown to be very stable, whereas intergeneric fusion products after few passages segregate to one of the auxotrophic parental strains[21].

Table 2. Relative Colony Number from Fused Prototrophic Cells on Minimal Medium in Dependence of E

$E/kV \cdot cm^{-1}$	S.cerev. (ade, his, lys, thr, arg)+ (his, thr, leu)	S.lip. (met, ade)+ (ile, lys)	S.lip+L.el. (arg, ade)+ (ade, his)
0 (control)	1	1	1
1.25	122	1	–
2.5	233	8	–
3.75	50	36	–
5.0	30	10	–
10.0	–	–	78
15.0	–	–	20
20.0	–	–	8

Fusion of Protoplasts of B.thuringiensis

The conventional fusion procedure with PEG as chemical fusogen in the
case of the given strains of B.thurinigiensis did not result in any colony
formation on the selective medium. This means that B.thuringiensis is an
organism for which the usual method of genetic transfer by protoplast is
difficult or, as under the given experimental conditions, fails completely.
Consequently, the additional applications of the electrostimulation technique was highly desirable in order to enable the protoplast fusion, the
frequency of which in the above control experiments was too low to be
detected.

With three successive pulses of 20 kV/cm, 5 µs, we succeeded in
obtaining the desired recombinants, namely colonies which were kanamycine
resistant and were able to form the brown pigment. The fusion frequency,
defined as the ratio of recombinant colonies to regenerated protoplasts,
was found to be 10^{-3} (recombinants/regenerated colonies) under these
electrostimulated conditions, i.e. at least one order of magnitude larger
than that reported in the only paper on PEG-induced protoplast fusion of
strains of B.thoringiensis, were for increasing the efficiency of the
fusion UV irradiation had to be used[19]

Our recombinants were stable even after 30 passages on selective as
well as on non-selective media and they did not segregate into one of the
parental strains. Microscopically it was checked that they were able to
form the typical crystalline protein which is responsible for the toxic
action of B.thuringiensis on insects. As it was pointed out this protein
was not found by plasmid DNA transformation via electroporation.

Fusion of Plant Protoplasts after Dielectrophoretic Collection

In contrast to technique A the pulse effect on the protoplasts collected as "pearl necklaces" can be observed and documented by photos or by
a video camera. The high pulse causes oscillations in field direction
during the first second followed by membrane flattening between the cell
(Figure 5).

The membrane material in this region migrates into the inner parts
during the fusion process. This can be measured two-dimensionally by
changes in the optical density in an image analyzer (Quantimet) and finally
the new membrane surface is lowered by about 26% compared to the surfaces
of both original cells.

The pulse parameters necessary for fusion of a certain kind of protoplasts[6] are shown in Figure 6.

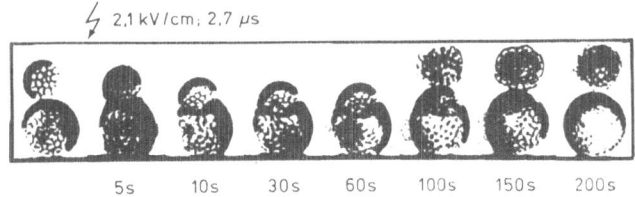

Fig. 5.   Microscopic observation of the fusion process of protoplasts of
          barley (hordeum vulgaris) before and after a single retangular
          pulse.

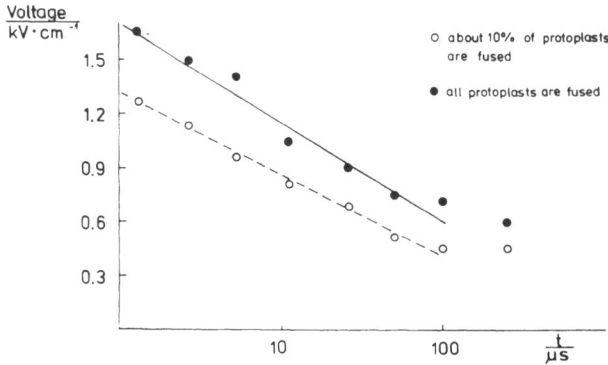

Fig. 6. Relation between field strength E and pulse length for
- all protoplasts are fused, - only 10% of protoplasts are
fused, in the case of barley.

Similar experiments were done for:

- mesophyll protoplasts or vacuoles of Kalanchoe daigremontiona[20]
  exhibiting intermediate vesicles of membrane material,
- Avena sativa[7]
- Saccharomyces yeast[22]
- Petunia inflata[20], Vicia faba[7], Lycopersion esculentum and
  Nicotina tabacum[25], Rauwolfia serpentina[26].

Fusion of animal cells proceeds principally according to the same pathway,
however, some peculiarities have to be taken into account:

- Shortly after the pulse the fusion process should be finished by
  rounding the cells in nutrient medium e.g. leukaemie L1210 cells.
- Red blood cells have to be treated with neuraminidase for removing the
  glycocalix[27], especially in order to get giant cells, consisting of
  up to about 1000 cells, of >100 µm in diameter.
- Before fusion of sea urchin eggs, the vitellin layer has to be destroyed enzymatically (pronase 1 mg/ml)[28]. Fused cells can be fer-
  tilized and will divide. Simultaneously the membrane becomes stabilized by addition of pronase for higher pulse energies[29].
- Optimal conditions have to be ruled out for fusion in monolayer
  cultures of fibroblasts[28].
- Difficulties arise in the case of hydbridoma fusion for antibody
  production from myeloma and spleen cells, because the difference in
  diameter is 3:1 and consequently the field strength for breakdown is
  different, too[31,32].

As a compromise the 0.3 M glucose solution must contain pronase (1 mg/ml)
before the pulse of 3 kV/cm, 50 µs. Five minutes afterwards the nutrition
medium has to be added slowly and for further cultivation the heterocaryons
have to be transmitted into appropriated walls[32]. Recently an electro-
hydraulic procedure for more continuous two-cell production (60-80% 1:1
fusion) has been described[31].

Fusion of Blastomers of Murine and Rat Oocytes, Zygotes etc.

In the case of two blastomers (Figure 7a, b) the electrodes are in
slight touch to the zona pellucida parallel to the adhering membranes
separating both cells. Some minutes after the pulse (4 kV/cm and about
50 µs) the adhering membrane parts are melted and interaction between the

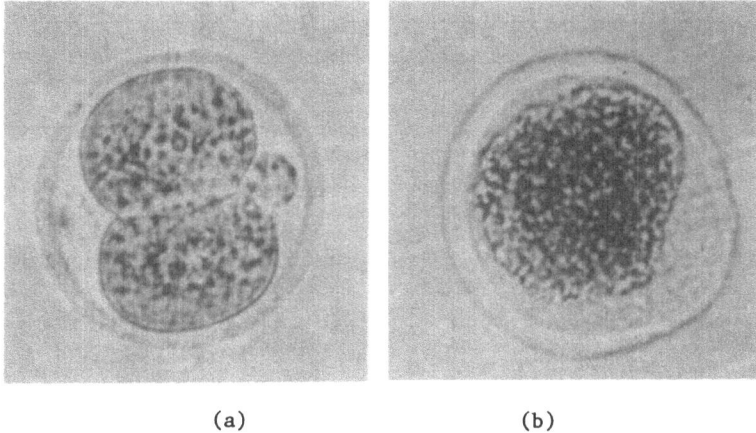

(a)   (b)

Fig. 7. High energy electrofusion of two-cell stage with needle
         electrodes not in touch with the cells: with zona pellucida
         a) before, b) after a pulse of 5 kV/cm 200 μs. (Note:
         electrodes in close contact with the cell surface need lower
         pulse energy e.g. <2 kV/cm for pulse length >100 μs).

nuclei may occur[10,11]. Higher pulse energy causes disruption of the
zona pellucida and destruction of the protoplasma observable as granul-
ation. On the contrary, the main aim is to obtain a high viability rate
of the fused products, tested by the subsequent ability of cell division
either within an incubator or after implantation[33]. In order to obtain
at least one division in the incubator a lowered field strength <2 kV/cm
applied during a pulse width >100 μs is suitable. The fused zygote implanted in a mouse was found alive after 11 days, however, this test is
still in progress.

For the purpose of fusion of genetically different zygotes in the
early stages of embryonic development the zona pellucida have to be removed
by pronase treatment. In contact with the nutrient medium after fusion the
membrane forms spheres quickly. Consequently fusion is now possible with
lower pulse energies, however, the naturally adherent membranes fuse
firstly followed by the outer artificially contacted membranes compressed
only by the tops of both needle electrodes. Therefore it is more difficult
to find out the optimal pulse parameters. On the other hand there may be
similarities with sea urchin eggs behavior[31].

THE ELECTROSTIMULATION OF METABOLISM PATHWAYS

The PEMIC application has been used widely causing a lot of surprising
effects quite different in nature as can be seen from Table 3. Starting
with basic studies of electrical potentials in skeletal tissues[23] in the
sixties it is nowadays a successful clinical therapy for the repair of
pseudoarthroses and in fracture healing[24] besides the techniques with
surgically-implanted electrodes[23].

The further types of applications (Table 3) are more or less primary
results of laboratory experiments showing first possibilities for biotechnological purposes. Anyway this will be the beginning of a new era of
electromagnetic field effects of importance in the future. (Table 3).

The nature of PEMIC effects have been addressed by many investigations, especially Pilla[55], taking into account the facilitation of ion

Table 3. Biological Effects of PEMIC

- Therapy : pseudoarthrosis	about 80% healing[23,24]
- Therapy : cancer	murine ascites tumor in combination with cytostatic agents increases life span up to 50% [41]
- Regeneration : amphibians	acceleration of limb healing[42]
- Permeation : membranes	enhancement of transport of Na^+, K^+ [43,44]
- Differentation : nerves	stimulation of neurite growth regeneration etc.[45,46,47,50]
- Dedifferentation : erythrocytes	of amphibian to juvenile structure[48,49]
- Stimulation of metabolism : rat embryos:	DNA + synthesis (enhancement of 3H-thymidin incorporation)[44,49]
rat skin:	protein synthesis[51]
bone cells:	ATP, cAMP activation[30,52]
microorganism:	influence of phosphate consumption[53]
enzymes:	kinetic changes[54]

transport across membranes, with resultant activation of a variety of enzyme systems and altered biochemical activity by energy uptake (excitation) on the reactive sites. However, there must be a lot of work done for elucidation of the various kinds of mechanisms.

DISCUSSION

Using the three techniques described one might assume that they are suitable for all kinds of membrane fusion, however, there are two principal difficulties: at first the relative broad range of the break-down potential of cells of different size and membrane ingredients. Since the cell diameter in our examples varies between 10 and 100 μm, the fusion to hybridoma for instance requires the admixture of membrane stabilizing substances for the larger cells (and vice versa), otherwise the myeloma cells may undergo lysis.

Secondly higher pulse amplitudes may cause damage inside the cell and can decrease its viability. From this point of view the admixture of tensides labilizing the membrane stability is useful in order to decrease the pulse energy. In this respect some polymers (PEG, PVA) have been tested successfully[35]. Concerning the influence of the alternating current on membranes causing dielectrophoresis of animal cells (method B) some similarities seem to exist to the pulsating electromagnetic induced current (PEMIC) according to Pilla[36].

The information system of the cell will be changed and processes will be accelerated comparable with the effects of light excitation or heating. In future closer relations between both methods will be found and their future development will be as fast as that of magnetic field effects and nuclear magnetic reasonance on living beings. Some trends of electroporation and electrofusion are recognizable and even in progress:

- application of a series of (weaker) pulses instead of a strong one,

- after cell fusion furthermore fusion of eukaryotic nuclei, too,
- large scale continuous fusion, e.g. for hybridona production,
- 1:1 fusion of different cells different in diameter, too,
- effective separation of fused products from the bulk of the cell suspension,
- enhancement of the viability (division) of fused cells by the synergism of a lower field strength with membrane melting tensides,
- cultivation of fused cells to whole new plants and animals.

To realize these aims it is necessary to develop appropriate models. Finally some current ideas will be described briefly.

With respect to complex biological membranes electric field-induced conformational changes of macromolecular structures and different orientation effects on the various molecular constituents must be taken into account. Such effects have been discussed in several theoretical papers[37,38,39]. Conformational changes of, e.g. embedded lipoproteins can be brought about by the primary action of the electric field on charges and dipoles present, favoring configurations associated with larger overall dipole moments.

An attempt to indicate the variety of such molecular events which will all contribute to the interesting electric field effects on membranes described in this survey is shown schematically in Figures 8 and 9.

The electroporation (Figure 8) for ions, drugs of low molecular weight and biopolymers is facilitated by the enlargement of pores by repulsion of induced protein dipoles etc. or by local destruction of the lipid bilayer. At higher pulse energies proteins can be oriented suddenly into the field direction causing also destruction and intermingling of the adjacent membranes of two or more cells starting the fusion process (Figure 9).

Despite the several lines of explanations given here, there are today more open problems and unexplained phenomena in this fast growing field of bioelectrochemistry, which will evolve to a new powerful tool in gene- and biotechnology as well as in medicine.

CONCLUSIONS

Mainly in the last ten years a rapid evolution of electric and magnetic field effects has occurred in biology and medicine. Nowadays high single pulses are used on the one hand and pulsating currents on the other

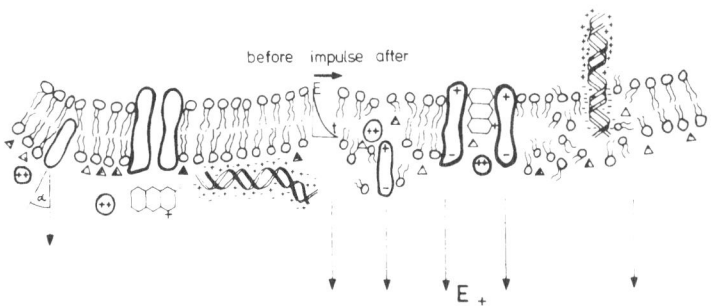

before impulse after

E +

Fig. 8. A membrane before the pulse with outside concentrations of not easy permeable cations, drugs, DNA etc. and after the pulse showing schematically the enlargement of pores, the disturbance and the breakdown of bilayer for electroporation.

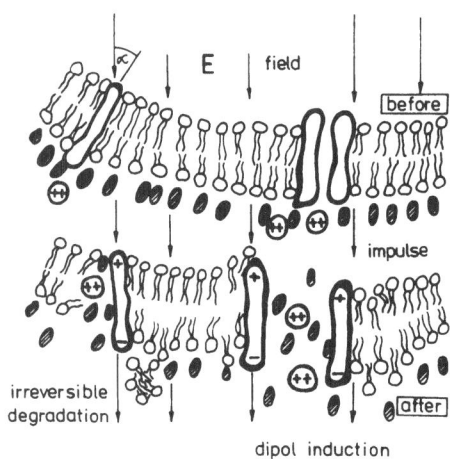

Fig. 9. Breakdown of membrane by sudden protein orientation and movement in the field direction before electrofusion of cells can occur.

linked by the dielectrophoretic effects and modified by chemical agents. The main targets are membranes and biopolymers and the responses are irreversible degradation or reversible conformational transitions causing a changing biochemical activity. In this way we now have a powerful tool for interference in the processes of living beings[29].

REFERENCES

1. J. W. Ritter, "Beweiß, daß ein ständiger Galvanismus den Lebensprozeß im Thierreich begleitet," Weimar (1798).

2. E. Du Bois-Reymond, "Untersuchungen über thierische Elektrizität," G. Reimer, Berlin (1848).

3. H. Berg, Historical roots of bioelectrochemistry, Experientia, 36:1247-1249 (1980).

4. R. Pethig, "Dielectric and Electronic Properties of Biological Materials," John Wiley, New York (1979).

5. M. Senda, General Discussion of the 5th Intern. Symp. Bioelectrochem. Bionerg., Weimar, Sept. 3-9 (1979).

6. H. Berg, K. Augsten, E. Bauer, W. Förster, H. -E. Jacob, P. Mühlig, and H. Weber, Possibilities of cell fusion and transformation by electrostimulation, Bioelectrochem.Bioenerg., 12:119 (1984).

7. U. Zimmermann, Electric field-mediated fusion and related electrical phenomena, Biochim.Biophys.Acta, 694:227-277 (1982).

8. H. -E. Jacob, W. Förster, and H. Berg, Microbiological implications of electric field effects. II. Inactivation of yeast cells and repair of their cell envelope, Z.Allgem.Mikrobiol., 21:225-233 (1981).

9. M. Senda, J. Takedo, Sh. Abe, and T. Nakamura, Induction of cell fusion of plant protoplasts by electrical stimulation, Plant Cell Physiol., 20:144 (1979).

10. H. Berg, Biological implications of electric field effects. V. Fusion of blastomeres and blastocysts of mouse embryos, Bioelectrochem. Bioenerg., 9:223-228 (1982).

11. H. Berg, A. Kurischko, and R. Freund, Biological implications of electric field effects. VI. Fusion of mouse blastomeres without and within zona pellucida, Studia biophysica, 94:103-104 (1983).

12. E. Neumann and K. Rosenheck, Permeability changes induced by electric impulses in vesicular membranes, J.Membr.Biol., 10:279-290 (1972).

13. U. Zimmermann, J. Schulz, and G. Pilwat, Transcullular ion flow in Escherichia coli.B and electrical sizing of bacterias, Biophys.J., 13:1005-1013 (1973).

14. K. Kinosita and T. Tsong, Voltage-induced pore formation and hemolysis of human erythrocytes, Biochim.Biophys.Acta, 471:227-242 (1977).

15. P. Lindner, E. Neumann, and K. Rosenheck, Kinetics of permeability changes induced by electric impulses in chromaffin granules, J.Membr.Biol., 32:231-254 (1977).

16. P. Mühlig, W. Förster, H. -E. Jacob, and H. Berg, Cell membrane permeation of the anthracycline violamycin BI induced by an electric field pulse, Poster on the X Jena Symp., Molecular-biological Mechanisms of Antitumor Antibiotics Actions, Weimar (1984), Studia biophysica, (1985) in preparation.

17. N. Shivarova, W. Förster, H. -E. Jacob and R. Grigorova, Micro-biological implications of electric field effects. VII. Stimulation of plasmid transformation of Bacillus cereus protoplasts by electric field pulses, Z.Allg.Mikrobiol., 23:595-599 (1983).

18. E. Neumann, M. Schaefer-Ridder, Y. Wang, and P. H. Hofschneider, Gene transfer into mouse lyoma cells by electroporation in high electric fields, The EMBO Journal, 1:841-845 (1982).

19. N. Shivarova, W. Förster, H. -E. Jacob, and R. Grigorova, Z.Allgem. Mikrobiol., 23:595 (1983).

20. U. Zimmermann and J. Vienken, in: "Cell Fusion, Gene Transfer and Transformation," R. Beers and E. Bassett, eds., p.171, Ravens Press, New York (1984).

21. H. Weber, W. Förster, H. Berg, and H. -E. Jacob, Parasexual hybrid-ization of yeasts by electric field stimulated fusion of proto-plasts, Current Genetics, 4:165 (1981).

22. H. J. Halfmann, C. C. Emeis, and U. Zimmermann, Electro-fusion and genetic analysis of fusion products of haploid and polyploid Saccharomyces yeast cells, FEMS Microbiology Letters, 20:13-16 (1983).

23. C. A. L. Bassett, A. A. Pilla, and R. J. Pawluk, A nonoperative salvage of surgically resistant pseudarthroses and nonunions by pulsing electromagnetic fields, Clin.Orthop., 124:128 (1977).

24. A. Pilla, Bioelectrochemistry, ions, surfaces, membranes, Adv.Chem., 188:126 (1970).

25. H. -E. Jacob, F. Siegemund, E. Bauer, and P. Mühlig, Fusion of plant protoplasts by dielectrophoresis and electric field pulse technique, Studia biophysica, 94:99-100 (1983).

26. M. Senda, H. Morikawa, and J. Takeda, Proc. 5th Internat. Congr. Plant. Tissues Cult., p.615 (1982).

27. G. Pilwat, H. -P. Richter, and U. Zimmermann, Giant culture cells by electric field-induced fusion, FEBS Letters, 133:169 (1981).

28. J. Teissie, U. P. Knutson, T. Y. Tsong, and M. D. Lane, Electric pulse-induced of 3T3 cells in monolayer culture, Science, 216:537-538 (1982).

29. U. Zimmermann and G. Küppers, Cell fusion by electromagnetic waves and its possible relevance for evolution, Naturwiss., 68:577 (1981).

30. J. Teissie, B. E. Knox, T. Y. Tsong, and J. Wehrle, Synethesis of adenosine triphosphate in respiration-inhibited submitochrondria particles induced by microsecond electric pulses, Proc.Natl.Acad. Sci.USA, 78:7473-7477 (1081).

31. H. -P. Richter, P. Scheurich, and U. Zimmermann, Electric field-induced of sea urchin eggs, Develop.Growth and Differ., 23:479 (1981).

32. D. Berg, I. Schumann, and A. Stelzner, Electrically stimulated fusion between myeloma cells and spleen cells, Studia biophysica, 94:101-102 (1983).

33. A. Kurischko and H. Berg, 2nd Seminar. Electrostimulated Cell Fusion and Transformation, Jena, Sept. (1984).

34. S. Walliser and K. Redmann, Membrane polarization as a candidate for signalling altered cell functions, Studia biophysica, 94:105-106 (1983).

35. M. Senda, Discussion at the UNESCO Forum Electrochemistry in Research and Development, Paris, June 1984.

36. A. A. Pilla, P. Sechaud, and B. McLeod, Electrochemical and electrical aspects of low-frequency electromagnetic current induction in biological systems, J.Biol.Physics, 11:51 (1983).

37. G. Schwarz, On the physico-chemical basis of voltage-dependent molecular gating mechanisms in biological membranes, J.Membrane Biol., 43:127-148 (1978).

38. E. Neumann, Electric field effects in biopolymer structures and electrical-chemical memory recording, in: "Ions in Macromolecular and Biological Systems," D. H. Everett and B. Vincent, eds., Scientechnica, Bristol (1978).

39. S. I. Sukharev, L. V. Chernomordik, I. G. Abidor, and Yu. A. Chizmadzhev, 466-Effects of UO_2^{2+} ions on the properties of bilayer lipid membranes, Bioelectrochem.Bioenerg., 9:133-140 (1982).

40. J. Weaver and R. Mintzer, Decreased bilayer stability due to transmembrane potentials, Physics Letters, 86A:57 (1981).

41. S. D. Smith and J. M. Feola, Effect of repetition rate and duty cycle on pulsed magnetic field modulation of LSA tumors in mice, J.Electrochem.Soc., 130:1210 (1983).

42. S. D. Smith and A. A. Pilla, Modulation of new limb regeneration by electromagnetically induced low level pulsating current, in: "Mechanisms of Growth Control," R. O. Becker, ed., Thomas Springfield, 11:137 (1981).

43. A. A. Pilla, Electrochemical information transfer and its possible role in the control of cell function, in: "Electrical Properties of Bone and Cartilage," C. T. Brighton, J. Black, and S. R. Pollack, eds., Grune and Stratton, New York, p.455 (1979).

44. L. A. Norton, L. A. Bourett, R. J. Majeska, and G. A. Rodan, Adherence and DNA synthesis changes in hard tissues cell culture produced by electric perturbation, in: "Electrical Properties of Bone and Cartilage," C. T. Brighton, J. Black, and S. R. Pollack, eds., Grune and Stratton, New York, p.443 (1979).

45. B. Sisken, B. McLeod, and A. A. Pilla, PMEF, DC and neuronal regeneration: Effect of electric field geometry, in: "Electrochemistry, Membranes, Cells and the Electrochemical Modulation of Cell and Tissue Function," A. A. Pilla and A. Boynton, eds., Springer Verlag (1984).

46. M. Schwartz and D. Neumann, Neuritic outgrowth from regenerative goldfish retina is affected by pulsed electromagnetic fields, Trans. Bioelectr.Repair and Growth Soc., 1:55 (1981).

47. A. A. Pilla, Electrochemical information transfer at living cell membranes, Ann.NY.Acad.Sci., 238:149 (1974).

48. A. A. Pilla, Mechanism of electrochemical phenomenon in tissue repair and growth, Bioelectrochem.Bioenerg., 1:227 (1974).

49. A. Chiabrera, M. Hinsenkamp, A. A. Pilla, J. Ryaby, D. Ponta, A. Belmont, F. Beltrame, M. Grattarola, and C. Nicolini, Cytofluorometry of electromagnetically controlled cell differentiation, J.Histochem.Cytochem., 27:375 (1979).

50. H. Murray, W. J. O'Brien, and M. Oregel, Pulsed electromagnetic fields and peripheral nerve regeneration in the cat, Anat.Rec., 205:137A (1983).

51. P. H. Delpert, N. Cheng, M. J. Hoogmartens, J. C. Mulier, W. Sansen, and W. De Loecker, The effects of pulsed electromagnetic fields on metabolism in rat skin, 3rd Annual BRAGGS, San Francisco, California, p.67, Oct. (1983).

52. R. Korenstein, D. Somjen, F. Laub, H. Fischler, and Y. Binderman, Electric stimulation of bone cells in culture, VII. Intern. Symp. on Bioelectrochem., Stuttgart, July (1983).
53. H. P. Große and E. Bauer, unpublished results.
54. E. Selegny, J. M. Valleton, and J. C. Vincent, Monitoring of mono-stable systems and memory in multistable systems, Bioelectrochem. Bioenerg., 10:133 (1983).
55. A. A. Pilla, The rate modulation of cell and tissue function via electrochemical information, in: "Mechanisms of Growth Control," R. O. Becker, ed., Charles C. Thomas, Springfield, p.211 (1981).

DISCUSSION

QUESTIONS AND REMARKS

H. W. Nürnberg: The transmembrane potential imposed by the external field pulse is in the order of the natural potential difference in the double layer!

A. Bard: What are the relations between the single pulses and PEMIC? (See answer in the text).

E. Gileadi: What is the relation between the necessary fusion field strength and the cell diameter? (See formula page 229).

M. Senda
Kyoto University, Japan (communicated):

I would like to make a comment about some interesting features of the electrofusion, which were observed with plant protoplasts[1,2].

Stimulation Strength-duration Relationship

When an electric field was applied for a short duration (Δt) to point-adherent (agglutinated) protoplasts by passing an electric current pulse (I) through protoplast suspension, fusion was immediately induced. The susceptibility of protoplasts to electrofusion can be evaluated by the product of $I \times \Delta t$ (at larger I) to attain a specified fusion percentage, and the relationship between electric impulse strength and duration can be quantitatively expressed by Weiss' formula (or its modified forms) $I - I_r = a/\Delta t$, where I_r and a/I_r are the rheobase and chronaxie, on the strength-duration relationship in biological stimulation.

Decay of Fusion-susceptible State of Membrane

When each of two point-adherent protoplasts was stimulated separately and successively with a time interval t_{ab}, the percentage of fusion decreased with increasing t_{ab} and eventually no fusion was induced if, for instance, $t_{ab} > 30s$ for R.serpentina protoplasts. This result indicates that the fusion-susceptible state, to which the membrane or protoplast is transferred by electric impulse, decays within about 30 s in this case.

Effect of Polymers

When a dilute amount (a few percent) of certain polymers, such as polyethylene oxide, polyvinyl alcohol, polyvinyl pyrrolidone or dextran is added the $Ix\Delta t$ values were greatly reduced. Also, addition of polymers increased the fusion percentage.

REFERENCES

1. M. Senda, H. Morikawa, and J. Takeda, Proc. 5th Intl. Cong. Plant Tissue and Cell Culture, p.615 (1982).
2. M. Senda, H. Morikawa, and J. Takeda, SEIBUTSU BUTSURI (Biophysics) 22:198 (1982).

SECTION IV
ELECTROCHEMISTRY IN TECHNOLOGY

POTENTIALITIES OF NEW TECHNOLOGIES

E. B. Budevski

Central Laboratory of Electrochemical Power Sources
Bulgarian Academy of Science
Sofia 1040, Bulgaria

Exactly one year ago I attended a Workshop in Rolla, Missouri, organized by the National Science Foundation and the Bureau of Mines of the US Department of the Interior, devoted to the Research Needs in Electrochemistry for mineral and Primary Material processing. In the Workshop the present situation of Electrochemical Technologies were discussed, as well as what we can expect from electrochemistry for an improvement of these technologies. I feel that a discussion, based partially on some of the conclusions of this Workshop, would serve as a good introduction to the deliberations scheduled by the organizers of the Forum for this session.

In a world of ever increasing demands for larger quantities and better quality by an uncompromizing consumer society on the one hand and the inexhaustible energy and material resources on the other, the problem of efficiency and quality cannot be overestimated.

Both terms, quality and efficiency, have to be understood in a very broad sense.

Quality includes not only the performance characteristics of a product but also in general new products having an impact on the quality of life. In a competitive world of information exchange, quality change and development can hardly be controlled or predicted.

As "efficiency" we have to understand not necessarily the profit but much more the efficiency of use of energy and raw materials, as well as the problems connected with worthless byproducts and wastes contributing significantly to the contamination of the environment. With respect to energy and raw material efficiency, electrochemistry offers unique possibilities which in very few cases have been realized to a satisfactory extent. With respect to waste and environmental pollution electrochemical gross technologies are especially dangerous, but in many cases have been brought under complete control.

Electrochemical technologies having a high impact on the world goods production are not so many. Table 1 shows the percentage of the production by electrolysis of several of the most important metals and non-metallic materials and the distribution of electric power and energy consumption in the production process of these materials. It is evident that the main

247

interests must be and are in reality directed towards metal electrowinning, especially aluminium and to chlor-alkali electrolysis.

In the metal electrowinning two problems can be identified. The first one is connected mainly with the chemistry of the process and the second primarily with cell construction and design. The classical parallel plate cell construction suffers from several drawbacks: high capital and operating costs due to low productivity per unit cell volume, high electric energy costs and intensive labor activities, as discussed in the Rolla Workshop[1]

There are not many improvements that can be introduced aiming at a current density, i.e. productivity, increase. Although a number of techniques have been studied and suggested in the last decades, very few of them have found their way to the production lines. These techniques involve either the time profile of the current, i.e. use of periodic current reversal[2,3], or the agitation of the electrolyte, e.g. use of the convection action of the anodically generated oxygen[4], forced electrolyte circulation[5,6] including high speed electrolyte flow parallel to the electrodes[7,8], ultrasonic agitation[9], etc.

A reduction of the total energy consumption can be achieved by the introduction of new electrolytes, particularly chlorides, or novel anode electrochemistry, e.g. electroleaching as used in the Dextec process, use of low energy anodes such as metal sulfides or fuel cell electrodes, etc.

A fuel powered fuel cell electrode can be used to replace the gas evolving anode in a number of processes decreasing the energy consumption significantly. Hydrogen, methanol or sulfur dioxide have been studied as fuels[10] Hydrogen and sulfur dioxide deserve a special attention. In a hydrogen economy this gas can easily be used to save electric energy in metal recovery plants. Figure 1 shows polarization characteristics of hydrogen gas diffusion electrodes consisting tungsten carbide as catalyst in $ZnSO_4$ solutions as used in the zinc recovery cells[11]. As seen the

Table 1. Percentage of the Production by Electrolysis, out of the Total, of some Materials and Electric Power and Energy Distribution for their Electrowinning

Material	Percentage of electrolytic production	Percentage of production power	Percentage of production energy*
Metals			
Al	100	13	58
Cu	85	12	1
Zn	60	8	5
Pb	10	6	1
Total metal-winning		39	64
Non-metals			
Cl_2	100	27	26
NaOH	100	31	
Total chlorine-caustic		59	26
Others		3	10

*Out of total of roughly 200.10^{12} Wh annual energy consumption for electro-chemical technologies.

energy saved corresponding to roughly 1 V decrease of the anode potential, can be expected to be significant. The main problem is connected with the life expectancy of this electrode. Figure 2 shows the recent progress made in the field.

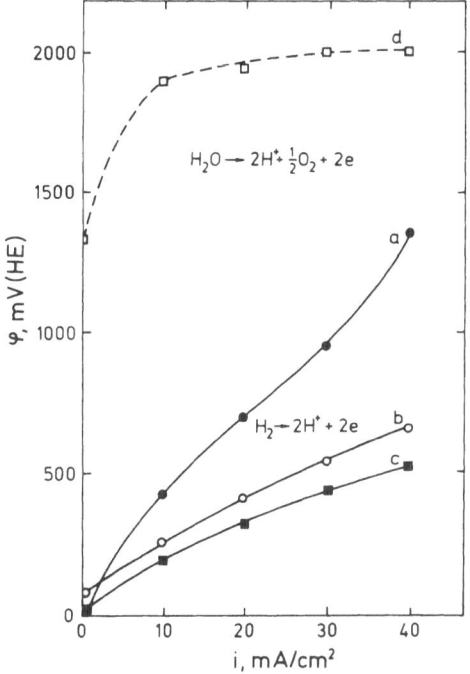

Fig. 1. Polarization curves of gas diffusion H_2-electrodes catalyzed by WC in a $ZnSO_4$ solution (Zn 130 g/l, H_2SO_4 30 g/l at 40°C). (a) 60 mg WC/cm^2; (b) 180 mg WC/cm^2; (c) 300 mg WC/cm^2; (d) Pb($\sim$1% Ag) electrode in the same solution.

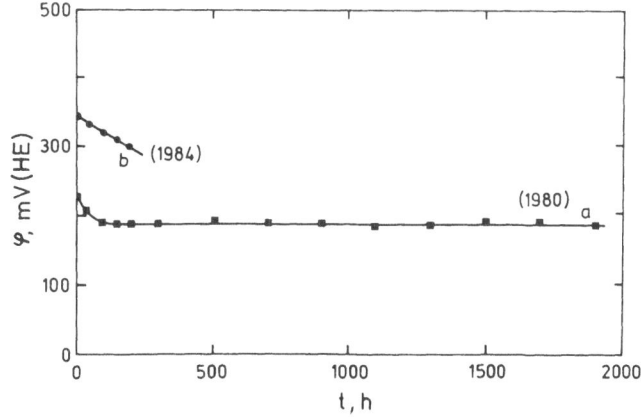

Fig. 2. Potential-time (life test) curves of WC-catalyzed H_2-gas diffusion electrodes. (a) 180 mg WC/cm^2, solution 240 g/l H_2SO_4, 60°C at 50 mA/cm^2; (b) 300 mg WC/cm^2, solution 30 g/l H_2SO_4, 130 g/l Zn, 40°C at 30 mA/cm^2.

Sulfur dioxide is also a useful depolarization. Figure 3 shows performance characteristics of sulfur dioxide gas diffusion electrodes[12]. The sulfur dioxide electrodes were developed for the sulfuric acid hydrogen production cycle. With respect to an oxygen evolving anode the decrease of electrode potential is significant especially when iodine is used as mediator. Life expectancy of the electrodes as well as the removal of the continuously produced sulfuric acid are still unresolved problems of this process.

A number of novel cell and electrode designs have been suggested in recent years. Only a few of them can be mentioned and shortly discussed here in general following the listing given by Flett and Evans in their paper presented at the Rolla Workshop[1]:

- The Duval's CLEAR cell design producing copper powder from cuprous chloride solutions[13,14].

- The Swiss-Roll cell design introduced by Robertson, Scholder, Theis and Ibl[15] for dilute solutions. This cell construction is characterized by an extremely high electrode surface to cell volume ratio amounting to 200 cm² per cm³. Additionally turbulence promoters are used in the cell to increase the current density.

- The Chemelec cell[16] using mesh electrodes in a nonconducting fluidized bed compartment, with continuously upwards flowing electrolyte.

- The fluidized bed electrode invented by Fleischmann et al.[17] has found a recent development in works by Evans and coworkers[1,18]. This development eliminates some early restrictions of the original design by introduction of a smooth diaphragm, appropriate particle size, bed expansion, etc. At anodic current densities of up to 5000 A m$_{-2}$ these authors claim an energy consumption as low as 1 kWh/kg copper[1]

High temperature electrolysis in molten salts is confined primarily to electronegative high reactive and refractory metal winning. The most important process is evidently the high energy consuming aluminium elec-

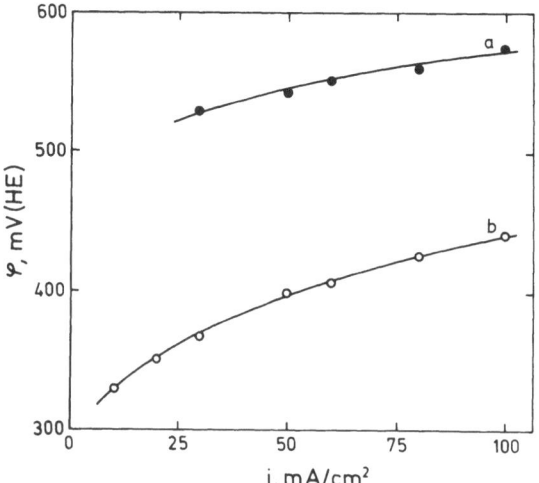

Fig. 3. Polarization curves of SO_2-gas diffusion electrodes with active carbon as a catalyst. (a) 20 wt% H_2SO_4 at 25°C[12]; (b) 20 wt% H_2SO_4 + 1.7 wt% KI at 25°C (K. Petrov, I. Nikolov, T. Vitanov, private communication).

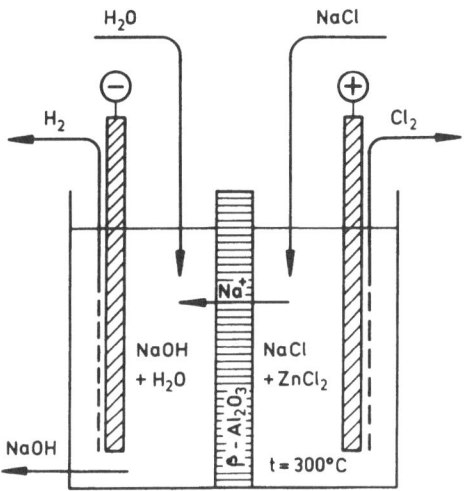

Fig. 4. Principle of operation of a chlor-alkali electrolytic cell with a β-alumina solid electrolyte separator.

trolysis. It consumes more than 50% of the energy used for electrochemical technologies - plating excluded. As discussed at the Rolla Workshop[19-21] progress in the field is expected to come by the introduction of non-reactive cathodes, nonconsumable anodes and, as an ultimate goal by the introduction of bipolar cell designs. The road to it may be long and requires patience. Otherwise, molten salt electrometallurgy seems to be progressing well. More than 30 metals and metal products have been prepared by molten salt electrorefining and electrowinning. Many of these materials have unique properties which are required in the development of rockets, space craft, atomic reactors and high temperature specialty alloys. Referring to Lei et al.[22] we can mention among them: electro-refining of titanium and vanadium, electrolytic separation of nickel and cobalt from their alloys, electrowinning of yttrium-aluminium alloys, electrowinning of zirconium, electrowinning of tungsten, etc. Due to the unique size and shape of electrowon tungsten particles, new technologies were developed for producing high density sintered material.

Another industry which deserves attention is the chlorine caustic electrolytic production with a consumption of ca. 26% of the energy spent for electrochemical processes. Innovations can be found in the replacement of the hydrogen evolving cathode with a fuel cell type oxygen (air) electrode. Modern air electrode technologies based on teflon-bonded gas diffusion layers as developed in our Laboratory (Iliev[123]) could be modified to the hard working conditions of the chlor-alkali cell. How much sense it makes to save energy by not producing hydrogen in the face of a coming hydrogen economy is another question.

To come to the end I would like to mention a revolutionary new process for chlor-alkali electrolysis which I saw working at a larger Laboratory cell in Japan. The process uses the sodium conducting solid electrolyte β-aluminium as a separator of the anodic and cathodic compartments[24]. The principle of operation is illustrated on Figure 4. The fast ion conductor β-aluminium has been developed as the basic component of the sodium sulfur battery cell. Whether it will give birth to a new technological process is too early to predict.

REFERENCES

1. D. S. Flett and J. W. Evans, Report of Workshop, in: "Electrochemistry Research Needs for Mineral and Primary Materials Processing," T.J. O'Keefe and J.W. Evans, eds., p.39, Rolla, Missouri, (1983).
2. D. A. Petrov, L. T. Lachev, and Y. D. Popov, Method for electrolytic refining of copper at higher current density, Bulg.pat.No. 10188 (1962).
3. R. G. Bautista and D. S. Flett, Warren Spring Laboratory, Report No. LR 221 (ME) (1976).
4. V. A. Ettel and A. S. Gendron, Chem.Ind.London, 9:376 (1975).
5. T. Balberyszski and A. K. Andersen, Proc.Aust.Inst.Min.Met., 244:11 (1972).
6. P. T. W. Stub and E. J. Clugston, jr., TMS paper selection AIME, New York, Paper No. A 72:76 (1972).
7. L. S. Krushev, Jahrbuch des Wissensch. Inst. fur Technologie des Maschienenbaues, ZNIIT Masch, Sofia, 4:11 (1969).
8. S. J. Wallden, S. T. Henriksson, P. G. Arbstedt, and Th. Mioen, J.Metals, 11:528 (1959).
9. G. Eggett, W. R. Hopkin, J. W. Garlick, and M. J. Ashley, in: "I. Chem. E. Symposium Series No. 42, Hydrometallurgy," G.A. Davies and J.B. Scuffham, eds., London (1975).
10. P. Duby, Report of Workshop "Electrochemistry Research Needs, etc.," T.J. O'Keefe and J.W. Evans, eds., Rolla, Missouri, p.91 (1983).
11. T. Vitanov, I. Nikolov, and V. Nikolova, private communication.
12. K. Petrov, I. Nikolov, and T. Vitanov, Int.J.Hydrogen Energy, in press.
13. G. E. Atwood and R. W. Livingston, Erzmetall., 33:251 (1980).
14. M. S. Cook and G. E. Atwood, US Pat. 4,025,400 (1977).
15. P. M. Robertson, B. Scholder, G. Thies, and N. Ibl, Chem.Ind.London, 13:459 (1978).
16. C. L. Lopez-Cacicedo, Trans.Inst.Met.Finish., 53:75 (1975).
17. M. Fleischmann, F. Goodridge, and J. M. Backhurst, UK Pat. 1,194,181 (1970).
18. M. Dubrovsky and J. W. Evans, Metall.Trans., B,13B:293 (1982).
19. N. Jarrett, Report of Workshop, "Electrochemistry Research Needs, etc.," T.J. O'Keefe and J.W. Evans, eds., Rolla, Missouri, p.117 (1983).
20. R. Keller, Report of Workshop, "Electrochemistry Research Needs, etc.," T.J. O'Keefe and J.W. Evans, eds., Rolla, Missouri, p.139 (1983).
21. J. Thonstad, Report of Workshop, "Electrochemistry Research Needs, etc.," T.J. O'Keefe and J.W. Evans, eds., Rolla, Missouri, p.145 (1983).
22. K. P. V. Lei, E. Morrice, J. E. Murphy, and G. M. Matinez, Report of Workshop, "Electrochemistry Research Needs, etc.," T.J. O'Keefe and J.W. Evans, eds., Rolla, Missouri, p.167 (1983).
23. I. Iliev, Bull.Soc.Chim.Beograd., 48(Suppl.):317 (1983).
24. Y. Ito and S. Yoshizawa, "Advances in Molten Salt Chemistry," G. Mamantov and J. Braunstein, eds., Plenum, New York, p.319 (1982).

INDUSTRIAL DYNAMICS INVOLVING ELECTROCHEMICAL PROCESSES

G. Silvestri

Istituto di Ingegneria Chimica
Universita di Palermo
Italy

INTRODUCTION

Many industrial activities are based on electrochemical technologies, and several other receive from electrochemical methodologies useful or even decisive contributions for their development. There are important electro-chemical industries in the fields of inorganic and organic chemical pro-duction, the treatment of metals, such as extraction, refining, finishing and processing, and the production of the various battery systems. Furthermore some electrochemical technologies are dependent on, and inte-grated with other industrial activities, and are related to corrosion control, water treatments, and environmental protection. Like physicians with their patients, these last applications have a prosperous future, based on the problems raised by the industrial activities to which they are connected! Many of the above listed branches are very active, and hold out promising prospects of application. In the introduction to his recent book on industrial electrochemistry, Pletcher[1] emphasizes the brightness of the future of electrochemical technologies, and the favorable expectations for their wider applications, despite the still scarce diffusion of an electrochemical knowledge in the industrial chemical world.

Before entering, with the lecture of Prof. Fleischmann, into the specific electrochemical field, it seems appropriate to give a concise survey of the present state and trends of some branches of the chemical industry in some way connected with electrochemical processes. Therefore, our attention will be focused on those aspects of industrial productions which are external to the plant, but have decisive importance in condition-ing the rise, the development and the disappearance of a given technology of production. Factors able to exert a decisive influence on the indus-trial strategies are found in many apparently disparate fields such as the opening or the closure of some market areas, alterations in the avail-ability of raw materials, local protectionism or support to selected industrial production, and, of course, even if much rarer, scientific innovations which find their proper technological application.

THE PETROCHEMICAL INDUSTRY

Present State

It seems appropriate that this short excursus on the chemical industry involving electrochemical processes should start with petrochemistry. In fact, a predominant part of the existing chemical processes, both in the commodity field and in the speciality chemicals field takes its feedstocks from oil derived materials. First of all, as far as the lower olefins industry is concerned, feedstock prices have assumed, in the last decade, such a predominant part in the production costs[2,3] that companies have to be ready to utilize the most widely differing raw materials for their productions, being conditioned by a market which is strongly affected by the enormous fuel business. It is compulsory to gain flexibility and the ultimate goal of a modern lower olefins producer should be the possibility of utilizing feedstock ranging from 100% ethane to 100% vacuum gas oil. And it is a general problem, for all chemical companies, to have a series of requirements which must be fulfilled[4], concerning energy and materials consumption, waste production and safety of operations. To complicate this pattern further, the present market situation is influenced by the appearance, among the world producers, of Middle-Eastern companies which have created new, efficient and strongly competitive production units.

What about the electrochemical industry involved in petrochemical processes? By far the most important electrochemical industry, chlor-alkali, has several close connections with petrochemical industry. Sodium hydroxide is employed in a multitude of steps in many chemical processes in the manufacture of synthetic and artificial fibers, in oil refining, and so on. Chlorine enters more specifically the commodity field: about 17% of its production goes to the manufacture of polyvinylchloride, 18% to various chlorinated solvents, among which polychlorinated ethane and ethylene, and 10% to the production of propylene oxide[5].

Vinyl chloride is a monomer of extreme importance for the production of a large series of homo-, co-, and terpolymers of wide commercial diffusion. This monomer was involved in a severe storm, due to the discovery, in 1974, of its carcinogenic properties. The severe restrictions imposed by many governments, gave small and out-of-date plants no further possibility of survival. Only modern large-scale plants could introduce the modifications imposed by the new restrictions without a prohibitive incidence of installation costs on production costs[6]. The rationalization currently taking place among the major chemical producers has had the effect of closing down some ethylene dichloride/vinyl chloride units in Europe, but elsewhere other units are being enlarged, and some other about to enter into production (see, for example the huge plant about to come onstream for a total 450,000 t yr^{-1} of ethylene dichloride in Saudi Arabia[4]).

Among propylene derivatives, propylene oxide and adiponitrile, the former has undergone major changes in the last tow decades. Propylene oxide via chlorohydrin was the sole production route in use until the end of the '60s. This process has the limitation of a wasteful production of a dilute aqueous solution of alkaline calcium chloride, through which the chlorine employed to produce the intermediate chlorohydrin is thrown away (ca 40 t of aqueous salt solution with 5 - 6% wt $CaCl_2$ and 0.1% wt $Ca(OH_2)$ result per t of propylene oxide[7]). Different processes, which use as epoxidizing agents some hydroperoxides, such as those derived from isobutane or ethylbenzene, spread quite rapidly in the US, but with greater difficulty in Europe. Indeed, the competition between the chlorohydrin and peroxide processes is far from being at an end. The main disadvantage of the hydroperoxide route, say the upholders of the chlorohydrin process, lies in the fact that at the best equimolar quantities of the alcohol corresponding to the hydroperoxide are obtained, and this, in terms of quantity of products, has heavy repercussions on the overall economy of production. For example, when propylene oxide is produced through the hydroperoxidation of ethylbenzene, 2.5 t of styrene are produced for each t of propylene oxide. In a situation of production overcapacity affecting styrene, the peroxide route might present more problems than those which were presented by the chlorohydrin process!

An electrochemical route for the production of propylene oxide has been extensively studied, has reached the pilot stage in several laboratories of applied electrochemistry. It consists of a chlor-alkali modified process in which propylene is fed to the electrolytic cell and reacts with HOCl arising from the anodically formed chlorine. The chlorohydrin thus formed reacts with OH$^-$ ions in the cathodic region, giving rise to the epoxide. The NaCl solution is continuously recycled to the cell, and the problem of waste disposal is therefore solved. According to current economic evaluation, this route is near to breakthrough, the investment costs and the large quantity of energy required for the recovery of propylene oxide from the dilute aqueous solution effluent from the cell, the heaviest items in the evaluation of the production costs.

Modern propylene oxide plants in which the chlorohydrin route is followed have reached a close integration of the chlorine cycle with a conventional chlor-alkali plant.

Another important process connected with propylene derivatives is the hydrodimerization of acrylonitrile, which has been developed commercially by Monsanto, up to a total installed capacity exceeding 2000.000 t y^{-1}, with plants located in the US and in Europe. This process easily withstands the competition with the concurrent chemical processes by using cheaper raw materials presenting minor problems of pollution and affording higher operational safety. Adiponitrile is hydrogenated to hexamethylenediamine, which is used in the manufacture of polyamides, and, to a lesser extent, of polyurethanes.

In conclusion, almost all the elements so far considered are concordant about the fact that, as regards the petrochemical industry, the present situation of electrochemical processes is very positive, and no danger of any fall in industrial demand in the short term can be forecast.

Future Trends

Apart from the present situation, briefly summarized in the preceding chapter, great concern is focussed on what will happen when oil derived feedstocks are replaced by other raw materials. Crude prices are decisive for the economy of the petrochemical industry and their incidence is bound to increase in the future. The limit of replacing oil fractions as a feedstock for the chemical industry was reasonably fixed at a crude price near to 50 US$ per barrel. Beyond this price it is most likely that new productions with alternative feedstocks will dominate the scene. This limit seemed to be very near many years ago, but the current projections postpone this by almost two decades[8]. Even if it seems possible to breathe freely for a little longer, everybody is well aware that the internation situation poses continuously new dramatic problems with heavy repercussions on the oil market. In any case, sooner or later, oil based feedstocks will become antieconomical, so that the challenge for exploiting alternative feedstocks has already started, involving the Research and Development programms of many petrochemical firms.

Among the alternative raw materials most liely to be used on a large scale, synthesis gas, its main derivative, methanol, and methane, receive the larger attention[9,10]. Apart from methanol production, synthesis gas has various industrial application, among which the carbonylation of methanol to acetic acid, the hydroformylation process, and, introduced very recently, the production of acetic anhydride. This last process is particularly interesting, because it is based on coal derived synthesis gas. And the incidence of coal derived syngas on the overall production, presently lower than 5%, will increase in the future with the prices of crude oil.

Actually, almost all the oxygenated compounds synthesized from ethylene can be produced from syngas, many of them are commercially attractive and some are already commercialized[10]. No electrochemical processes are involved in the production of those commodities. In fact, electrochemical technologies can find some application in large-scale productions only if they are particularly adapted to solving some of the critical problems of a synthetic procedure: reducing the number of stages from the starting building block to the final commodity, avoiding the use of toxic or unsafe reactants reducing substantially the production of increasing selectivity and yields, or, the crucial point, reducing the energy consumption of the synthetic procedure. Apart from the above mentioned production of adiponitrile from acrilonitrile, which meets the requirements for being commercialized on a large scale, up to now none of the several other interesting syntheses of commodities proposed even on a pilot scale has passed the severe test of economic comparison with existing conventional processes[11].

The connections of chlorine production which petrochemistry, as already said, concern mainly the ethylene derivatives, and particularly the production of vinyl chloride. Therefore, looking into the far future, it is of vital importance to find alternative feedstocks either for the direct production of vinyl chloride or for the production of ethylene. Vinyl chloride could be produced, according to the so called "Lummus-Armstrong Transcat" process, by high temperature direct chlorination of ethane[12] or, following a modification of the so called Benson process[10], by high temperature chlorination and dehydro-halogenation of methane. The Benson process is at present studied in view of its utilization for the production of ethylene. Both these processes have up to new been studied just on a pilot scale, and need to be tested in a first commercial scale demonstration unit, to obtain reliable figures for a correct evaluation of their possibilities of industrial application.

FINE CHEMICALS

Large tonnage processes, like those connected with commodities, such as polymers or fertilizers, were involved, until last year in a deep crisis which seems to have been partially overcome just in the last period. But even if the managements of chemical companies feel a bit more optimistic about the future of their large scale productions, there is the diffuse awareness that the field leaves room just for moderate profits, and that it is necessary to look for further diversification in their activities. The speciality market seems to be a promising direction for the expansion strategies of many chemical firms, even for those who have not previously specialized in small scale productions. This move which started in the 1970's, is involving more and more chemical companies which enter the field either revolutionizing their research and development sectors, or simply by acquiring already staffed small plants or even small companies already active in the field. Industrial sectors involved in small-volume production, range from pharmaceuticals to agrochemicals, cleaners, additives

for petroleum and for plastics and food industries, dyes, products for
water and oil treatment. The US market alone for those speciality
chemicals amounted to 27.4 billions of dollars in 1981, and about 45% of
those sales were due to pharmaceuticals[13]. There is, of course, a great
concern about the possible evolution of a market experiencing deep changes
in such a short time, and there have been authoritative warnings in the
specialized literature about the need to be cautious in entering this
business[14,15].

Although the two fields of fine chemicals and speciality chemicals are
strictly interrelated[14], there is a different approach to the two kinds
of production; whilst fine chemicals have to meet stringent chemical
specifications, such as chemical composition, purity and so on, speciality
chemicals have to solve only performance needs, in some way regardless of
the chemicals used for preparing the commercial product.

Specialized productions may need new, and perhaps more sophisticated,
technologies. The most promising areas for technological developments
appear to be biotechnologies, advanced catalytic applications, and electrochemical technologies. As far as electrochemistry is concerned, the lower
the production scale, the more appropriate its application becomes. The
data of Table 1, published in 1979[16], show how the incidence of the
energy consumption due to the electrochemical stage, on the overall production cost, decreases markedly with the capacity of the plant. Furthermore, electrochemical technologies offer, in several cases, considerable
advantages with respect to the chemical processes. But a correct evaluation of the actual possibilities of application of a laboratory reaction
should take into account a series of elements such as the current density
at the working electrode, the nature of the solvent, the supporting electrolyte and the electrodes, the solubility of the substrate and of the
products of the synthesis and the way of separating the products from the
electrolytic solution.

Table 2 reports the electroorganic processes currently applied on a
commercial scale: apart from the electrohydrodimerization of acrylonitrile,
already seen in the petrochemical context, a series of processes, is
listed, among which only the lead alkyls and the polyfluorinated productions are applied on a medium scale. Several of the process listed in
the table are realized by indirect synthetic methods, in which a selected
inorganic oxidizing agent is continuously regenerated in an electrochemical
cell, and continuously fed to a chemical reactor in which the organic
substrate is oxidized. In many respects, these indirect processes appear
to be more promising than the direct electrochemical reactions[17]. In
fact, indirect processes benefit from the advantages of an easier and less
expensive electrochemical step, as the inorganic reagent is regenerated in

Table 1. Capacity and Operating Costs of Electroorganic Processes[16]

Operating costs	capacity (t y^{-1})		
	40 5 x 100	400 x 100	4000 0.5 x 100
Personal costs*	55	30	15
Energy costs*	2	12	25
Capital costs*	6	12	13
Electrolytic cell*	1.5	5	6
Maintenance*	10	14	16
Others*	27	32	31

* calculated as % of operating cost.   2 e –process, divided cell.

Table 2. Industrial Electroorganic Processes[19]

Starting material	Electrolyte	Products
anthracene (1)	H_2O, Cr^{6+}, H_2SO_4	anthraquinone
furan	CH_3OH, Br	dimethoxydihydrofuran
glucose	H_2O, NaBr, $CaCO_3$	calcium gluconate
starch (1)	H_2O, NaOH, IO_4^-	dihaldehyde starch
various organics	HF, KF	perfluoroorganics
acrylonitrile	H_2O, quaternary salts	adiponitrile
α-methylindole	H_2O	α-methyldihydroindole
pyridine	H_2O, H_2SO_4	piperidine
salicylic acid	H_2O, $NaHSO_3$, H_3BO_3	salicylicaldehyde
tetrahydrocarbazole	H_2O, C_2H_5OH, H_2SO_4	hexahydrocarbazole
cyanide waste stream	H_2O, various	CO_2, N_2 (2)
montan wax (1)	H_2O, H_2CrO_4	purified montan wax

Notes: (1) The process is a two stage operation, in which the redox reagent is regenerated electrochemically and passed into a second reactor where the oxidant and the organic feed are contacted. (2) The initial products are cyanate and cyanogen.

aqueous media at high current densities, with higher current efficiencies and lower energy consumption. Further advantages come from the possibility of making use of the well assessed technology of chemical processes both in homogeneous two-phase liquid systems.

The presence of electrochemistry in the fine chemistry field is due not only to the electrorganic processes listed in Table 2, but also to a certain number of inorganic processes, affording auxiliary chemicals and reactants for a large variety of industrial activities. Among the products, there are strong oxidizing agents, such as fluorine, chlorates, chromates, permanganates, peroxodisulphates and hydrogen peroxide. Very pure hydrogen from water electrolysis, and some metal oxides (MnO_2, Cu_2O) complete this list.

REFERENCES

1. D. Pletcher, Preface, in: "Industrial Electrochemistry," Chapman and Hall, London, New York (1982).
2. B. F. Greek, Natural Gas Liquids Remain Strong Petrochemical Feedstocks, Chem.Eng.News, 62(41):11 (1983).
3. A. R. Hirsig and M. O. Schlanger, How can the Olefins Industry Survive in the 80's? Chem.Eng.Progress, 80(2):24 (1984).
4. Petrochemicals, ECN Chem-Scope, suppl. Eur.Chem.News., 41(1112): (1983).
5. C. J. Harke and J. Renner, Report on the Electrolytic Industries for 1974, J.Electrochem.Soc., 125:455c (1978).
6. U. S. Environmental Protection Agency, "Emission Standard for Vinyl Chloride," EPA, Research Triangle Park, USA (1975).
7. K. Weissermel and H.-J. Arpe, Oxidation Products of Propene, in: "Industrial Organic Chemistry," Verlag Chemie, Weinheim, New York (1978).
8. Staff Report, World oil prices expected to fall, Chem.Eng.News, 61(41):14 (1983).
9. G. F. Pregaglia, Which technologies of tomorrow will be successful? Chim.Ind.Milan, 66:105 (1984).
10. Staff Report, The C_1 way to petrochemicals, ECN Chem-Scope, suppl. Eur.Chem.News, 41(1108): (1983).

11. K. Weissermel and H.-J. Arpe, Basic products of industrial syntheses, in: "Industrial Organic Chemistry," Verlag Chemie, Weinheim, New York (1978).

12. K. Weissermel and H.-J. Arpe, Vinyl-halogen compounds, in: "Industrial Organic Chemistry," Verlag Chemie, Weinheim, New York (1978).

13. P. B. Godfrey, Speciality and fine chemicals, a panacea for profits? Speciality Chem., 4(1):4 (1984).

14. Speciality Chemicals, ECN Chem-Scope, suppl. Eur.Chem.News, 41(1104): (1983).

15. W. J. Storck, Speciality chemicals push raises new management problems, Chem.Eng.News, 61(31):8 (1983).

16. H. Nohe, Comparison of conventional and electro-organic processes, AIChE Symp.Series, 75(185):69 (1979).

17. R. Clarke, A. Kuhn, and E. Okoh, Indirect electrochemical processes, Chem.Brit., 11(2):59 (1975).

18. G. Silvestri, S. Gambino, G. Filardo, C. Cuccia, and E. Guarino, Electrochemical Processes in Supercritical Phases, Angew.Chem.Int. Ed.Engl., 20(1):101 (1981).

19. Electrochemical Technology Corporation, A survey on organic electrolytic processes, US National Technical Information Service, Argonne Ill (1979).

INDUSTRIAL APPLICATIONS

M. Fleischmann and D. Pletcher

Department of Chemistry
The University
Southampton

The diverse applications of electrochemistry and its many-sided interactions with other branches of science and technology (Figure 1), have been
illustrated by other papers contributed at this symposium. Table 1 lists
some specific examples of present day applications and shows that these
range from large scale syntheses (tens of millions of tons per annum, e.g.
Cl_2/NaOH, metal winning) to devices using electrochemistry on a very small
scale, for example for on-line analysis or to batteries to power electronic
circuits. Current trends in research are strongly influenced by these
applications and are illustrated in Table 2; the increasing need for
selective, non-polluting and energy efficient processes has led to a strong
emphasis on research on kinetics (and especially on electrocatalysis), the
study of new reaction systems (including organic electrode reactions and
primary and secondary battery couples), the investigations of new solvent/
electrolyte combinations as well as to new developments in electrochemical
engineering.

In view of the wide range of applications, this paper will necessarily
be restricted to a small number of topics. We first consider research on
cell design and electrochemical engineering since the interface between
electrochemistry and chemical engineering is of crucial importance to the
development of the applications of electrochemistry. Next we consider
trends in the development of two large scale industrial processes; this is
followed by a brief discussion of the present status of organic electrosynthesis. In view of the present day rapid development of the characterization of electrochemical systems at the molecular level (not covered
elsewhere in this symposium) we conclude by illustrating the application
of these techniques to systems of practical importance.

ELECTROCHEMICAL ENGINEERING

Cell Design

In view of the wide range of electrochemical processes[1,2,3]
(Figure 1 and Table 2), there is a diversity in the objective underlying
cell design. For example, the aim in electroplating[4] or electrophoretic
painting[5] (see e.g. Figure 2) is the production of a uniform coating
frequently at relatively low rates whereas in electroforming and electrochemical machining[6] (Figure 3), the objectives are respectively the

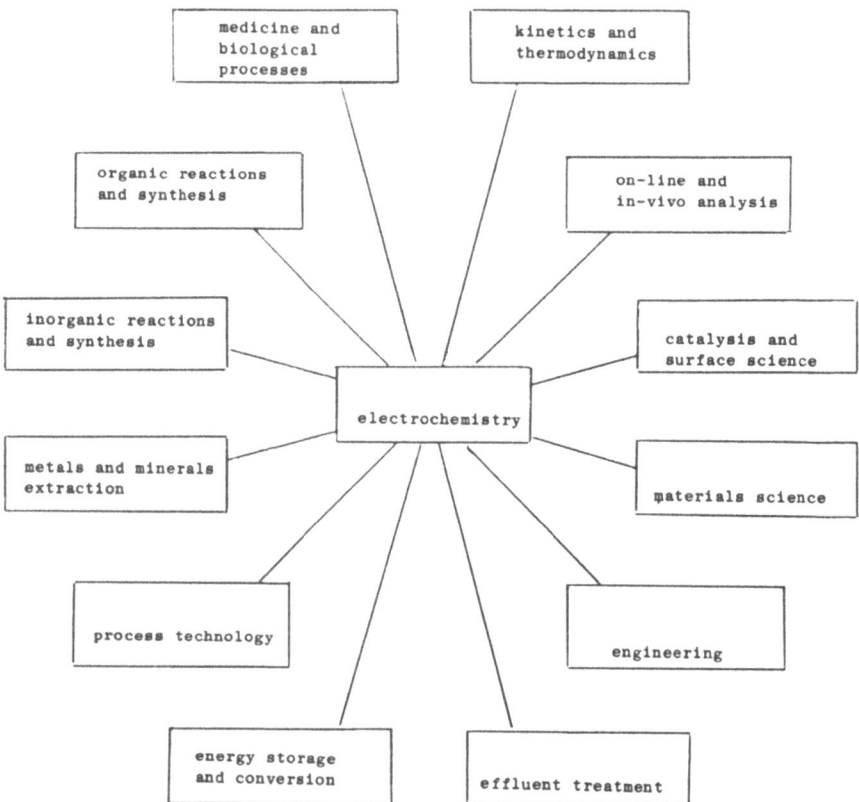

Fig. 1. The applications of electrochemistry and its interactions with other branches of science and technology.

Fig. 2. An example of a product of electrochemical machining. An engine casing and the cathode tool used to form one of the features.

Table 1. Examples of the Industrial Applications of Electrochemistry

1. Synthesis of Inorganic and Organic Chemicals

$Cl_2/NaOH$, H_2, F_2, MnO_4^-

$NCCH_2CH_2CH_2CH_2CN$, PbR_4 ($R=CH_3$, C_2H_5), ⬡$\begin{smallmatrix}COOH\\COOH\end{smallmatrix}$, $C_8F_{17}COOH$

2. Metal Extraction, Refining and Recovery

   extraction - Al, Mg, Zn, Cu, Na, K, Li
   refining   - Cu, Sn, Pb, Al
   recovery   - Cu, Pb, Au, Ag

3. Metal Finishing

   electroplating - Zn, Sn, Cu, Ni, Fe, Cr, Au, alloys, composites
   anodising   - Al, Cu, steels
   electrophoretic painting
   electrochemically initiated polymerization
   electropolymerization

4. Electrochemical Process Technologies

   electrochemical machining, grinding, electroforming

5. Energy Storage and Conversion

   conventional primary and secondary batteries
                - Pb/Acid, Ni/Cd, Leclanché cells
   novel systems - ambient T Li cells, Li/FeS, Na/S, Al/air solid
                   state batteries
   fuel cells

6. Effluent Treatment

   metal recovery, hypochlorite cells, CN^- removal,
   electroflotation, electroosmosis, electrodialysis

7. Corrosion Protection

   anodic and cathodic protection, sacrificial anodes

8. Analysis

   laboratory and on-line analysis, medical sensors, process
   control, ion selective electrodes, gas sensors

cathodic deposition and anodic removal of metal (usually from high duty alloys) at high rates while achieving a desired shape to close levels of tolerance. Inevitably therefore the primary preoccupation is the design (now the computer aided design) of cells and electrodes to achieve these sizes.

On the other hand the major objective in inorganic and organic synthesis, in hydrometallurgy and, indeed, in efficient treatment is the efficient conversion of process streams. Although there is more common ground in these applications, the associated objectives vary considerably. A wide range of cell design has therefore been investigated in recent years. Indeed, one objective of recent research has been to delineate and exploit the flexibility in design. In parallel plate filter press cells (Figure 4) (which remain the favored cell in industrial practice) the aim has been the reduction of the interelectrode gap and the simplification of cell design, manifolding pipe work and electrical connections. Many other cell designs although apparently radically different from filter press cells, retain the essentially parallel plate geometry. Such designs

Table 2.   Trends in Fundamental Electrochemical Research

1. Precise and detailed kinetic and mechanistic studies of surface changes, electron transfer processes and of coupled homogeneous chemical reactions.

2. The development of models for electron transfer, nucleation and growth of new phases and the structure of interfaces.

3. Characterization of electrode-solution interfaces at the molecular level; comparisons of metal-electrolyte with metal-gas and metal-UHV systems.

4. The application of modern techniques to the study of technologically important systems.

5. The development of new coatings: electrocatalysts, composite coatings, polymer films.

6. Novel baths for electrodeposition.

7. Intensification of extraction metallurgical processes; reduction in energy consumption.

8. Development of realistic synthetic reactions.

9. New feedstocks e.g. CO_2, biomass and coal.

10. New energy sources: solar energy, fuel cells, high energy and power density batteries.

11. The application of electrochemistry to the manufacture of electronic devices.

12. The study of biological systems.

13. On-line and in-vivo analysis e.g. CHEMFETS, solid state gas sensors, biomedical sensors.

14. Electrochemical engineering - the development and characterization of novel cell designs and electrode structures: simplification of traditional cell desighs.

15. Computer design of cell configurations.

16. Control of electrochemical processes.

---

illustrated in Figure 4 (for detailed references see e.g.[7,8]) include the Swiss Roll cell (containing expanded mesh cathodes, separators and anodes wound into a cylindrical configuration); the Chemelec cell (containing expanded mesh planar cathodes in a bed of inert particles fluidized in the electrolyte); the ECO cell (a rotating cathode inside a cylindrical anode); the capillary gap cell (a bipolar stack of disc electrodes with radical outflow of electrolyte in the capillary gaps between the electrodes) and the Colloid Mill or Pump Cell (a rotating disc opposite a disc stator or a stack of such discs).   In the cells having a cylindrical configuration the radius of curvature is large compared to the interelectrode spacing so that these cells can still be classified as belonging to the category of plane parallel cells.   A major objective for most of these designs has been the enhancement of mass transfer to the electrodes at acceptable pumping costs (by inducing turbulent flow at low solution flow velocities).

The major objective in other designs such as packed beds of particles, fibers, reticulated foams, bipolar trickel towers and fluidized bed electrodes has been the extension of the electrode area in the direction of current flow (the so-called 'three-dimensional electrodes') thereby increasing the specific area and hence the space-time yield.

264

Fig. 3. The bath for the automatic electrophoretic painting of a car body.

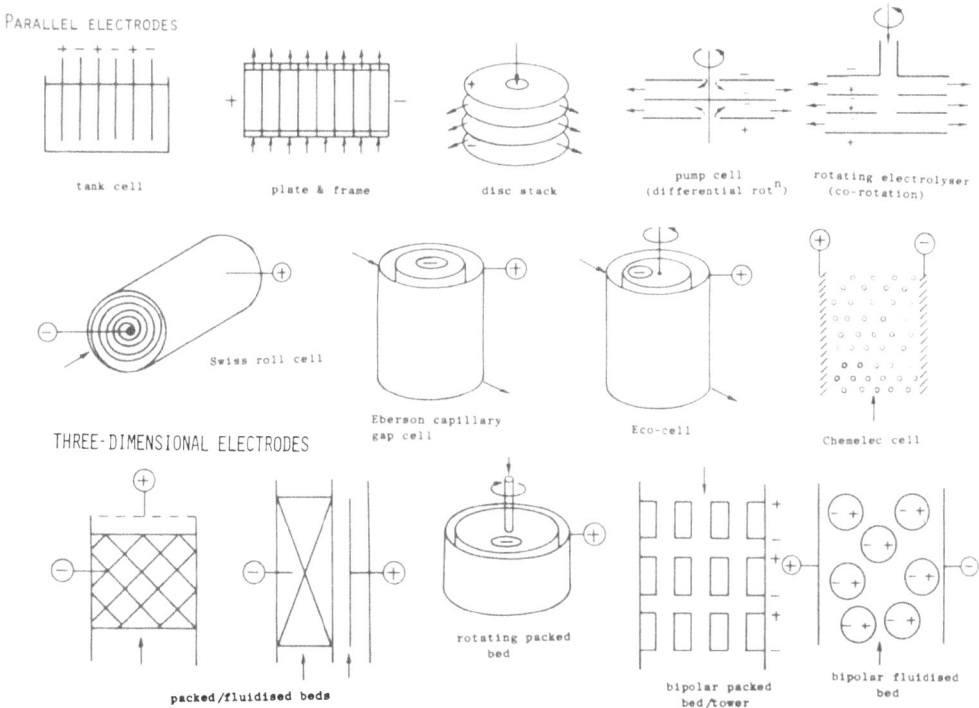

Fig. 4. Schematic diagrams of some electrochemical reactors.

There is considerable flexibility for detailed changes in design for all cell types. For example, fluidized bed electrodes[8] have been investigated in plane parallel vertical and inclined plane configurations with current flow either parallel or perpendicular to solution flow, (the so-called flow-by configuration) (Figure 4), in cylindrical form with radical outflow of current to the counter electrode and with cylindrical arrays of counter electrodes having radical inflow of the current. Cells have been made with membranes or diaphragms as well as without any separator.

Changes in cell geometry have usually been made to meet a number of subsidiary objectives, e.g. to distribute the current or to achieve a variation in bed conductivity and thereby of the current distribution. Bipolar fluidized beds and trickle towers offer yet further flexibility in design: in the first case, for a sufficient bed expansion and low electrolyte conductivity each element of the fluidized bed becomes a bipolar cell; the same objective is achieved in the second case (containing typically arrays of conducting Raschig rings separated by insulating meshes) by maintaining a sufficient voltage gradient in the thin film of electrolyte flowing over the array.

In addition to the increase in the specific surface area these cell designs also meet to varying extents a number of further objectives including:

(a) achievement of high mass transfer coefficients at relatively low solution flow velocities i.e. at low pumping costs (e.g. the pump cell, fluidized bed electrodes)
(b) low hold up of electrolyte and consequent increase in conversion per pass
(c) combination of unit operations (e.g. of pumping and reaction in the pump cell; of absorption/disengagement of reactants/products in the trickle tower)
(d) the ability to handle multiphase streams (e.g. the ECO cell, pump cells)
(e) continuity of operation and the ability to recycle products to different parts of the flow sheet
(f) control over the mixing history and the fine tuning of the design to fit the contacting patterns required for a given reaction (e.g. independent control over the residence time and local mixing in the pump cell).

The relative importance of the various factors naturally depends on the nature of the application. For example the yield and selectivity of many organic electrosynthesis are dependent on the contacting pattern so that (f) is important while (d) is the key factor for reactions using solid-liquid or liquid-liquid dispersions. By contrast (e) is important in metal winning and in the removal of metals from effluents and both these applications also require control over the selectivity (purity of product); efficient treatment also requires (a) and, in general, the use of high specific surface area electrodes. In each application the cell designs must satisfy the set economic criterion of capital and operating costs and optimization must be carried out within the constraints posed by a particular duty[9].

Characterization and Modelling

The brief description of cell design indicates that electrochemical reaction engineering has much in common with chemical reaction engineering except that we have to consider in addition the distribution of potential within the cells (and within three-dimensional electrodes) since this controls the rate of reaction at the metal/solution interfaces.

The characterization of reactors can be carried out at the first level of detail using conventional measurements of overall performance such as of concentration changes in single pass operation or measurements of polarization curves, see Figure 5 [10]. The interpretation of such data naturally requires appropriate models which are also required for the prediction of cell performances. Multivariate analysis of even such complex structures as fluidized bed electrodes show that although there are complex interactions between parameters, these instructions are weak (apart from certain self-evident interactions such as between bed expansions and flow rate)[11]. It is therefore expected that the behavior of electrodes and cells should be predictable by relatively simple models; the correlations between strongly related variables will in any event be built into such models. At the first level of approximation cells can be described by appropriate lumped parameter models. Although the flow is certainly dispersive (see further below) it is sufficient to model simple processes such as metal deposition by plug flow in the axial direction coupled to radical flow of charge[10,12-14]. If the conductivity of the electrode is much higher than that of the solution we have in the steady state

$$\bar{u} \frac{dc}{dx} + r(c,\eta) = 0 \tag{1}$$

$$\frac{1}{\rho_s} \frac{d^2\eta}{dy^2} - nFr(c,\eta) = 0 \tag{2}$$

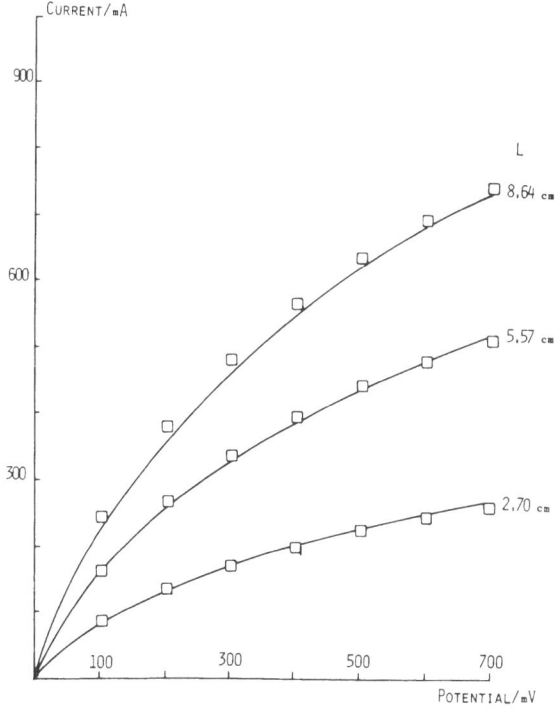

Fig. 5. Comparison of theoretical (solid lines) and experimental (points) polarization curves for electrodes of various lengths of nickel foam[10]. Solution 10^{-3}M $K_4Fe(CN)_6$ + 10^{-3}M $K_3Fe(CN)_6$ + 0.1M KOH. Specific area of Ni foam 56 cm^{-1}.

Here $\bar{u}$ (cm s^{-1}) is the rate of solution flow, ρ_s (ohm cm) is the solution resistivity, η(V) is the local overpotential, $r(c, \eta)$ (mol cm^{-3} s^{-1}) is the local rate of reaction and nF (C mol^{-1}) is the charge transferred per mole of reactant. Although the model is therefore simple, the solution for any particular cell nevertheless poses difficulties in view of the coupling of the equations through the non-linear reaction terms, e.g. see[14]. It has been shown, however, that even further simplifications can be made in predicting a set of 'operating local' in the radial direction at different axial positions within the bed: the procedure[10,15] has much in common with the prediction of the performance of non-isothermal chemical reactors[16] (except that cells cannot store charge in the same way as reactors can store heat). The overall cell performance is obtained by summing along all sets of operating loci. Figure 5 shows that such a simple model does indeed account for the behavior of a relatively complex system; evidently further elaboration of the model would not be justified for the interpretation of the behavior of simple processes such as metal deposition reactions. The behavior of systems such as fluidized and packed beds is determined by single length scales

$$A = \bar{u}/k_m a \tag{3}$$

where a (cm^{-1}) is the specific area and k_m (cm s^{-1}) the mass transport coefficient and by a polarization parameter which governs the distribution of reaction in space

$$B \cong \rho_s nF\bar{u}c_i \tag{4}$$

where c_i (mol cm^{-3}) is the inlet concentration[10].

At the next level of characterization measurements can be made of local reaction rates, mass transfer coefficients, turbulence intensities, potentials, e.g. by using segmented electrodes. Alternatively, the lumped parameter description can be extended to include for example radial and axial dispersion in the cells or electrodes. Electrochemical analogues of the tracer techniques of chemical reaction engineering have been found to be particularly useful for such measurements e.g.[17,18]. It is found that in general electrochemical reactors can be modelled as 'black boxes' containing stagnant and fast moving phases. However, in contrast to conventional chemical reaction engineering, reaction takes place in the stagnant phase. Figure 6 illustrates such measurements for a reticulated foam electrode[19]. Consideration of these more detailed models is important in the evaluation of the influence of cell design on reactions more complicated than metal deposition processes e.g. on organic electro-syntheses.

It can be seen that electrochemical reaction engineering has much in common with chemical reaction engineering. Scaled-down versions of full-sized cells are readily constructed and the system behavior can be characterized in detail using electrochemical techniques. One can anticipate therefore that there will be a valuable interplay between electrochemical and conventional chemical reaction engineering. It is relevant in this context that whereas some reactions are carried out on very large scales, many desired products are made on the scale of a few tons to a few thousand tons per annum. The investigation of the chemical engineering of 'small' systems is a prerequisite for the successful implementation of such process, a field which can be readily explored in the area of electro-chemistry.

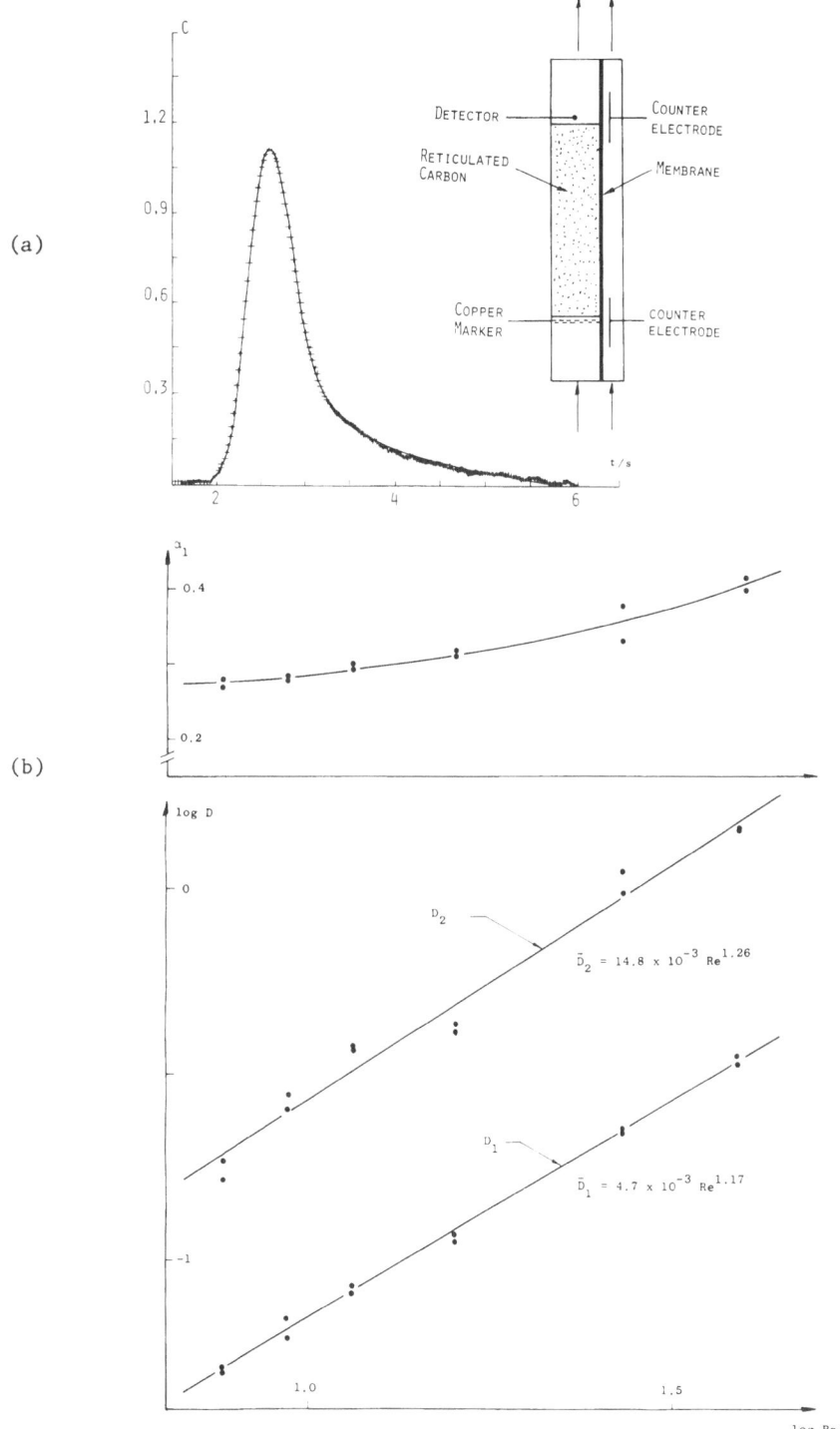

Fig. 6. (a) Response of detector electrode to a square pulse at a copper marker electrode to form Cu^{2+}. Solution 1M H_2SO_4. Inset shows the cell configuration. (b) Derived parameters for fast and slow moving phases; average diffusion coefficients for the two phases and the fraction of slow moving phase.

EXAMPLES OF LARGE SCALE SYNTHESIS

1. The Manufacture of Chlorine and Sodium Hydroxide

Electrolysis has been used for the production of chlorine and sodium hydroxide for over a hundred years. Even so, the technology has developed very rapidly during the last decade and the trends indicate well the application of the principles of electrochemical engineering. Of course, the driving force for these developments is the massive scale of the process, approximately 3.3×10^7 ton Cl_2/year and the consequent energy consumption, some 10^{11} kWh/year. In addition, concern over mercury pollution has led to a need to develop cells which do not use the amalgam system.

For the past ten years, three technologies have computed within the chlor-alkali industry[1,3,20,21]. The oldest is the mercury cell where 35% brine is electrolyzed in an undivided cell with a mercury cathode. The electrode reactions are

$$2Cl^- - 2e^- \longrightarrow Cl_2$$

$$Na^+ + Hg + e^- \longrightarrow NaHg$$

and the sodium amalgam is reacted with water in the presence of a catalyst in a separate reactor

$$2NaHg + 2H_2O \longrightarrow 2NaOH + 2Hg + H_2$$

This process has the advantages that it may be operated at a very high current density (because of the absence of separator and the small inter-electrode gap) and that it produces pure H_2, Cl_2 and NaOH in the appropriate form for sale (generally a 50% aqueous solution). For many years its only disadvantage was the need to handle very large volumes of mercury. Recently it has been under pressure for purely economic reasons; its energy consumption is too high. The second process technology to be developed was the diaphragm cell. In this cell the electrode reactions are

$$2Cl^- - 2e^- \longrightarrow Cl_2$$

$$2H_2O + 2Na^+ + 2e^- \longrightarrow 2NaOH + H_2$$

and a porous diaphragm, usually asbestos is used to separate the gaseous products. Although this technology avoids the use of mercury, it has never been entirely satisfactory because of the low current density which must be used, the relatively impure gases formed and the very poor quality caustic soda produced (this is dilute, 12% and contains a very high chloride ion contamination). The third process is based on the membrane cell wherein the separator is a high performance, selective ion exchange membrane which allows the transport of Na^+ but not other ions or H_2 and Cl_2 gas. Materials of sufficient stability have only been available for 10-15 years and are based on perfluorinated resins. In spite of the high cost of these membranes, it is now clear that the future of the industry lies with membrane cells.

Before outlining the advantages of these cells in greater detail, it is appropriate to examine the advances which have contributed to the revolution in the chlor-alkali industry.

Electrocatalysts. Most will be aware of the success of the dimensionally stable anodes (DSA), the anodes based on titanium coated with RuO_2. These anodes combine excellent electrocatalytic properties with long life-times and perhaps most importantly can be readily fabricated in many shapes (cf. C as was previously used as the anode). This has greatly enhanced the possibilities for improved cell design.

More recently the problem of overpotential for H_2 evolution on the cathodes of membrane (and diaphragm cells has been attacked. Traditionally the cathode is made of steel (overpotential $\sim$ 400 mV) but coatings based on high area nickel alloys have been developed which reduce the overpotential to 80–120 mV.

Membranes. The properties required from the separator are clear. It should only permit the transfer of Na^+ from anolyte to catholyte without allowing movement of OH^-, Cl^- or H^+ between the two compartments. It must be stable to Cl_2 and concentrated NaOH with great physical stability and a low electrical resistance. As always the final design requires compromise, for example, over thickness.

The breakthrough came with the design of the membranes based on perfluorinated resins which combine stability and low resistance. In the early materials the transport was based on sulphonate groups. These strong acid resins had low resistance but only withstood the large pH gradient across them in the working cell – pH 4 in the anolyte and concentrated NaOH in the catholyte – to a limited extent. The maximum caustic soda concentration was 12%. Later resins based on the carboxylate group allowed more concentrated caustic soda to be produced (30–40%) but being weak acid resins had a higher electrical resistance particularly at the anolyte interface where the active groups protonate at pH 4.

Modern materials (e.g. Nafion 901) make use of the best features of both materials, see Figure 7. Most of the membrane is fabricated from the sulphonate resin to obtain the low resistance but the catholyte surface is coated with a thin film of carboxylate resin which permits the high caustic soda concentration to be produced. The plastic net gives the membrane additional mechanical and physical strength.

Cell Design. Clearly, low energy consumption requires the inter-electrode gap to be minimized. The extreme is the 'zero gap cell'. In practice, in membrane cells, this means that the two electrodes are in contact with the membrane, and hence can also act as support for

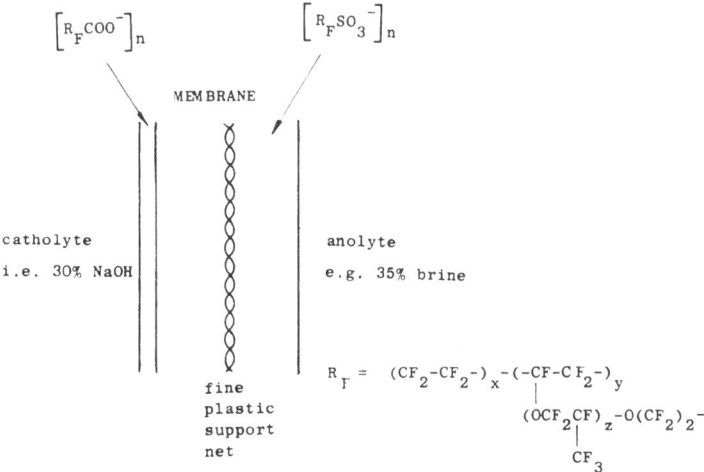

Fig. 7. Structure of modern chlor-alkali membranes.

the membrane. In such arrangements it is of course essential that the gases are released quickly from the electrode surface and away from the inter-electrode gap. This can be achieved by appropriate design of the electrode structures.

The cost of the fabrication of components and the manufacture of cell units is another critical factor. As a result the design of simpler cells with few fluid and electrical connections and less components is a clear modern trend.

Control. The age of the computer is allowing significant improvements in performance. In a modern plant, the performance of each cell can be monitored and it is possible using on-line analytical devices to watch and optimize all the process streams. The latter is important since low concentrations of impurities can adversely affect the performance of many of the cell components.

The FM21 Membrane Cell, produced by ICI Ltd, is an example of a cell unit which takes advantage of many of these modern concepts and components[22]. The unit shown in Figure 8 is based on a modular concept; the number of monopolar cells within the filter press is adjusted to give the required capacity. Each cell makes use of the modern concepts and components described above. They are essentially zero gap cells where the cost of components is minimized by producing the electrodes by a pressing process whereby a sheet of metal is converted into two sets of active ribs, with the gas disengagement space between, and using rubber gasketing between electrodes and membranes which also fulfils the role of cell walls. The cells also make use of the latest membranes and both catalytic cathodes and anodes. It is claimed that the cell combines low capital cost for both electrolyses and cell room, low maintenance costs with reliable and low cost operation. Recent commercial plants have been operating with an

Fig. 8. The FM21 cell used by ICI plc for the manufacture of Cl_2 and NaOH.

272

energy efficiency of 2150 kWh/ton NaOH (at 0.3A cm^{-2}); this compares with more than 3000 kWh/ton for a typical mercury plant.

As the world economy recovers, the outlook for chlor-alkali is again improving and a real change of attitude to cell technology is now to be seen. Both for conversion of existing cell rooms and greenfield sites, the choice is expected to be membrane cells. A belief that investment in mercury cells might be difficult because of legislation over pollution has been replaced by a confidence in the concept of membrane cells and evidence that they can give real economic advantage.

The Monsanto Adiponitrile Process

The history of the Monsanto adiponitrile process demonstrates well the development of a synthetic process[1,23-25]. The first electrolytic process for the hydrodimerization of acrylonitrile was introduced on a 14,000 tons/year scale in 1965. The late 1970's saw a large expansion of this process both in the UK and USA and it now produces some 200,000 tons adiponitrile/year. At least the recent cell houses are based on quite different technology to that used in the first plant and this has allowed large reductions in both capital investment and energy consumption.

The essential feature of the two technologies are compared in Table 3. In 1965 it was believed that a high yield of adiponitrile was dependent on using catholyte which contained a high concentration of both acrylonitrile and a quaternary ammonium salt and to obtain a homogeneous solution a complex low water content solution was devized This was not stable to anodic oxidation and hence a membrane had to be used to separate anolyte and catholyte. The low conductivity of the catholyte and the presence of the membrane inevitably led to a high cell voltage and hence energy consumption. Moreover the use of a membrane and the need for turbulence promoters (to avoid base catalyzed reaction leading to by-products) determined the choice of a plate and frame cell and it was also necessary to have complex and capital intensive work-up units to separate product and recover both acrylonitrile and the quaternary ammonium salt.

Several papers in the following years[26-28] pointed the way to carrying out the reaction with a saturated solution of acrylonitrile in an aqueous electrolyte containing only a low concentration of a quaternary ammonium salt; the process no longer required a separator and by having acrylonitrile present in the electrolyte as a second phase, the product could be extracted in the cell. Hence it was no surprise when it was confirmed[24] that recently installed Monsanto plant used such technology (see Table 3). The cell is based on a simple stack of carbon steel plates whose cathodic faces are electroplated with cadmium, although additives, borax + EDTA, are necessary to reduce the rate of anode corrosion. The large gain in energy efficiency is obvious from Table 3 but because of it simplicity the cell is also very much cheaper to install; this explains the reduction in current density to further reduce the energy consumption. The use of the two phase electrolyte also reduces the investment in extraction processes since the aqueous and non-aqueous streams may be treated separately.

ORGANIC ELECTROSYNTHESIS

It is now established that electrolysis can make a contribution to organic synthesis both in the laboratory and on an industrial scale. The size of the new edition of Baizer[29] (well over 1000 pages) testifies to the number and variety of reactions which have now been reported. However, the chief lesson of recent years has been that commercial processes are dependent on the design of a total system. Thus in addition to finding

Table 3.

Cell type	Plate and frame cell with membrane in filter press	Undivided bipolar stack of plates
Cathode	Pb	Cd
Anode	PbO_2(+ 1% AgO)	Steel
Separator	Ionics CR 61	None
Electrode gap (mm)	7.1	1.8
Catholyte	$Et_4N^+EtSO_4^-$ (40%) H_2O (15%) CH_2=CHCN + $(CH_2CH_2CN)_2$ (50%)	$EtBu_2N^+(CH_2)_6N^+Bu_2Et$ HPO_4^{2-} (0.4%) Na_2HPO_4 (10%) $Na_2B_4O_7$ (2%) Na_4EDTA (0.5%) Sat. CH_2=CHCN all in H_2O
Anolyte	H_2SO_4 (1M)	as catholyte
Temperature (°C)	50	55
Catholyte velocity (m s^{-1})	2	1–1.5
Catholyte resistivity (ohm cm)	38	12
Current density (A m^{-2})	4500	2000
Voltage distribution (V)		
Estimated reversible cell voltage	2.50	2.50
Overpotentials	1.22	0.87
Electrolyte IR	6.24	0.47
Membrane IR	1.69	–
Cell voltage (V)	11.65	3.84
Energy consumption (kWh/lb)	3.0	1.1

conditions where the chemical conversion of interest occurs with the required selectivity and rate (usually a current density above 100 mA cm^{-2}), it is also necessary (i) to design a counter electrode reaction which does not add substantially to the cost of the system: the reaction product should be stable at the counter electrode (thereby avoiding the use of diaphragms or membranes); (ii) to have a system where the product may be extracted and maybe solvents and/or electrolytes recovered; (iii) to design a cell appropriate to the chemistry. As a result, the success of processes has commonly been due to the careful design of all parts of the process and/or the combination of several unit processes within the cell; examples include the use of a distillable electrolyte[30], $R_3NH^+OAc^-$ and the use of a second phase within the cell for product extraction[31]. Good engineering plays an essential role and the reaction conditions must be adjusted to the demands of the total process.

To evaluate the role of electrolysis in the commercial manufacture of organic compounds it is necessary to understand the nature of the industry itself. Relatively few compounds (maybe 300)[32] are made on a very large scale and then mostly by gas phase catalysis. On the other hand, a very large number (maybe 10^6) are made to an annual tonnage between 1 and 10^4 tons/year; many are high cost materials whose market may not be secure or invariant with time. It is in this latter market that electrolysis must compete and hence, product selectivity, low cost cell design, the cost of solvents and product isolation, the generality of the plant for other chemicals are features much more likely to be important than energy consumption. Electrochemical processes are, perhaps, most intensively studied when (a) they allow chemistry not otherwise possible e.g. oxidations at very positive potentials, (b) they replace reaction with hazardous chemicals or reagents which lead to a pollution problem, (c) there are families of products which may be made by electrolysis.

Another general feature of the low and medium tonnage organic compounds industry is that reaction routes are seldom announced and thus our knowledge of electrolytic processes in current commercial operation remains far from complete. It is interesting to note that a US Chemicals Supply House is reported to use electrolysis routinely while a range of diverse processes have been operated at least on a pilot scale. These include[33-36,30].

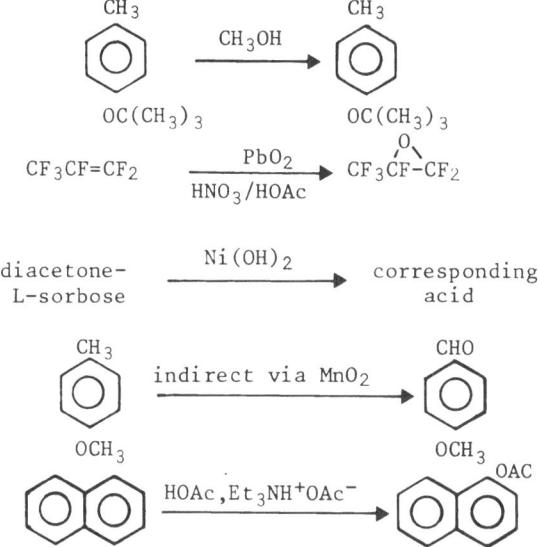

Many such processes continue to use parallel plate reactors, a number of which are now available for purchase (e.g. the SU cell[37], the dished electrode cell[38] and the FM21 cell[22]) but many syntheses fail to be applied because economic assessments are based on unnecessarily complex designs and expensive cell components developed for other application e.g. membranes, Ti cell bodies, electrode materials. Hence it is particularly good to see the development and application of very cheap configurations, for example the new Monsanto process[24] or the Swiss Roll cell[35]. It is to be hoped that more companies give consideration to cell systems specifically designed for particular processes.

We would like to highlight one laboratory development which we believe could have considerable impact on industrial practice. That is multi-phase electrolysis where the electrolyte is an emulsion of an aqueous phase and an immiscible organic solvent. Advantages of such techniques are (i) they provide a feasible way to apply the many reaction described in aprotic media in industry, (ii) by continuous extraction into the organic solvent of the product its isolation can be greatly simplified, (iii) very high current densities can be achieved because it is possible to decouple the link between maximum current density and solubility of the organic compound, (iv) it is possible to design the system so that the working electrode reaction occurs in an organic medium and the counter electrode is simply water electrolysis without recourse to a separator, (v) the cell voltage can be greatly reduced by the aqueous phase in most of the inter-electrode gap, (vi) the aqueous solution may be used to buffer the organic medium minimizing acid or base catalyzed reactions which reduce selectivity in many aprotic electrolyses. To give three illustrations of multiphase electrolysis:

(a) It is possible [39] to reduce nitrobenzene with a current of 1A cm^{-2} by electrolyzing an emulsion of nitrobenzene and an aqueous solution, 0.4M ZnCl$_2$ + 1M HCl. The cathode reaction is

$$Zn^{2+} + 2e^- \longrightarrow Zn \text{ (powder)}$$

and the high area, oxide free surface, zinc powder then reacts rapidly

This reaction is quite general for substituted nitrobenzenes and the current density is many orders higher than that achieved in direct reductions.

(b) The corresponding oxidation reactions employ the anodic generation of an oxidizing agent in an aqueous phase. If the oxidizing agent is an anion, e.g. hypobromite or dichromate, a possible route to carrying out reactions in the organic medium is to employ phase transfer catalysis. For example [40], the anode reaction in H$_2$SO$_4$/H$_2$O

$$2Cr^{3+} + 7H_2O - 6e^- \longrightarrow Cr_2O_7^{--} + 14H^+$$

can be coupled to the oxidation

in dichloroethane by simply adding a phase transfer agent, e.g. Bu$_4$N$^+$.

(c) The anodic substitution of aromatic hydrocarbons by electrolysis of an emulsion of the substrate in methylene chloride and the nucleophile and a phase transfer catalyst in water. For example, the cyanation[41], chlorination[42] and acyloxylation[43] of naphthalene have all been reported in good yields.

DEVELOPMENTS IN BASIC ELECTROCHEMISTRY

The successful exploitation of many ideas based on electrochemistry requires a detailed understanding of the kinetics of the processes and of the structure and properties of electrode-solution interfaces: the purely empirical approach to the solution of problems and to optimization of processes will in future rarely prove to be adequate.

The development during the last three decades of laboratory methods for investigation of the kinetics of electrode reaction[44,45,46] e.g. of relaxation techniques (potential step, current step, A.C. impedance methods, etc.), of cyclic voltammetry and, most recently of the use of electrodes of very small dimensions, has led to a marked increase in our level of understanding of these processes. The value of these techniques has been greatly enhanced by the development of computer based methods of data analysis.

Inevitably, however, these methods cannot give us a detailed insight into the structure of the electrode-solution interface and our understanding of the relationships between structure, energetics and kinetics is therefore still rudimentary. Research in numerous groups is therefore turning increasingly to the simultaneous examination of electrode processes by conventional means combined with appropriate in situ and ex situ (especially U.H.V.) structural probes[46,47]. Table 4 gives a listing of these techniques; those marked * are at this time especially useful while those marked + hold considerable promise.

Investigation of electrode solution interfaces by in situ vibrational spectroscopy has two principal advantages: firstly the species present and their structures are directly characterized by their spectra and, secondly, these spectra are sensitive to the environment and therefore can be used to probe complex interactions. Raman spectroscopy is particularly well suited to the investigation of aqueous systems and in certain cases the adsorption of neutral species, of anions in the double layer and of the solvent (as well as interactions between these species) can now be characterized[48]. Vibrational spectroscopy of systems of practical importance is illustrated by the Surface Enhanced Raman Spectra (SERS) of the corrosion inhibitor thiourea adsorbed at silver and copper electrodes[49]; it should be noted that inhibitors such as thiourea are also used as plating additives. Figures 9A and B compare the SER spectra in neutral and acid solutions. We

Table 4. Techniques Used for the Study of the Structure of Electrode-solution Interfaces and of Adsorption at these Interfaces

In situ methods	Ex-situ methods
*1) Raman vibrational spectroscopy	*1) X-ray photoelectron spectroscopy (XPS)
*2) Infra-red vibrational reflectance spectroscopy	2) Ultra violet photoelectron spectroscopy (UPS)
3) Ultra violet-visible reflectance spectroscopy; ellipsometry as a spectroscopic tool	*3) Auger spectroscopy
4) Transmission through optically transparent electrodes	4) Electron energy loss spectroscopy
5) Photoelectrochemistry	5) Inelastic electron tunneling
6) Mössbauer spectroscopy	6) Ion scattering spectroscopy
7) Electron spin resonance	7) Secondary ion mass spectroscopy
8) Acoustoelectrochemical methods	*8) Low energy electron diffraction (LEED)
*9) X-ray diffraction	9) Reflection high energy electron diffraction (RHEED)
	10) Transmission electron diffraction microscopy

Being developed

*10) EXAFS

+11) Neutron diffraction (and dynamic neutron scattering?)

Promising methods

+12) γ-ray induced X-ray fluorescence

+13) NMR broadline and high resolution

+14) Positron annihilation

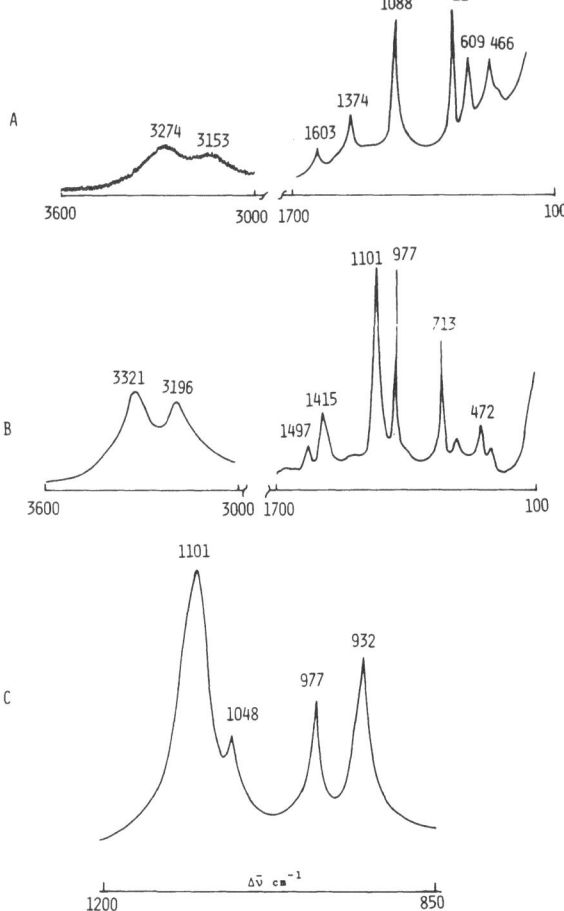

Fig. 9. SERS of thiourea (0.1M) at a silver electrode. (A) in 0.1M K_2SO_4 at -1.2V; O-H and N-H stretch region at -0.6V; (B) in 2M H_2SO_4 at -0.6V; (C) in acidic solution of 0.1M ClO_4^-, 0.01M NO_3^- + 0.002M SO_4^{--} at -0.6V.

observe that for neutral solutions the bands at 1088 and 711 cm^{-1} (assigned as mainly CN symmetric stretch and CS stretch respectively) are shifted from the values 1094 and 729 cm^{-1} for the solution-free species. This is consistent with an increase of the CN and a decrease of the CS bond order for the adsorbed species which would be caused by adsorption via the sulphur atom (compare electrochemical data); a Ag-S stretching mode is observed at 242 cm^{-1}. The strong band at 611 cm^{-1} (which is not observed in solution) is close to an infra-red active band (at 629 cm^{-1}) and is assigned to an SCNN out of plane bending mode. We are led to a model

$$S = C \begin{smallmatrix} \ddot{N}H_2 \\ \\ \ddot{N}H_2 \end{smallmatrix} \longleftrightarrow S - C \begin{smallmatrix} \overset{+\cdot}{N}H_2 \\ \\ \ddot{N}H_2 \end{smallmatrix} \longleftrightarrow S - C \begin{smallmatrix} \ddot{N}H_2 \\ \\ \overset{+\cdot}{N}H_2 \end{smallmatrix}$$

(i)   (ii)   (iii)

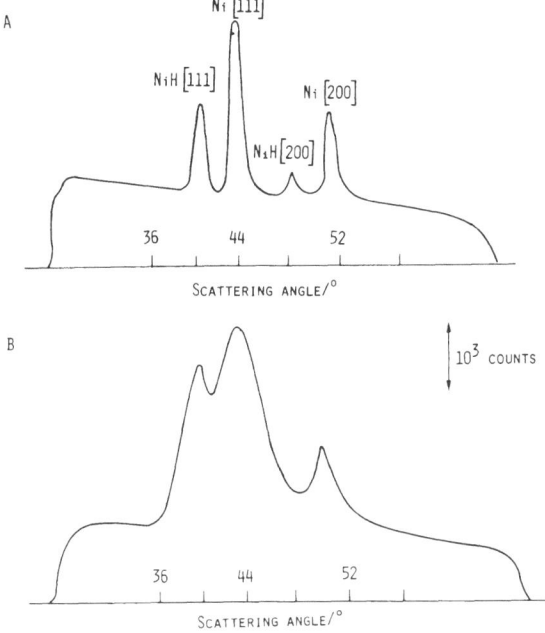

Fig. 10. X-ray diffractograms of Ni (A) bulk Ni cathodically polarized in aqueous acid (B) Ni depositing from a Watts bath containing thiourea.

in which the canonical forms (ii) and (iii) of the adsorbed species contribute more to the adsorbed state than does the form (i). The observation of the out of plane bending mode as a strong Raman active mode however suggests that the thiourea molecular must be inclined to the surface.

The spectra in acidified solutions are illustrated in Figure 9B. Adsorbed thiourea remains unprotonated even in strong acid solutions; the spectra increase in intensity and coadsorbed anions are observed, Figure 9C. It should be noted that the sulphate and not the bisulphate ion is coadsorbed even in 2M H_2SO_4 and that coadsorption of a number of oxyanions can be detected. The relative amounts of the various ions depend on the concentration ratios and the potential, i.e. there is a potential dependent anion exchange of the species. We assume that anions are coadsorbed via hydrogen bonding to one or more $-NH_2$ groups of the adsorbed thiourea (indicated by the broad and shifted NH bands, Figure 9B). OH^- ions coadsorbed in neutral solutions (not detected in the spectra) are replaced by the oxyanions in acid solutions.

Whilst vibrational spectroscopy provides information about the nature, energetics and environment of adsorbed molecules it does not yield direct information about long range order in the interface. In situ X-ray diffraction (INSEX), both by transmission and reflection[50], is one such method capable of giving information about order parameters. Although the scattering of X-ray photons is extremely weak, the use of multiplex single photon count detection coupled to modulation of the electrode potential and computer based signal averaging techniques provide sufficient sensitivity to allow measurements on monolayer amounts of materials at electrode-solution interfaces and of changes in structure of these interfaces.

Measurements of this type are illustrated by changes in structure of cathodically polarized nickel surfaces. In this case the changes are sufficiently marked that they can be directly detected by reflection diffraction (i.e. without using modulation techniques). Figures 10A and B show that following cathodic polarization of a bulk nickel electrode diffractions of βNiH_x can be observed at $2\theta = 42.0°$ and $49.1°$ (corresponding to the (111) and (200) reflections) in addition to the (111) and (200) reflections of nickel at $44.5°$ and $52.0°$. Diffractograms showing broad peaks in the same regions can be observed for electrodeposits formed in the presence of thiourea i.e. mixtures of finely crystalline nickel and βNiH_x are formed under these conditions.

It is evident therefore that it is now possible to characterize electrode-solution interfaces at the molecular level: fundamental measurements can be made on systems of practical importance. The significance of the development of these methods (as well as of other new methods in the 'pipeline') lies at least in part in the scope this gives for the design and selection of appropriate conditions for electrocatalysis, synthesis, corrosion inhibition, metal plating, operation of battery systems etc.

CONCLUSIONS

Electrochemistry already has an essential role in a wide range of industrial activities and has a very healthy interaction with many areas of science, medicine and technology. It is to be expected that these contributions of electrochemistry will grow in importance during the coming years and we believe that more fundamental studies, both of electrode reactions themselves and in engineering, have an important role to play. It is particularly encouraging that the science of electrochemistry has developed to an extent where it can contribute to technology by the study of real systems.

REFERENCES

1. D. Pletcher, "Industrial Electrochemistry," Chapman and Hall (1982).
2. A. T. Kuhn, ed., "Industrial Electrochemical Processes, Elsevier (1971).
3. J. O'M. Bockris, B. E. Conway, E. Yeager, and R. E. White, eds., "Comprehensive Treatise of Electrochemistry, Vol. 2, Electrochemical Processing," Plenum (1981).
4. D. R. Gabe, "Principles of Metal Surface Treatment and Protection," Pergamon (1978).
5. G. E. F. Brewer, J.Appl.Electrochem., 13:269 (1983).
6. P. J. Boden, Phil.Trans.R.Soc.London, A302:297 (1981).
7. M. Fleischmann and G. Kelsall, Proceedings of "Electrochemistry in Minerals and Metal Processing," Electrochemical Society, Vol. 84-10, 572.
8. F. C. Walsh and R. J. Marshall, Surface Technol., 24:45 (1985).
9. R. E. W. Jansson, Phil.Trans.R.Soc.London, A302:285 (1981).
10. M. Fleischmann and R. E. W. Jansson, Electrochim.Acta, 27:1023,1029 (1982).
11. M. Fleischmann and G. Kelsall, J.Appl.Electrochem., 14:269 (1984).
12. A. J. S. Walker and A. A. Wragg, Electrochim.Acta, 22:1129 (1977).
13. R. E. Sioda, J.Appl.Electrochem., 7:135 (1977).
14. J. S. Newman and W. Tiedemann, Adv.Electrochem. and Electrochem Eng., 11:353 (1978).

15. M. Fleischmann and R. E. W. Jansson, J.Chem.Technol. and Biotechnol., 30:351 (1980).
16. O. Levenspiel, "Chemical Reaction Engineering," John Wiley and Sons (1972).
17. M. Fleischmann and Z. Ibrisagic, J.Appl.Electrochem., 10:151,157,169 (1980).
18. R. E. W. Jansson and R. Marshall, Electrochim.Acta., 27:823 (1982).
19. A. Bejerano and M. Fleischmann, to be published.
20. M. O. Coulter, ed., "Modern Chlor-alkali Technology," Vol. I, Ellis Horwood Ltd. (1980).
21. C. Jackson, ed., "Modern Chlor-alkali Technology," Vol. II, Ellis Horwood Ltd. (1983).
22. C. Jackson, I. M. Girvan, and R. A. Harrison, Electrochemical Society Meeting, Cincinnati, May (1984).
23. M. M. Baizer and D. E. Danly, Chem.Ind., 435,439 (1979).
24. D. E. Danly, Hydrocarbon Processing, 161 (1981).
25. D. E. Danly, J.Electrochem.Soc., 131:435C (1984).
26. A. P. Tomilov, S. L. Varshavskii, and I. L. Knunyants, British Pat. 1,089,707 (1967).
27. W. V. Childs and H. C. Walters, Hydrocarbon Processing, 139 (1978).
28. C. A. van Eugen, C. A. Hendricks, J. Romioulle, J. Walravens, and A. Verheyden, Chemie et Industrie, 104:71 (1971).
29. M. M. Baizer and H. Lund, eds., "Organic Electrochemistry," Marcel Dekker (1983).
30. D. Degner, "Technique of Electroorganic Synthesis - Part III," N.L. Weinberg and B.V. Vilak, eds., John Wiley (1982).
31. S. Torii, H. Tanaka, and M. Mishima, Bull.Chem.Soc.Japan, 51:1575 (1978).
32. T. Beck, "A Survey of Organic Electrolytic Processes," Electrochemical Technology Corporation, ANL/OEPM-79-5 (1979).
33. D. Degner, "Electroorganic Synthesis," RSC, Wrexham (1982).
34. H. Millauer, Electroorganic Synthesis," RSC, Wrexham (1982).
35. P. M. Robertson, P. Berg, H. Reimann, K. Schleich, and P. Seiler, J.Electrochem.Soc., 130:591 (1983).
36. P. Millington, "Electroorganic Synthesis," RSC, Wrexham (1982).
37. SU cell, Swedish National Dev. Corp., PO Box 34, S-18400, Akersberga, Sweden.
38. DEM Cell, Steetley Eng. Ltd., PO Box 20, Brierley Hill, West Midlands EY6 8XA, England.
39. N. E. Gunawardena and D. Pletcher, Acta Chem.Scand., B37:549 (1983).
40. L.-C. Jaing and D. Pletcher, J.Electroanal.Chem., 152:157 (1983).
41. L. Eberson and B. Helgee, Chem.Scripta., 5:47 (1974).
42. S. R. Ellis, D. Pletcher, W. N. Brooks, and K. P. Healy, J.Appl. Electrochem., 13:735 (1983).
43. S. R. Ellis, D. Pletcher, P. H. Gamlen, and K. P. Healy, J.Appl. Electrochem., 12:693 (1982).
44. A. J. Bard and L. R. Faulkner, "Electrochemical Methods," John Wiley (1980).
45. P. T. Kissinger and W. R. Heinman, eds., "Laboratory Techniques in Electroanalytical Chemistry," Marcel Dekker (1984).
46. R. Greef, R. Peat, L. M. Peter, D. Pletcher, and J. Robinson, "Instrumental Methods in Electrochemistry," Ellis Horwood Ltd. (1985).
47. E. Yeager, J.Electrochem.Soc., 128:161C (1981).
48. R. K. Chang and T. E. Furtak, eds., "Surface Enhanced Raman Scattering," Plenum (1981).
49. M. Fleischmann, I. R. Hill, and G. Sundholm, J.Electroanal.Chem., in press.
50. M. Fleischmann, P. Graves, I. R. Hill, A. Oliver, and J. Robinson, to be published.

DISCUSSION

G. Silvestri

Università di Palerneo
Italy

CHAIRMAN'S SUMMARY

The notes to the discussion which followed the lecture of Professor
Fleischmann find their proper place here, as they are connected with the
preceding introduction, being focussed essentially on the general strategies to be developed with a view to widening the field of application of
electrochemistry. The contribution from the audience stimulated the
discussion on the ways of increasing the overall energy efficiency of
existing industrial processes (Professor Wiesener), on the incidence of
capital costs on the economy of electrochemical processes (Professor
Murray), and on the possible advanced technological applications of
electrochemical methodologies (Professor Bard).

How to increase the generally low overall energy efficiency of large
electrochemical processes, such as chlor-alkali or aluminium production?
Among the two possibilities of improvement, concerning the performances of
electrochemical cells or the energy integration of all the steps of production, the most provising way appears to be the second. In fact,
electrochemical steps have achieved in the last decades, substantial
improvements in the electrode materials and in cell engineering by lowering
both the electrode overvoltages and the internal resistance of the cells,
but, as for the rest of the chemical industry, a great deal of work must be
done to minimize energy and waste products. Looking at aluminium production for example, one of the main causes of inefficiency lies in the
need to produce pure Al_2O_3 from bauxite and has nothing to do with the
efficiency of the electrochemical step. Furthermore, in unintegrated
processes, a great deal of the heat which issues from the stacks is bought
as thermal energy; in fact, it could be used in many of the process operations, to give rise to an integrated, and much more efficient process.

One of the most common prejudices concerning applied electrochemistry
is that the crucial factor for the production costs of an electrochemical
process is the cost of electricity. But capital costs are almost in all
cases the dominant factor, and this should be taken into account when an
electrochemical reaction is studied with a view to its industrial application. In fact, the possibilities of applications are connected with the
improvements which can be achieved in the overall synthetic sequence, from
the raw material to the final product. The most important goal is to
cumulate several reaction steps in one: this immediately improves energy

efficiency, and allow a large capital cost saving. Further improvements may be obtained via a simplification of the cell design: the best solution, if possible, is to work with undivided cells and with extraction procedures built into the process.

Looking particularly at the fine chemicals field, many of the conventional processes currently in operation use inorganic chemicals, 50% of which are made by electrochemical techniques, and this is a strong argument in favor of indirect electroorganic processes.

The best field for the introduction of electrochemical technologies, in order to substitute the conventional ones, lies in the range of compounds having neither too high nor too low production costs. In fact, in these two extreme cases, for opposite obvious reasons, nobody is going to bother to install a cheaper process, replacing the existing one.

As far as the electrolytic cells are concerned, there are at present on the market some rather specialized plate and frame cells. Therefore, if there is the opportunity of introducing a new technology in some synthetic procedure, it is necessary to base it on a plate and frame design, with relatively small modifications. Several other reactors have been proposed in the patent literature, and may be adopted as they can be developed by any reasonably competent engineering firm. The ultimate goal must be, in any case, to reach the simplest and cheapest solution of the problem.

Organic electrochemistry has made great progress in the use of non-aqueous aprotic media, but it is not clear whether this will continue to be a productive area, at least in terms of industrial applications, or whether organic solvents will only be of laboratory interest. In fact, after the first period of euphoria, there is now much more prudence is scaling up processes in which organic solvents are involved. Of course, in the case of small scale processes affording products with high added value, organic solvents will continue to be used, but as far as large scale processes are concerned, industrial management feel very reluctant to go into the handling of large tonnages of polluting non-aqueous media. A possible suggestion is to continue to use non-aqueous electrochemistry as a guide for the study of electroorganic processes, but, whenever possible, one should try to move to a more acceptable form, such as, for instance, the multiphase electrolyses.

Another possibility might be to make use of completely different systems, in which liquid ammonia or sulfur dioxide is involved, as these solvents are potentially cheaper and easier to remove. Furthermore, there are other fields to be explored, in which electrochemistry can do things which are much harder to do by other techniques: very high temperatures and pressure electrolyses, up to the critical conditions of the system. This kind of electrolytic system has already been experimented, but nothing is known about the possibility of using it in electroorganic processes.

POLYMER ELECTRODES

R. W. Murray

Department of Chemistry
University of North Carolina
North Carolina, USA

INTRODUCTION

This article has the following purposes:

- to delineate the various existing types of polymer electrodes in a broad sense, and
- to illustrate the contributions that some of the newer varieties of polymer electrodes have made to electrochemistry and to broader segments of chemistry.

Not all types of polymer electrodes will be given equal attention by the author's choices, and apologies are made in advance for the necessary selective referencing which will omit significant papers of many authors. A complete bibliography of this field would be well in excess of 500 citations. Detailed reviews[1-10] should be consulted for more comprehensive treatments of various polymer electrode types. See also a similarly aimed article by Faulkner[11].

CLASSES OF POLYMER ELECTRODES

The subject of polymer electrodes is taken to include those electrodes where a polymeric material exerts some specific direct or indirect control over an electrochemical reaction or response. In polymer electrodes, the polymeric material is usually employed as a film or membrane in contact with (or coating) the electrode. In some electrodes, the membrane is separated from the electrode by a layer of reactant. (Excluded from this discussion are thus structural applications of polymers in electrochemical cells, such as battery casings, Teflon electrode shrouds, capillary barriers in gas-consuming cells, etc.).

Polymeric films which contain no electroactive groupings have been used with electrodes for many years and continue to find new uses. These nonelectroactive films are typically used as membrane barriers to transport, often in some permselective way. Permselective membrane barriers are used:

- in electroanalytical sensors,
- for corrosion inhibition, and as
- battery separators.

Polymeric films which do contain electroactive (and chemically reactive) groupings have given rise to many new kinds of polymer electrodes.
The membrane electroactivity is turned on by electron transfers to/from the
underlying electrode, and dominates its response. These electroactive
polymer electrodes are often termed "chemically modified electrodes", and
they follow the theme initiated in the mid-1970's of covalently binding
monolayers of electroactive species to electrodes for the purposes of
predictively eliciting some interesting or desired electrochemical or
electrocatalytic behavior[1]. Electroactive polymer coatings, because of
their manipulable thickness (5 mm to 10 μm), can additionally be specially
arranged in space relative to the electrode(s), so that the electrode is
not only chemically but also "specially modified". Special modification is
presently an active research subject.

Electroactive polymer films on electrodes fall into the categories:

- polymers with fixed redox sites,
- ion exchange polymers with electroactive counterions,
- polymers with fixed coordinating sites for redox metals or ligands,
- and polymers with significant electronic conductivity.

The special modifications which have been made include:

- bilayer electrodes, with two (or more) overlaid layers of different
  electroactive polymers,
- sandwich electrodes, where the film(s) of electroactive polymers are
  contracted at different places with two (or more) different working
  electrodes, in a closed or "open" face sandwich.
- ion gates, electroactive polymer membrane with embedded porous
  electrodes, and
- composite films, electroactive polymers with interseparated
  particles of conductor, semiconductor, or other catalytically active
  materials.

The next section will illustrate these various polymer electrodes.

ELECTROACTIVE POLYMER FILMS ON ELECTRODES

Polymers with Fixed Electroactive Groupings

Considerable research interest accompanied the discovery by several
laboratories[12-16] in the late 1970's, that electrodes could exchange
electrons not only with bonded monolayers[1] of electroactive sites, but
also with the equivalent of multiple monomolecular layers of redox groupings in polymeric films. In one of the initial publications, Merz and
Bard[12] coated Pt electrodes with poly(vinylferrocene) by electrochemically oxidizing a CH_2Cl_2 solution. The less soluble (more absorbing?)
poly(vinylferricenium) product of this reaction coated the Pt electrode,
which could then be transferred to an CHCN solvent for stable observation
of $Fer^{+/o}$ cyclic voltammetry. Miller and Van de Mark[13] concurrently
described the electrochemical reactivity of poly(p-nitrostyrene) films
applied to Pt electrode by the simple process of dip coating. Wrighton
and coworkers[14] then described a hydrolytically reactive organosilane
ferrocene derivative which would both bond to an oxide layer on the electrode and cross-link the polymeric film, and the authors' coworkers[15,16]
showed that electroactive poly(vinylferocene) films could be deposited on
electrodes from RF plasmas into which vinylferrocene was introduced.

In the above polymers, the electroactive site was either part of or
pendant to the polymeric chain and was applied to the electrode as part of

the polymeric film. Subsequent research has expanded the range of electro-active groupings incorporated into redox polymers; these now include[3] ferrocene and alkylated 4,4'-bipyridines (viologens) in a variety of poly-mers $-PyFe(CN)_5^{3-}$, tetrathiafulvalene, anthraquinone, pyrazolines, dopa-mine, cobaltocene, and a number of Fe, Ru, and Os polypyridine complexes. Three illustrative examples are shown in Figure 1: I–III. The films have been formed[3] by a great diversity of methods; electrochemical and photo-chemical precipitation (ferrocenes and viologens), dip coating (adsorbable high molecular weight polymers), RF plasma (from monomer, usually with some chemical degradation), condensation of hydrolytically unstable organosilane derivatives, and electrochemical polymerization onto the electrode from monomer solution (metal polypyridines, ferrocenes, and porphyrins).

Ion Exchange Polymer Films with Electroactive Counterions

Oyama and Anson[19] initiated another now popular route to electro-active polymer films by the simple expedient of ion exchanging ferrocyanide ion into a protonated (anion exchanger) film of poly(vinylpyridine) which had been adsorbed onto a carbon electrode (Figure 1: IV). Because of the favorable (high charge) partition coefficient for the $Fe(CN)_6^{4-}$ ion enter-ing the polycationic film, the cyclic voltammetric response of a $PVPyH^+$ coated electrode placed in a very dilute $Fe(CN)_6^{4-}$ solution could be observed to grow steadily, as the $Fe(CN)_6^{4-}$ concentration in the ion exchanger polymer film "solution" next to the electrode was increased in-partitioning.

The ion exchange procedure is appealing on several accounts, and other ion exchanger[3], both cationic and anionic (Figure 1: V, VI) have now been applied, including sulfonated poly(styrene) and analogs, alkylated polypyridines, the perfluorinated polymer Nafion, and several of the fixed redox site polymers which were charged and thus also ion exchangers. Electroactive ions incorporated have included metal cyanide, chloride, bipyridine, EDTA, and oxalate complexes, tetrathiafulvalenium, bromide, and a cationic ferrocene derivative. The diversity of accessible redox chemicals is not as great as that in the fixed site redox polymers due to the requirement of a significant partition coefficient into the film. However not all of the partitioning forces need be electrostatic; Nafion for example (Figure 1: V) is a clustered polymer with hydrophobic (fluoro-carbon) regions into which hydrophobic materials can partition in a sense more related to solvent extraction. Consequently, it may well be that this category of electroactive films will in the future be termed simply "films which (reversibly) partition electroactive ions and molecules". The author thinks it probable that even if the film partitions (dissolves) only neutral electroreactants, to be successful it should nonetheless contain some fixed ionic sites simple for the purpose of allowing low ionic resist-ance for the flow of change across the polymer membrane. Note that all of the resistance of a polymer film to current flow appears, even in a three electrode cell, as "uncompensated resistance".

Polymers with Fixed Coordinating Sites for Redox Metals or Ligands

Another approach to an electroactive polymer membrane is coating the electrode with a coordinatively reactive polymer film and binding the redox group to it in a second step. The first example of this was presented by Anson and coworkers[26], who reacted $Ru(EDTA)(H_2O)$ solutions with adsorbed poly(vinylpyridine) films, incorporating the complex via -Py-Ru bonding. Pentacyanoferrate could be similarly incorporated[22] (Figure 1: VII). Utilizing the polymer film reactivity in this way is coordinatively versa-tile, and avoids the task of synthesizing the metallated polymer in a monomer suitable for film-making. The metal complex sites may not however, be uniformly dispersed in the film[22].

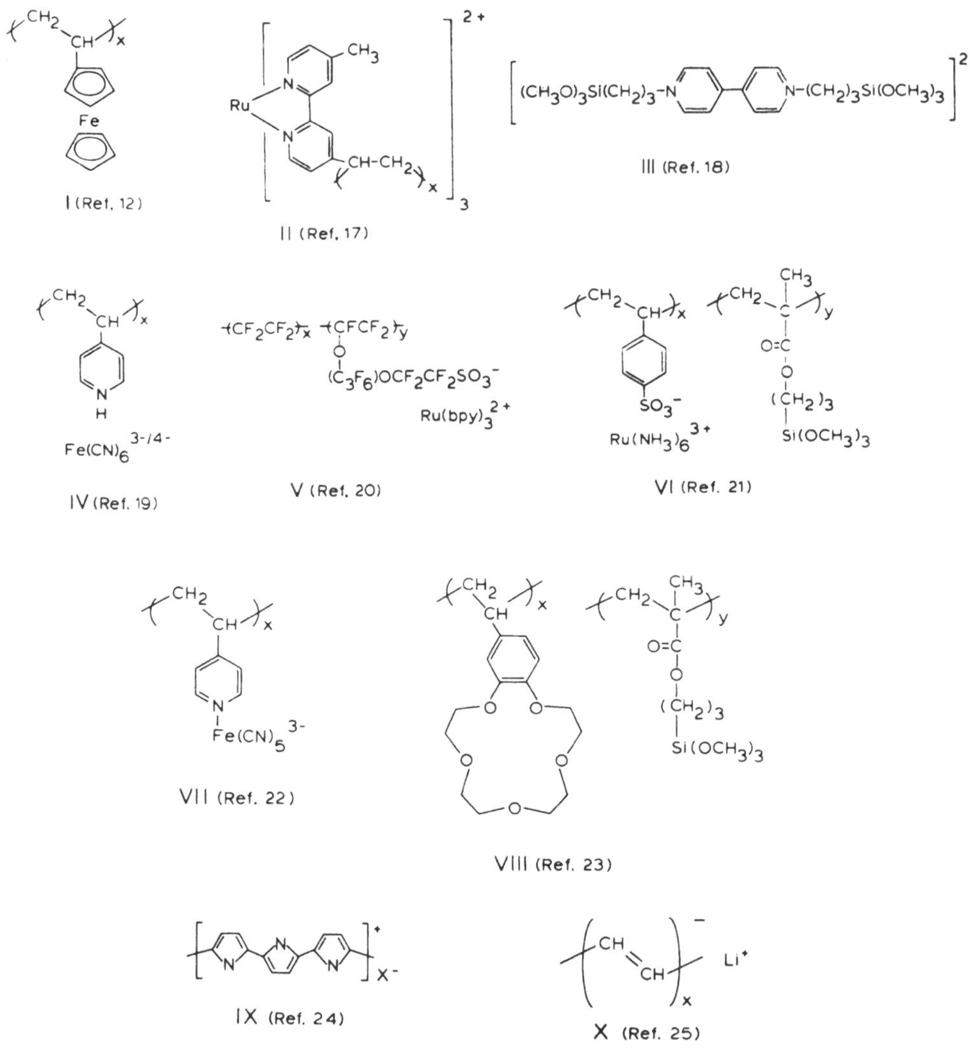

Fig. 1. Examples of Electroactive Polymers

The polymer films can also be coordinately reactive in a more reversible way, so that they can be regarded as metal ion partitioning films. Polymeric films containing macrocycles ("crown ethers") are of this class, and polymer VIII shown in Figure 1 can be used to partition and concentrate, for example Tl^+ from its acetonitrile solutions.

Polymers with Significant Electronic Conductivity

The electroactive polymers in the preceding sections conduct electrons by some combination of electron self exchange reactions between neighbor oxidized and reduced sites (electron hopping) and, for non-polymer bound redox groups, physical diffusive notions. These are electron conduction mechanisms that require and are primarily driven by concentration gradients of oxidated and reduced sites, not gradients of electrical potential. Because of this basically localized state electron occupancy character, the effective electron mobilities in the preceding electroactive polymers are small relative to those in materials with well developed electronic band structures and long electron mean free paths. Such electronically conducting materials have been fashioned from organic polymers in recent years[5,6,10] and are variously termed "organic conductors, organic metals, or conducting polymers".

Electronically conducting polymers include (Figure 1) poly(acetylene), poly(pyrrole), poly(thiophene), poly(phenylene), poly(phenylene sulfide), and O-bridged metallophthalocyanines. These have been prepared by procedures ranging from thermal to electrochemical polymerization. Conducting polymers are often used in film thicknesses larger than typical for the localized state electroactive polymers, since their electron conductivity allows communication between electrode and electroactive polymer over larger distances, and the potentialities for charge storage (batteries) favor large masses of electroactive material. However, charging/discharging reactions of these materials involve a change in chemical state and ionicity, which demands influx/egress of charge-compensating counterions just as in the case for the localized state electroactive polymers. The film thicknesses optimum for rapid, low ohmic loss, charge storage will with conducting polymers probably be lowered with further research on physical design of the films.

SPECIAL MODIFICATIONS WITH ELECTROACTIVE POLYMER FILMS

Bilayer Electrodes

The physical thicknesses of electroactive polymer films allow their special structuring relative to the electrode and each other. Recognition of this led to a series of investigations in the author's laboratory, beginning with the so-called bilayer electrode. In a bilayer electrode, the electrode is first coated with one ("inner") electroactive polymer and a film of a different ("outer") electroactive material is then overlaid on this (Figure 2). Because the inner polymer film is of the localized state electroactive variety (either fixed polymer site or ion exchanged), a chemical potential gradient can exist between it and the outer electroactive film, at the film/film interface. This chemical potential gradient causes the film/film boundary to act as a current rectifying junction[17,27], without invoking the electrostatic space charge effects of a conventional semi-conductor. The outer electroactive film was, in early bilayer assemblies, a redox polymer film, but the outer film can in principle be an electroactive, electronically conducting material, such as silver[28] or poly(pyrrole).

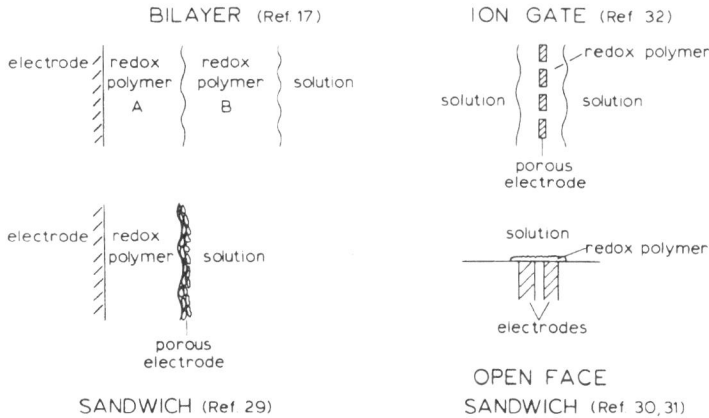

Fig. 2. Examples of spatially modified electrodes.

Sandwich Electrodes

Another special arrangement contacts the electroactive polymer film in two (or more) positions with controlled-potential working electrodes. The initial version[28,29] of this was an electrode coated first with the electroactive polymer, followed by evaporation of a thin Au film onto the outermost surface of the thin polymer film. When used, the Au film is in contact with the solvent/electrolyte solution of the cell (Figure 2). The Au film must have electrical continuity of contact with the polymer yet also have sufficient porosity that charge compensating counterions can migrate in/out of the film as its oxidation state is changed by manipulations of the two working electrode potentials. Steady state currents flow across the polymer film when one electrode's potential is set so as to reduce the polymer at its interface and the other's so as to oxidize the polymer at the opposite electrode-polymer interface. The currents serve to measure the electron mobility in the polymer film, both for localized state electroactive polymer such as the metal polypyridines (Figure 1: II) and for conducting polymers like poly(pyrrole) (Figure 1: IX). The sandwich electrode response also serves to demonstrate explicitly that potential gradients per se are not required for electron transport through the metal polypyridines, and that oxidized poly(pyrrole) in a solvent wetted state acts like an ohmic conductor just as it does in the more conventionally studied dry state.

Instead of contacting the opposite faces of the electroactive polymer film, one can alternatively contact opposite edges in an "open face" sandwich as illustrated in Figure 2. Open face sandwich electrodes require an insulating gap between the two conductor electrodes which is large compared to the film thickness but small enough to yield current flow which is measureable given the electron mobility of the particular polymer studied; this translates to a several micrometer gap. While no published versions of open face sandwich electrodes have appeared, successful experiments have been conducted in both the author's laboratory[30] and in Wrighton's laboratory[31] at MIT, using metal polypyridine polymers and poly(pyrrole), respectively. The open face sandwich should be more broadly applicable to different polymers than the early, "closed face" version.

Ion Gates

Redox polymers with fixed redox groupings and conducting polymers usually undergo a change in charge as a result of oxidation or reduction by the electrode. Thus, poly(vinylferrocene) is neutral as such but becomes poly-cationic and an anion exchanger when oxidized. Likewise, poly(pyrrole) is poly-cationic in its (conducting) oxidized state but neutral (and less highly conducting) when reduced. The "ion gate" exploits this electrode-controlled change in polymer film ionicity. An electroactive polymer such as poly(pyrrole) is coated onto a porous (Au minigrid) electrode so as to fill all pores, and this membrane is then used to separate two electrolyte solutions (Figure 2). By means of electrical control of the electrode potential, the ionic permeability of the membrane to ions moving between the two electrolyte solutions can be dynamically, reversibly, and remotely controlled. This is a unique property of both natural and man-made membranes.

The ion gate was demonstrated initially[32] with poly(pyrrole) as the electroactive polymer. (In this application, the electronic conductivity of poly(pyrrole) is strictly incidental). When the embedded porous electrode potential is positive, the poly(pyrrole) is oxidized and poly-cationic, and anions flow freely between the two solutions. When the electrode potential is negative, the reduced, more neutral poly(pyrrole) presents a much higher impedance to ion flow. The ion gate is "on" and "off" in these two states.

Composite Films on Electrodes

This term is used to include a variety of electrodes where particulate material of some sort is interspersed within or on a polymer film on an electrode. The first example of a composite film was described by Wrighton and coworkers[33,34] who electroreductively generated small $Pt°$ particles within viologen polymer films on an electrode. When the viologen film was charged to its reduced (V^+) state, the $Pt°$ particles became poised at a potential sufficient to evolve H_2 by reduction of the water medium. Thus the $Pt°$ particles served as heterogeneous catalysts; this early experiment was important in illustrating this novel idea. Frank[35] has since described an analogous application of RuO_2 particles as oxygen-evolving catalyst on a photoanode.

A completely different form of particle-in-film was devised by Bard and coworkers[36] and by Rolison and coworkers[37]. In both cases, some means was taken to immobilize small adsorbant particles within a polymeric matrix on an electrode: clay particles by former and zeolitic particles by the latter authors. The adsorbant particles in both cases were used to incorporate an electroactive species into the film ($[Ru/bpy)_3]^{2+}$ and O_2, respectively). The incorporated substances are, by some as yet unclear mechanism, able to undergo electron transfer reactions with the electrode. This author regards these developments as forerunners of electrodes capable of capturing the powerful catalytic selectivities known for zeolites and clays for electrocatalytic uses. More research can be expected on this subject.

A third category of composite films incorporates conducting particles into the polymer. In one case, the conducting particles were carbon whiskers[38], in a study designed to show how high surface area polymer coated electrodes could be assembled to avoid difficulties with electron transport rates in electrocatalytic application. In another study[39], conducting $TTFBr_{0.7}$ needles were grown by a novel ion exchange/electrochemical precipitation experiment. Lastly, highly reducing polymer films were used to reduce and electrodeposit metal films[40] onto the outer surface of the polymer, in an illustration of a new approach to metallization of plastics, an important technological topic.

NON-ELECTROACTIVE POLYMER FILMS

Electroanalytical Sensors

Perhaps the oldest applications of polymers to electrochemistry are found in the covering of electrodes with permselective membrane transport barriers. Clark and coworkers[41] in 1953 covered a Pt electrode with a poly(ethylene) membrane for measurements of blood O_2; the membrane is permeable to O_2 but not to other physiological components which would poison the Pt electrode response. Mancy, Okun, and Reilley[42] later incorporated this idea into a galvanic oxygen analyzer for waste and neutral water analysis which became a widely used device. In the latter electrode, both working and counter electrodes were encased in the membrane along with a thin film of electrolyte solution, so that the ionic resistivity of the membrane was of no consequence. The Clark and Mancy electrodes responded amperometrically. The field of gas-responsive membrane and electrolyte solution film-covered electrodes is now well developed[4,8,9] and also includes potentiometric arrangements such as (pH) glass electrodes covered with a bicarbonate solution film and membrane, for CO_2 analysis.

Another well developed electroanalytical topic is "enzyme" electrodes[4,9,43], where enzymes are immobilized on electrode surfaces, often within a polymeric matrix. These polymeric films are designed more for their specific chemical (enzymic) reactivity toward a substrate than for permselectivity, but are mentioned here for completeness.

Corrosion Inhibition

The coating of polymeric films on conducting surfaces (such as metals) would, intrinsically, appear attractive to seek inhibition of corrosion of the metal, and indeed, polymer coatings are widely employed in commerce. The corrosion applications appear to outstrip by far the published research activity aimed at a fundamental understanding of how such polymer coatings actually work to reduce corrosion in a variety of environments. Most coatings sought are non-ionic, dense, pinhole-free, unreactive materials. For an example of potentially useful materials, see the electrochemically formed phenolic studied by Dubois and associates[44,45]. Certainly part of the protective action of these films must be their action as transport barriers to water and ionic substances (they would not be expected to be effective barriers to gases). However, a more reactivity-oriented type of corrosion protective is possible with films that combine electroactivity with a transport barrier, as illustrated in the photoanodic corrosion work of Wrighton and coworkers[46,47] on n-type Si° photoanodes. It would appear that this latter approach could be applied to conductor materials as well, and that the corrosion field might profit from attention to fundamental membrane transport investigations.

Battery and Electrochemical Cell Separators

For isolation of electrode reactants in either an energy storage cell or an electrochemical cell employed for electrosynthesis, a chemically and physically stable, permselective separator membrane is generally required. Two recent classes of polymeric materials are of interest for such application: perfluorinated ionomers of which Nafion[48,49] is an example, and alkali metal-coordinating polyethers and polyesters of which polyethylene oxide-LiCF$_3$SO$_3$ mixture is an example[50-52].

As a membrane, Nafion has many desirable separator characteristics, and its analogs are attractive for technological application. Nafion, because of its mixture of anionic-polar and fluorocarbon-non polar functionalities, also has fundamentally interesting partition characteristics which encompass both electrostatic (ion exchange) interactions with hydrophilic cations and hydrophobic dissolution interactions with hydrophobic cations[53-55]. It has been proposed that Nafion exists as a biphasic polymer via clustering of the hydrophilic and hydrophobic groupings[56,57]. Both partitioning and transport characteristics of electroactive materials through Nafion films on electrodes are being studied[58,59]; these investigations have direct relevance to understanding the transport permselectivity which has utility in the separation membrane application.

(Polyethers like poly(ethylene oxide) (PEO) when mixed with alkali metal salts, serve as effective complexing media to yield, often in amorphous form, ionically conducting polymeric solids. The considerable potential permselectivity and excellent redox stability of these newer processible solids is attractive for battery separator applications, and many research groups have been attracted to development of this subject. Armand[60] has published a useful overview of the available polyether conductivity, stability, interfacial kinetic, and ionic transference number literature.

This section presents some examples of the lines of research which have developed using electrodes coated with electroactive polymer films, stressing those which contain contributions or potential contributions beyond electrochemistry.

Basic Electron Transfer Chemistry

The propagation of oxidized and reduced redox states through many monolayers of electroactive sites (charge transport) was the object of intense curiosity and probing following the initial reports of electroactivity for poly(vinylferrocene) and other redox polymers. The general understanding of charge transport for fixed site redox polymers, following a proposal by Kaufman[61,62], measurement approaches by the author's associates[16,63] and Oyama and Anson[64], theory by Andrieux and Saveant[65] and Laviron[66], and numerous measurements of charge transport rates[3], is now reduced to a picture of electron self-exchange reactions between neighbor oxidized and reduced redox sites which serve to propagate the electron (or, equivalently, the hole) in space relative to the electrode. The electrochemical experiment has proven to be an effective probe of the rates of these reactions. The flow of electroactivity in ion exchange polymer films containing redox counterions has in general terms also been explained; publications from the Anson[59] and Bard laboratories[58,67] show that physical diffusion of the redox counterion through the film is involved in addition to electron self-exchange between its oxidized and reduced forms.

These studies of charge transport have broad consequences in polymer chemistry, electron transfer kinetics, and electrical conductivity. First, the paucity of existing electroactive polymers available to serve as objects of study drove electrochemists to unnatural but rewarding explorations into polymer synthesis, resulting in the emergence of entire new classes of localized redox state polymeric materials. Many of these materials are likely to have further non-electrochemical uses. Secondly, the kinetics of electron self-exchanges are in the polymer films studied, being investigated in new and novel (polymeric) media. In the polymer, the reaction barrier to electron transfer may not always (or even typically) be the same as that applicable to homogeneous solutions. Motions of charge compensating counterions and segmental motions of the polymer framework may be in control; the latter circumstance allows new insights into the microdynamics of polymeric chain mations. Thirdly, electrons flow through the polymers under concentration not voltage gradients, and at specific ($E°$) energies. This comprises a new form of electron conduction.

Mediated Electrocatalysis

Electroactive polymer films can serve electrocatalytic functions by acting as kinetically rapid electron transfer intermediaries between the electrode and a catalytic reaction substrate. Research in this area has aimed both at understanding the principle of mediation with electroactive polymer films, and at devising new chemical mediation systems. The literature is now extensive[3], but so are both the challenges to theory and to the chemical problem of driving reactivity of interesting reaction substances. Mediated electrocatalysis is a vigorous research field today and is far from maturity. The principal theoretical framework, due to Saveant and coworkers[68-72], shows that charge and substrate transport in the polymer film, plus steps having to do with film non-homogeniety, compete with the chemical (mediation electron transfer) step for catalytic rate control. Theory has not yet encountered and surmounted the catalytically significant category of mediation site-substrate adduct formation equilibria and kinetics, but surely will. In chemical terms, significant examples of substrate reactions that have been driven include alcohol and hydrocarbon oxidations[73], cytochrome c[74], NADH[75], O_2[76,77], and Marcussian electron transfer cross exchanges[78]. The four-electron, reversible reduction of O_2, the reduction of CO_2, and further oxidations of alcohols and hydrocarbons have not yielded to polymer film electro-analysis yet but the importance of these and related reactions, and the potency of the research groups interested in them, make it probable that they will.

It is in mediated electrocatalysis that polymer coated electrodes display some of the strongest advantages of modified electrodes, and perhaps the largest, but at present still long term, potential for impact on society. Because of the present versatility and reliability of pro-cedure for immobilizing mediator-catalysts in polymers, a researcher can invest in (easier) homogeneous solution experiments designing a mediation reaction to be fast, stable, and selective to the desired degree, and when satisfied proceed to the immobilization and mediated electrocatalysis format. Mediated electrocatalysis seems probable to make significant con-tributions to chemical energy conversion in second generation fuel cells, to electrosynthesis of speciality chemicals, and when used in conjunction with semiconductor electrodes, to radiant energy conversion and storage. The physical form of the electrodes used for these purposes may not be simple electrodes-plus-smooth polymer films; porous flow-through assem-blies, particulate electrodes, electrodes with designed binding sites[79], and electrodes with immobilized heterogeneous as well as mediator catalysts contain promising avenues to improved catalytic efficiencies and to dealing with stability requirements.

Sensors and Devices

This is a rapidly emerging subject, based substantially upon ideas of specially modifying polymer films on electrodes. The rectifying features of bilayer electrodes can for instance be employed to detect analytically low concentrations of redox substances in solutions and to render astable the switching of an electroactive electrochromic film between its color states[80]. Sandwich electrodes in closed face form have with redox polymers been designed as simple chemical diodes[28] and in open face form with those contacting electrodes and poly(pyrrole) films as a simple transistor[31]. Ion gate membranes[32] serve to control the flow of ionic species in an electrically controlled "on-off" monomer analogous to elec-tronic gates. Reagents can also be released from polymer films by chemical cleavage[81]. These various experiments can be viewed as the beginning of "molecular electronics" with polymer electrodes. It is far too early to forecast the utility of devices and sensors based on specially structured polymer electrodes, except that it seems clear that they are unlikely to compete directly with the electronic functions and speed of modern solid state microelectronics. The benefits are more likely to be in electronic behaviors not readily achievable in the conventional solid state world. To speculate on the future, promising application may be structured electrodes in contact with solutions as species selective analytical sensors (perhaps with built-in signal amplification), in color display formats, and in modeling of biological phenomena.

Acknowledgement

Preparation of this article was supported in part by a grant from the National Science Foundation.

294

REFERENCES

1. R. W. Murray, Chemically modified electrodes, Accts.Chem.Res., 13:135 (1980).
2. R. W. Murray, Chemically modified electrodes, in: "Electroanalytical Chemistry," Vol.13, A.J. Bard, ed., M. Dekker, New York (1984).
3. R. W. Murray, Polymer modification of electrodes, Ann.Rev.Mats.Sci., 14:145 (1984).
4. M. E. Meyerhoff and Y. M. Fraticelli, Ion-selective electrodes, Anal.Chem., 54:27R (1982).
5. R. B. Seymour, ed., "Conductive Polymers," Plenum Press, New York (1981).
6. D. Baeriswyl, G. Harbeke, H. Kiess, and W. Meyer, Conducting polymers: Polyacetylene, in: "Electronic Properties of Polymers," J. Mort and G. Pfister, eds., Wiley-Interscience, New York (1982).
7. J. S. Miller, ed., "Chemically Modified Surfaces in Catalysis and Electrocatalysis," ACS Symposium Series Vol.192 (1980).
8. W. E. Morf, "The Principles of Ion-selective Electrodes and of Membrane Transport," Elsevier Scientific Pub. Co., New York (1981).
9. D. W. Lubbers, ed., "Progress in Enzyme and Ion Selective Electrodes," Springer-Verlag, Berlin (1981).
10. G. Wegner, Polymers with metal like conductivity, Angew.Chem.Int.Ed., 30:361 (1981).
11. L. R. Faulkner, Chemical microstructures on electrodes, Chem.Eng. News., pp.28 Feb. 27 (1984).
12. A. Mertz, and A. J. Bard, A stable surface modified platinum electrode prepared by coating with electroactive polymer, J.Am.Chem.Soc., 100:3222 (1978).
13. L. L. Miller, and M. R. van de Mark, A poly-p-nitrostyrene electrode surface: Potential dependent conductivity and electroactive properties, J.Electroanal.Chem., 100:3223 (1978).
14. M. S. Wrighton, R. G. Austin, A. B. Bocarsly, J. M. Bolts, O. Haas, K. D. Legg, L. Nadjo, and M. C. Palazzotto, A chemically derivatized platinum electrode: Persistent attachment to an electroactive ferrocene derivative, J.Electroanal.Chem., 87:429 (1978).
15. P. Daum, J. R. Lenhard, D. R. Rolison, and R. W. Murray, Diffusional charge transport through ultrathin films of radiofrequency plasma polymerized vinylferrocene at low temperature, J.Am.Chem.Soc., 202:4649 (1980).
16. R. J. Nowak, F. A. Schultz, M. Umana, R. Lam, and R. W. Murray, Chemically modified electrodes. XX. Radiofrequency plasma polymerization of vinylferrocene on glassy carbon and platinum electrodes, Anal.Chem., 52:315 (1980).
17. H. D. Abruna, P. Denisevich, M. Umana, T. J. Meyer, and R. W. Murray, Rectifying interfaces using two-layer films of electrochemically polymerized vinylpyridine and vinylbipyridine complexes of ruthenium and iron on electrodes, J.Am.Chem.Soc., 103:1 (1981).
18. D. C. Bookbinder and M. S. Wrighton, Thermodynamically uphill reduction of a surface-confined N,N'-dialkyl-4,4'-bipyridinium derivative on illuminated p-type silicon surfaces, J.Am.Chem.Soc., 102:5123 (1980).
19. N. Oyama and F. C. Anson, Electrostatic binding of metal complexes to electrode surfaces coated with highly charged polymeric film, J.Electrochem.Soc., 127:247 (1980).
20. I. Rubinstein and A. J. Bard, Polymer films on electrodes. 4. NAFION-coated electrodes and electrogenerated chemiluminescence of surface attached ru(bpy)$_3$$^{2+}$, J.Am.Chem.Soc., 102:6641 (1980).
21. C. R. Leidner and R. W. Murray, Electron-transfer reactions of iron, ruthenium, and osmium bipyridine and phenanthroline complexes at polymer/solution interfaces, J.Am.Chem.Soc., 106:1606 (1984).

22. N. Oyama, K. Shigehara, and F. C. Anson, Electrochemical responses of electrodes coated with redox polymers: Evidence for control of charge transfer rates across polymeric layers by electron exchange between incorporated sites, J.Am.Chem.Soc., 103:2552 (1981).

23. J. Massaux, P. Burgmayer, E. Takeuchi, and R. W. Murray, An electroactive polymer film on Hg electrode, based on a thallium macrocyclic polyether complex, Inorg.Chem., in press.

24. K. K. Kanazawa, A. F. Diaz, R. H. Geiss, W. D. Gill, J. F. Kwak, J. A. Logan, J. F. Rabolt, and G. B. Street, Organic metals: Polypyrrole, a stable synthetic metallic polymer, J.Chem.Soc.Chem.Commun., 854 (1979).

25. A. G. MacDiarmid and A. J. Heeger, Organic metals and semiconductors: the chemistry of polyacetylene, $(CH)_x$, and its derivatives, Syn.Metal., 1:101 (1980).

26. N. Oyama and F. C. Anson, Facile attachment of transition metal complexes to graphite electrodes coated with polymeric ligands: Observation and control of metal-ligand coordination among reactants confined to electrode surfaces, J.Am.Chem., 101:739 (1979).

27. P. Denisevich, K. W. Willman, and R. W. Murray, Undirectional current flow and charge state trapping at redox polymer interfaces on bilayer electrodes: Principles, experimental demonstration, and theory, J.Am.Chem.Soc., 103:4727 (1981).

28. P. G. Pickup and R. W. Murray, Redox Conduction: Its use in electronic devices, J.Electrochem.Soc., 131:833 (1984).

29. P. G. Pickup and R. W. Murray, Redox conduction in mixed valent polymers, J.Am.Chem.Soc., 105:4510 (1983).

30. C. Chidsey, G. Lin, and R. W. Murray, unpublished work, Univ. N.C., Chapel Hill, N.C. (1984).

31. H. S. White, G. P. Kittlesen, and M. S. Wrighton, private communications (1984).

32. P. Burgmayer and R. W. Murray, An ion gate membrane: Electrochemical control of ion permeability through a membrane with an embedded electrode, J.Am.Chem.Soc., 104:6139 (1982).

33. D. C. Bookbinder, N. S. Lewis, M. G. Bradley, A. B. Bocarsly, and M. S. Wrighton, Photoelectrochemical reduction of N,N'-dimethyl-4,4'-bipyridinium in aqueous media at p-type silicon: Sustained photogeneration of a species capable of evolving hydrogen, J.Am.Chem.Soc., 101:7721 (1979).

34. D. C. Bookbinder, J. A. Bruse, R. N. Dominey, N. S. Lewis, and M. S. Wrighton, Synthesis and characterization of a photosensitive interface for hydrogen generation: Chemically modified p-type semiconducting silicon photocathodes, Proc.Nat.Acad.Sci., 77:6280 (1980).

35. A. J. Frank and K. Honda, Visible-light-induced water cleavage and stabilization of n-type cadmium sulfide to photocorrosion with surface-attached polypyrrole-catalyst coating, J.Phys.Chem., 86:1933 (1982).

36. A. J. Bard and P. K. Ghosh, Clay-modified electrodes, J.Am.Chem.Soc., 105:5691 (1983).

37. C. G. Murray, R. J. Nowak, and D. R. Rolison, Electrogenerated coatings containing zeolites, J.Electroanal.Chem., 164:205 (1984).

38. P. Burgmayer and R. W. Murray, Increasing the rate of charging of redox polymer films with extended surface electrodes, J.Electroanal.Chem., 135:335 (1982).

39. A. J. Bard, T. P. Henning, and H. S. White, Polymer films on electrodes. 10. Electrochemical behavior of solution species at nafiontetrathiafulvalenium bromide polymers, J.Am.Chem.Soc., 104:5362 (1982).

40. P. G. Pickup, K. N. Kuo, and R. W. Murray, Electrodeposition of metal particles and films by a reducing redox polymer, J.Electrochem.Soc., submitted.

41. L. C. Clark, R. Wolf, D. Granger, and Z. Taylor, Continuous recording of blood oxygen tensions by polarography, J.Appl.Physiol., 6:189 (1953).

42. K. H. Mancy, D. A. Okun, and C. N. Reilley, A galvanic cell oxygen analyzer, J.Electroanal.Chem., 4:65 (1962).

43. J. E. Davis, R. L. Solsky, L. Giering, and S. Malhotra, Clinical chemistry, Anal.Chem., 55:202R (1983).

44. M. C. Pham, P. C. Lacaz, and J. E. Dubois, Obtaining thin films of 'Reactive Polymers' on metal surfaces by electrochemical polymerization. Part I. Reactivity of functional groups in a carbonyl substituted polyphenylene oxide film, J.Electroanal.Chem., 86:147 (1978).

45. F. Bruno, M. C. Pham, and J. E. Dubois, Polarmicrotribometric study of polyphenylene oxide film formation on metal electrodes by electrolysis of disubstituted phenols, Electrochim.Acta., 22:451 (1977).

46. J. M. Bolts, A. B. Bocarsly, M. C. Palazzatto, E. G. Walton, N. S. Lewis, and M. S. Wrighton, Chemically derivatized n-type silicon photoelectrodes: Stabilization to surface corrosion in aqueous electrolyte solutions and mediation of oxidation reactions by surface-attached electroactive ferrocene reagents, J.Am.Chem.Soc., 101:6179 (1979).

47. A. J. Bard and M. S. Wrighton, Thermodynamic potential for the anodic dissolution of n-type semiconductors: A crucial factor controlling durability and efficiency in photoelectrochemical cells and an important criteria in the selection of new electrode/electrolyte systems, J.Electrochem.Soc., 124:1706 (1977).

48. Nafion is a trademark of E.I. du Pont de Nemours and Company.

49. W. Grot, Use of nafion perfluorosulfonic acid products as separators in electrolytic cells, Chem.Ing.Tech., 50:299 (1978).

50. R. D. Lunberg, F. E. Bailey, and R. W. Callard, Interactions of inorganic salts with poly(ethylene oxide), J.Polym.Sci., Part A4:1563 (1966).

51. J. Moacanin and E. F. Cuddihy, Effect of polar sources on the viscoelastic properties of poly(propylene oxide), J.Polym.Sci., C14:313 (1966).

52. B. E. Fenton, J. M. Parker, and P. V. Wright, Complexes of alkali metal ions with poly(ethylene oxide), Polymer, 14:589 (1973).

53. C. R. Martin and H. Freiser, Ion-selective electrodes based on an ionic polymer, Anal.Chem., 53:902 (1981).

54. P. C. Lee and D. Meisel, Luminescence quenching in the cluster network of perfluorosulfonate membrane, J.Am.Chem.Soc., 102:5477 (1980).

55. M. N. Szentirmay and C. R. Martin, Ion exchange selectivity of nafion films on electrode surfaces, Anal.Chem., in press.

56. A. Eisenberg and M. King, "Ion Containing Polymers," Academic Press, New York, Chap. 2 (1977).

57. H. L. Yeager and A. Steck, Cation and water diffusion in nafion ion exchange membranes: Influence of polymer structure, J.Electrochem. Soc., 128:1880 (1981).

55. C. R. Martin, I. Rubinstein, and A. J. Bard, Polymer films on electrodes. 9. Electron and mass transfer in nafion films containing Ru(bpy)$_3^{2+}$, J.Am.Chem.Soc., 104:4817 (1982).

59. D. A. Buttry and F. C. Anson, Effects of electron exchange and single file diffusion on charge propagation in nafion films containing redox couples, J.Am.Chem.Soc., 105:685 (1983).

60. M. Armand, Polymer solid electrolytes: An overview, Solid State Ionics, 9:745 (1983).

61. F. B. Kaufman and E. M. Engler, Solid-state spectroelectrochemistry of cross-linked donor bound polymer films, J.Am.Chem.Soc., 101:547 (1979).

62. F. B. Kaufman, A. H. Schroeder, E. M. Engler, S. R. Kramer, and J. Q. Chambers, Ion and electron transfer in stable, electroactive tetrathiafulvalene polymer coated electrode, J.Am.Chem.Soc., 102:483 (1980).

63. P. Daum, J. R. Lenhard, D. R. Rolison, and R. W. Murray, Diffusional charge transport through ultrathin films of radiofrequency plasma polymerized vinylferrocene on glassy carbon and platinum electrodes, J.Am.Chem.Soc., 102:4649 (1980).

64. N. Oyama and F. C. Anson, Factors affecting the electrochemical responses of metal complexes at pyrolytic graphite electrodes coated with films of poly(4-vinylpyridine), J.Electrochem.Soc., 127:640 (1980).

65. C. P. Andrieux and J. M. Saveant, Electron transfer through redox polymer films, J.Electroanal.Chem., 111:377 (1980).

66. E. Laviron, A multilayer model for the study of space distributed redox modified electrodes. Part I. Description and discussion of the model, J.Electroanal.Chemi., 112:1 (1980).

67. H. S. White, J. Leddy, and A. J. Bard, Polymer films on electrodes. 8. Investigation of charge transport mechanisms in nafion polymer modified electrodes, J.Am.Chem.Soc., 104:4817 (1982).

68. J. Saveant, F. C. Anson, and K. Shigehara, Self-exchange reactions at redox polymer electrodes: A kinetic model and theory for stationary voltammetric techniques, J.Phys.Chem., 87:214 (1983).

69. C. P. Andrieux, J. M. Dumas-Bouchiat, and J. M. Saveant, Catalysis of electrochemical reactions at redox polymer electrodes: Effect of film thickness, J.Electroanal.Chem., 114:159 (1980).

70. C. P. Andrieux, J. M. Dumas-Bouchiat, and J. M. Saveant, Catalysis of electrochemical reactions at derivatized electrodes: Kinetic model for stationary voltammetric techniques and preparative scale electrolysis, J.Electroanal.Chem., 123:171 (1981).

71. C. P. Andrieux, J. M. Dumas-Bouchiat, and J. M. Saveant, Catalysis of electrochemical reactions at redox polymer electrodes: Kinetic model for stationary voltammetric techniques, J.Electroanal.Chem., 131:1 (1982).

72. C. P. Andrieux and J. M. Saveant, Kinetics of electrochemical reactions mediated by redox polymer films: Irreversible cross-exchange reactions: Formulation in terms of characteristic currents for stationary techniques, J.Electroanal.Chem., 134:163 (1982).

73. G. J. Samuels and T. J. Meyer, An electrode supported oxidation catalyst based on Ru(IV): pH encapsulation in a polymer film, J.Am.Chem.Soc., 103:307 (1981).

74. N. S. Lewis and M. S. Wrighton, Electrochemical reduction of horse heart pericytochrome c at chemically derivatized electrodes, Science, 211:944 (1981).

75. A. Kitani and L. L. Miller, Fast oxidants for NADH and electrochemical discrimination between ascorbic acid and NADH, J.Am.Chem.Soc., 103:3595 (1981).

76. A. Bettelheim, R. J. H. Chan, and T. Kuwana, Electrocatalysis of oxygen reduction. Part II. Absorbed cobalt(III)tetrapyridyl-porphyrin on glassy carbon electrode, J.Electroanal.Chem., 99:391 (1979).

77. P. Martigny and F. C. Anson, Catalysis of the reduction of dioxygen by poly(xylylviologen) coatings on graphite electrodes, J.Electroanal. Chem., 139:383 (1982).

78. C. R. Leidner and R. W. Murray, Kinetics of electron transfer reactions between polymeric and solution redox couples: Cross reactions between polymeric Os mediator and Fe, Ru and Os bipyridine and phenanthroline complexes, J.Am.Chem.Soc., 106:1606 (1984).

79. T. Matsue, M. Fujihira, and T. Osa, Selective electrosynthesis on chemically modified electrodes. 3. Regio-selective anodic chlori-

nation of some benzene derivatives with a cyclodextrin chemically modified electrode, J.Electrochem.Soc., 128:1473 (1981).

80. K. W. Willman and R. W. Murrary, Viologen homopolymer, polymer mixture, and polymer bilayer films on electrodes: Electropolymerization, electrocatalysis, and spectroelectrochemistry, J.Electroanal.Chem., 133:211 (1982).

81. L. L. Miller, A. N. Lau, and E. K. Miller, Electrically stimulated release of neurotransmitters from a surface: An analogue of the presynaptic terminal, J.Am.Chem.Soc., 104:5242 (1982).

Volta, Alessandro, 9, 10, 17, 55, 212
Voltammetric electrodes, 190-191
Voltammetry 105, 152-154, 167, 177, 206
  advantageous features of, 123
  applications and potentialities of, 121
  determination limits, 126
  in ecochemistry of metals, 123-126

Waste chemicals, 108
Waste water treatment, 111-113
Waste waters, heavy metals in, 138-139
Water
  chemical analysis of, 106-108
  electrolysis of, 83, 94
  oxygen in, 106
  pH values of, 106
  purification of, 46
  splitting of, 46
  surface active compounds in, 107

Whewell, William, 11
Wind power, 65
World Calendar, 35

X-ray diffraction, 279

Yeast, fusion of protoplasts, 232

Zero gap cell, 271, 272
Zero separation cells, 95
Zona pellucida, 234, 235